HANDBOOK OF MECHANICAL SPECIFICATIONS FOR BUILDINGS AND PLANTS

OTHER McGRAW-HILL HANDBOOKS OF INTEREST

HANDBOOK OF MECHANICAL SPECIFICATIONS FOR BUILDINGS AND PLANTS

A Checklist for Engineers and Architects

ROBERT HENDERSON EMERICK, P. E.

Member, ASME

Member. NSPE

McGRAW-HILL BOOK COMPANY

New York San Francisco Toronto London Sydney

HANDBOOK OF MECHANICAL SPECIFICATIONS FOR BUILDINGS AND PLANTS

 Library of Congress Catalog Card Number 66-16332

19313

2345 VB 987

Preface

In addition to describing the forms of specifications and enumerating the largely standardized General Conditions and the particularly oriented Special Conditions, this *Handbook of Mechanical Specifications for Buildings and Plants* is intended to stimulate the imagination of the specification writer with reference to the six basic functions of the technical clauses of a mechanical specification.

These six functions are identified as:

1. Telling bidders exactly what kind of equipment is wanted;
2. Telling them how it should be made for this project;
3. Stipulating the desired performance;
4. Stating what the buyer will furnish in supplementary or auxiliary materials and services;
5. Pointing out what must be done to install the equipment properly;
6. Requesting bidders, finally, to submit in detail how they propose to meet the specified requirements.

Since no checklist can be complete to the last detail and yet be confined in anything less than a library, much of this Handbook's value will be found in its capacity to suggest ideas for inclusion in a particular specification. These ideas will appear when the imaginative writer asks himself, "Have I satisfied all six functions for this particular item? In what section am I telling bidders exactly the kind of equipment I want? Where do I describe how it is to be made?" and so on, right down through each of the six functions, until the writer is satisfied that he has covered every corner of his field, and that his wants will not be misunderstood.

A common practice in active engineering and architectural offices is to take from the file an old specification covering similar equipment on a past project, then change the figures, strike out inapplicable clauses, add a few new ones, and let it go out for bids. This writer has done that many times and expects to do it again, but there is danger in the practice unless certain safeguards are observed.

First, the old specification is quite likely to contain errors, some important, some not. Probably they were caught either before or during the installation in the field, but the specification was not corrected, and the errors themselves have been forgotten. Result, the latest specification repeats the errors and sets up a new cycle of discovery, embarrassment, and perhaps some financial penalties as well.

Second, there exists something we call the "state of the art," which is a manner of evaluating scientific and engineering progress as this relates to the equipment being specified. From this viewpoint, a specification that was quite acceptable a few years ago can be sadly inadequate for the present state of the art, and should be kept in the history files and not revamped in an effort to bring it up to date.

The safeguarding of a new specification against these hazards is the responsibility of the specification writer, and requires the exercise of both knowledge and visualization. Without these two faculties, the writer may go far wrong, unknowingly, and there is no "or equal" for them, nor for the sound judgment that is their product.

This Handbook is expected to suggest ideas to the writer preparing a specification, whether by original writing or by purging an old one of errors and obsolescent items, and to offer him a mechanical equipment checklist probably unmatched in any other compact work of reference. By means of the cross references in the text, the specification writer is guided in the assembly of components for a power plant, for an underground or in-plant steam distribution system, for air conditioning and heating systems, refrigerating, ice-making plants, absorption cycles, compressed-air distribution, swimming pools, skating rinks, and other complexes of mechanical machinery. The range of subjects will be particularly useful to specification writers who may be entering a new field for which they have limited backgrounds of experience.

Where materials, shapes, designs, and arrangements are identified by name, the intent is to assist the specification writer in more accurately visualizing his problem, and is not to promote any item in preference to another. Selection and evaluation in all cases remains the responsibility of the specification writer and the engineer with whom he is working, if these two have separate relationships to the project in hand.

Illustrations and drawings are presented only to illuminate particular aspects of a unit or system. The author assumes the specification writer has access to manufacturers' catalogs and other publications that will provide him with adequate general information.

ACKNOWLEDGMENTS

The author gratefully acknowledges the help of those manufacturers who on request, generously provided illustrations and valuable information concerning their products. Also, the help of the United States Atomic Energy Commission, which furnished information, criteria, and technical data of high value in the very special area of specifications for nuclear reactors.

Finally there remains the unpayable debt to those engineers and architects who, in the course of his long career, gave to the author the benefits of their own experiences and ideas in the building of a successful specification.

Robert Henderson Emerick

To Robert Hardeik Emerick,
a second generation engineer,
who is also my son

Contents

DIVISION 1

Specification Forms and Definitions

The term *standard specification* has two forms and meanings, both of which are illustrated by usage in this Handbook. In what might be called its first form and meaning, a standard specification is one that is prepared and promulgated by a technical society, institute, association, or other specialized and competent group, with the purpose of setting minimum levels of performance, quality, safety, or other attributes of clearly defined materials and items.

Examples of such groups, familiar to most engineering and architectural offices, include the American Society for Testing and Materials (ASTM); American Standards Association (ASA); American Society of Mechanical Engineers (ASME); American Gas Association (AGA); National Electrical Manufacturers Association (NEMA); National Fire Protection Association (NFPA); American Petroleum Institute (API); and the Commercial Standards (CS) of the U.S. Department of Commerce.

An example of such issues is the widely recognized ASME Code for Unfired Pressure Vessels, or one of the American Standards Association dimensional standards for pipe fittings, as B16.15 "Brass or Bronze Screwed Fittings for Maximum Working Saturated Steam Pressure of 125 psig."

The availability of such standards substantially reduces the labors of the specification writer. Simply by requiring suppliers to comply with the standards indicated by name of the issuing group and the publication number, the measure of their responsibility is fixed. Lengthy elaboration is unnecessary.

Standard specification in its other form and meaning refers to that compilation of clauses prepared by individual engineering and architectural offices and intended for issue, with little change, on project after project. This is practical because many of the clauses are legalistic and fairly inflexible; others simply reiterate already established and commonly recognized rules of conduct that all bidders understand, but which the specification writer prudently reminds them of.

Repetition of technical clauses as a *standard specification* is less frequently used, owing to the diversity of the equipment and its performance as required by various projects. For example, the technical specifications for a water-tube boiler are quite different from those of a fire-tube boiler, and if these are both power boilers, they are again different from low-pressure heating boilers.

Repetitive specifying is possible, of course, when the equipment needed by a new project is similar in type, form, and materials to that previously specified on another job. In these instances, perhaps only the figures will need changing, and specification writers commonly mark up such old specifications for rewrite and issue as new ones. This time-saver is practiced in all offices that have been in business long enough to possess a usable backlog of specifications from past jobs.

When this form of standard specification has been printed, changes in the various clauses which appear necessary for a particular job are accomplished by striking out words, phrases, or even whole clauses, and substituting revised wording. The stamped words "Not Applicable" make a familiar appearance on many of these specifications. Equally familiar is the practice of adding new clauses as addenda, where the basic specification is lacking in some respect.

The visual impact of a specification that is scarred by X marks, stricken words, and rubber-stamp modifications is not that of beauty, but since such creations of the specification writer are profit savers, most offices feel obliged to follow the practice. The problem here is to make sure that this form of specification adequately covers every aspect of the job, before it is sent out for bids. At this stage it is well to remember that the purpose of a specification is to obtain what the ultimate owner wants and needs. It is written for his prime benefit.

A specification so loose that it permits sneaked-in substitutions, or so incomplete that field omissions follow, reflects unfavorably on the skill of the specification writer. The practice of lifting a printed standard specification sheet from a shelf is easy and attractive; checking it for adequacy clause by clause, in some instance word by word, is hard. It is a chore in which this Handbook is designed to assist.

Closed Specifications. A closed specification is one that limits acceptable equipment or materials to name-identified models, types, and sizes of one, or sometimes two or three, suppliers. Products from other sources are not considered. The specification is inflexible.

The advantage of the closed specification is in the time it saves, both in obtaining bids and in awarding the contract. Also it enables the designer to obtain exactly what he wants in the components of his overall design.

As a disadvantage, particularly for the owner who is paying the bills, a closed specification by eliminating competition also may eliminate any chance of buying at a more favorable price. Furthermore, since few manufacturers will divulge details of their trade secrets or production procedures, the specifier has little or no control of the way in which the product is made. He takes what he specifies on faith, the reputation of the maker, and the known performance record of the model or material that the closed specification calls for.

Economy in the preparation of a closed specification can be anticipated only when it covers some item that is to be purchased retail. That is, the specification names the item, which is delivered by the supplier and is paid for, and the supplier has no further responsibility. The only difference between this purchase of mechanical equipment and the purchase of an item in a supermarket is in the written description of the former. The latter transaction is wholly oral, or else the buyer simply helps himself.

Closed specifications involving installation, testing, inspections, and perhaps instructions to the user, must of course be issued in completeness.

Open Specifications. An open specification is distinguished by its willingness to accept the products of various suppliers who may or may not be named, provided their products meet the requirements detailed in the specification.

As an example of what is wanted, a make and model may be specified, followed by the words "or equal." This is a common practice and immediately introduces the question of what is equal and by whose estimate. Answers by which the engineer, architect, owner, or other designated person is made the sole and final judge are beside the point. There is no measure of equality, just an opinion.

Perhaps more acceptable wording to follow a designated make and model would be "or approved alternate." Questions of equality are avoided, and the fact that the decision to accept or reject rests on somebody's opinion is clearly indicated.

As a variation of the "approved alternate" technique, the named make and model may be designed as a base bid, with other makes and models offered as alternates either at no change in the base price or at prices quoted. Should a specifier particularly desire a certain make and model, chances are he can get it without resorting to a closed specification, by employing this technique.

To illustrate, suppose there are a half-dozen pumps of one make in a plant and all are satisfactory. When the time comes to add another, nothing less than an unusually attractive alternate bid is likely to offset the plant's successful experience and familiarity with the existing pumps. Yet the fact that such an alternate could be possible, and would be accepted if offered, tells bidders they can break in here if they make it worthwhile. This keeps the specification open while rendering such a break-in extremely difficult.

The open specification can be employed for the simple purchase of equipment, a transaction in which the supplier's delivery to the purchaser ends his responsibility except for whatever guarantees or warranties are given or implied. As in the closed specification, any further action required of the supplier must be covered by an adequate specification.

Construction Specifications Institute. With a membership that includes engineers and architects, contractors and maintenance men—in fact just about all workers in the construction field who have responsibilities in the preparation and use of specifications—the Institute has produced a specification format that is being adopted by some important organizations, notably the U.S. Navy's Bureau of Yards and Docks.

This format has been established as the result of lengthy study and evaluation. It is designed to cover the entire scope of projects of all sizes, and it works. Sixteen major divisions are provided, plus four unnumbered preliminary headings. The complete list is:

Bidding Requirements
Contract Forms
General Conditions
Supplementary General Conditions
Division 1. General Requirements
Division 2. Site Work
Division 3. Concrete
Division 4. Masonry
Division 5. Metals—Structural and Miscellaneous
Division 6. Carpentry
Division 7. Moisture Protection
Division 8. Doors, Windows, and Glass
Division 9. Finishes
Division 10. Specialties
Division 11. Equipment
Division 12. Furnishings
Division 13. Special Construction
Division 14. Conveying Systems

Division 15. Mechanical
Division 16. Electrical

Behind this format is the intention of placing related subjects close together for ready reference; for example, conveying systems are largely mechanical. Thus there is no need to trace related data through numerous pages and clauses; they are next door to each other.

Under each division, appropriate sections are identified by capital letters, as A, B, C, and so on. The combination of division number and section letter becomes a locating reference. For example, in a building project, Mechanical Division 15 might include Section A, "Heating"; Section B, "Air Conditioning"; Section C, "Sewage Ejector System"; Section D, "Drinking Water System"—these lettered sections to continue in sequence until the entire mechanical requirements are covered.

For the purposes of this Handbook a single mechanical division as presented in the CSI format is much too broad. The many diverse forms of subjects covered require individual and detailed treatment, in accordance with their intended applications. Each might be considered a mechanical project in itself, and in consequence is assigned division status in the Handbook.

Specification writers desiring to issue a CSI specification will of course arrange the subjects and clauses in the order, and in the number-letter system, of that format. This Handbook number-letter system of subject identification is not directly transferable to the CSI arrangement.

Shall: When used in a specification, *shall* is to be interpreted in the mandatory sense. Compliance in full is required. *Shall not* is similarly a mandatory prohibition. Generally applicable to contractors.

Should: Should refers to a provision of the specification that is desired, but not mandatory. A similar interpretation applies to the words *recommended* or *it is recommended.*

Will: Usually *will* applies to the actions and responsibilities of the Owner, or his representatives the engineer and architect. For example, the Owner *will* provide light and heat, but the Contractor *shall* furnish his own water and sanitary facilities.

Either: This word grants somebody, usually the contractor or supplier, a choice. Not to be used in the sense of *both.* Example: A specification calling for railings on *either* side of the walkway can result in a single railing being provided. To be sure of a railing on both sides, the specification must say *both.*

Contract Documents. All directives issued by the Owner or his representative the engineer or architect, and which are aimed at describing the work to be accomplished by the Contractor and which require his compliance, are properly included in the contract documents. Conventional inclusions are:

1. *Advertisement* or *Notice to Bidders* and inviting bids.
2. *General Conditions,* usually bound in the specifications.
3. *Special Conditions* (or *Supplementary General Conditions*), also usually bound in the specifications.
4. *Contract* (or *Agreement*), the form signed by the Parties.
5. *Specifications.*
6. *Plans or Drawings* covering the work involved, issued by the Owner in conjunction with the specifications. Also those manufacturer's prints and drawings as furnished by the successful bidder and approved by the Owner or his authorized deputy.

Each of these is a *Contract Document.* Others are sometimes included, for example, a sheet of *Instructions to Bidders,* issued when adequate instructions are not present elsewhere.

Guarantee and *Warranty:* These words are often used interchangeably, but this is in error. Actually there exists a difference in the time factor which is extremely important to all concerned. The *guarantee* (or *guaranty*), once made, continues in effect for a clearly stated and definite time. This period of the guarantee may range from 90 days, as with a new automobile, to 10 years in some tanks for domestic hot water.

In contrast, time is not an important element in a *warranty;* instead, the item must meet all represented levels of quality and performance. To illustrate, if stainless steel piping marketed as 16 per cent chromium and 10 per cent nickel is discovered later to be actually 18 per cent and 8 per cent, the buyer can sue the seller for a breach of warranty. In such cases it is assumed that the plaintiff, had he been aware of the product's deviation from its qualities as represented, would not have made the purchase.

The specification writer might consider presenting his guarantee or warranty clause somewhat like this: "The Contractor (or seller) warrants all materials, items, and performances to be equal to, or better than as represented; and guarantees all materials and workmanship against existing or developing defects, for a period of one year."

A *guarantee bond* or a *warranty bond* provides support for prospects of recovery, particularly if the guarantor or the warrantor is thought to be financially weak.

Streamlined Specifications. In this form, developed in fairly recent years, long and wordy paragraphs are avoided. To some degree, a streamlined specification resembles a skeleton specification. The subject item is stated, then the modifiers are appended as briefly as practical.

Example of this format:

HEATING SYSTEM, HYDRONIC

(a) *Circulating Pumps*

- Centrifugal.
- All bronze.
- Motor-driven through flexible coupling.
- Capacity ________ gpm at ________ ft total dynamic head.
- As manufactured by (name) or approved alternate.

Although the requirements of this form are readily comprehended, and the staccato arrangement of the modifiers mitigates against the chances of omitting anything important, it makes for a longer specification.

The arrangement of the checklists in this Handbook closely resembles the streamlined specification form in many instances.

Owner: When referred to in a specification, this term is interpreted to mean the individual, firm, or corporation for which the projected design, installation, and construction is being carried on. By extension, Owner also may be used to mean the Owner's deputy, as the architect or engineer, who is in charge of the Owner's interests.

Contractor: The person or firm who agrees to undertake and complete the work as set forth in the Contract Documents. On a straight purchase specification, the Supplier may be termed the Contractor, since he undertakes to comply with the requirements of the specification and the purchase order.

DIVISION 2

General Conditions

The *General Conditions* of a specification, which become part of the Contract Documents, cover those overall aspects of the work that must be satisfied prior to its beginning and also during its progress.

General Conditions, being largely standardized, are frequently carried from job to job and thus become practical candidates for large printings as standardized clauses.

Common practice in engineering and architectural offices today is to incorporate the general conditions developed by either the American Institute of Architects or the Consulting Engineers Council, latest editions. These are quite complete, the results of years of evaluations and revisions to meet the changing conditions of changing times. They have been widely accepted by the contracting associations of the United States.

The relationship of a mechanical specification to the general conditions takes two forms.

First, if the mechanical specification is to be just one division of the overall project specification, the responsibility of the mechanical writer is to make sure that the general conditions as bound in the project specification cover adequately those items related to the mechanical factors involved. If omissions are found, appropriate clauses must be added; if wording is ambiguous or fuzzy, it must be clarified. And finally, the mechanical contractor is told in his division of the specifications that the general conditions apply to him also, wherever they fit his work.

In its second form, the mechanical specification is the whole project. Other trades are to be handled by forces of the mechanical contractor, or by subcontractors under him. The list of clauses in this division of the Handbook are considered to be the minimum for accompanying a mechanical specification that is issued as a prime project. Additional clauses, and established wording of all clauses, may be found in the general conditions as published by the American Institute of Architects and the Consulting Engineers Council. All specification writers should obtain and study these documents.

The National Society of Professional Engineers, through their functional section of professional engineers in private practice, has issued several valuable guiding forms for the successful execution of contracts. These are:

1. "Standard Form of Agreement between Owner and Engineer for Professional Services"
2. "Standard General Conditions of the Construction Contract"
3. "Change Order Form"
4. "Certificate of Substantial Completion"
5. "Approval of Payment"
6. "Standard Form of Agreement between Owner and Contractor"
7. "Commentary on Contract Documents"

These issues should be in the reference file of all specification writers.

When a project is completed, signed off, and paid for, each of these documents which performed a function in the conduct of the work will be properly included with the contract documents.

2-1. STANDARD CONTRACT FORM

(a) Indicate if to be AIA, CEC, or other.
(b) If deletions, changes, or additions are to be made, make note of articles affected.

2-2. ACCEPTABLE SIGNATURES

(a) Requires official position of individual or individuals authorized to legally commit their company.
  - For contractor or seller.

2-3. PERMITS AND LICENSES

(a) Stipulates those required of the contractor or seller.
  - As building and construction permits.
  - Qualification licenses required. Example: Contractors' licenses as issued by a state or municipality.

2-4. LAWS AND ORDINANCES

(a) Cautions contractor or seller to become familiar with and observe.
  - And save the owner harmless for any infringements thereof.

2-5. INSURANCE

(a) Designates forms and kinds required.
- Required amounts for each class.

2-6. CONTRACT DOCUMENTS

(a) Names documents that become part of the contract. (See Division 1 for listings.)

2-7. CHANGES AND ADDENDA

(a) Describes manner of handling.
- Conditions that impose no obligation on owner.

(b) Extra work.
- Determining costs incurred, methods, and inclusion.
- Who authorizes.

(c) Method of paying for extra work.
- Unit prices, materials, and labor.
- Cost plus a percentage. States percentage to be agreed on.

2-8. LIENS

(a) Discharge of.
- Stipulate release necessary before final payment is made.
- Or a bond provided by contractor large enough to protect owner against all liens.

(b) Gives priority of liens.

2-9. UNION-COMPANY RELATIONSHIP

(a) Bidder to indicate status and name union.
- Closed shop or open shop.
- Expiration date of current agreement.

(b) Bidder to state existence of any dispute that might culminate in a strike.
- Expected settlement of dispute, by arbitration or other.

2-10. BID GUARANTEE

(a) If required, sets percentage of bid.
- Submitted as cash or bond.
- States grounds for forfeiture.

2-11. INSPECTION BY OWNER OR DEPUTY

(a) Available for inspection at all times.
- Contractor to act promptly when unsatisfactory conditions are called to his attention.

2-12. FORFEITURE OF CONTRACT

(a) Reasons therefor, listed.
- Bankruptcy.
- Failure to adhere to specifications.
- Failure to provide competent workers and superintendents.
- Failure to meet completion dates and extensions.
- Other reasons.

2-13. PATENTS AND COPYRIGHTS

(a) Requires owner to be saved harmless for any infringement or alleged infringement.
- Court costs borne wholly by contractor or supplier, as appropriate.

2-14. DISCHARGE OF CONTRACTOR'S EMPLOYEES

(a) Gives reasons for which owner may request discharge.
- Incompetence, drunkenness, theft, etc.

2-15. DISCRIMINATION IN EMPLOYMENT

(a) Prohibited by law.
- Contractor to prove his position is lawful, if requested.

2-16. ACTS OF GOD

(a) Definitions of.

2-17. ORAL AGREEMENTS

(a) Must be put in writing for inclusion with the contract.
- Within time stipulated, hours or days.
- Signed by authorized parties.

2-18. INTENT OF DRAWINGS AND SPECIFICATIONS

(a) What they are to accomplish.
- Requirements of contractor.

2-19. CONFLICT OF DRAWINGS AND SPECIFICATIONS

(a) Notification to owner or his deputy the engineer or architect, by the contractor.
(b) Indicates which requirement overrides.

2-20. AUTHORIZATION OF OVERTIME WITHOUT COST TO OWNER

(a) Lists conditions to which this clause applies.

2-21. GENERAL SAFETY REQUIREMENTS

(a) Steps to be taken by contractor.
(b) By owner, if any.
(c) Lists signals, warning signs, etc. Kinds and where placed.

2-22. DISCREPANCIES OF MATERIALS AND EQUIPMENT

(a) Notification of Owner or his deputy.
- If materials and equipment show on either drawings or specifications but not on both.
- Unless directed otherwise, is contractor to furnish on basis of a single stated requirement?

2-23. ERRORS AND OMISSIONS BY BIDDERS

(a) Corrections thereof.
- At whose expense?

2-24. SIGNS ALLOWED, KIND AND PLACEMENT

(a) Lists.
- Other than safety warning signs.

2-25. CUTTING AND PATCHING

(a) Makes contractor responsible for restoring area after cutting and patching.
- Contractor's required action if cutting and patching introduces a hazard, as:
  - Structural collapse.
  - Atmospheric contamination.
  - Other hazards.

2-26. ARBITRATION

(a) Who may evoke and when.
(b) Selection of arbitrators.
- Selection of impartial chairman or single arbitrator.

(c) Award to be final.
(d) Costs to be divided equally.

2-27. CLEANING UP

(a) During course of work, how often, and how careful.
(b) After completion of work. Describe area degree of cleanliness.
(c) Disposal of scrap and debris, where, when, and how.

2-28. FIRE PROTECTION

(a) Makes contractor responsible for his materials and equipment.
(b) Precautions in prosecuting the work. Describe methods of controlling
- Welding sparks.
- Smoking.
- Extinguishment of salamanders and other spot heaters before leaving the site.

2-29. MANUFACTURER'S DIRECTIONS

(a) States to be followed in assembly and installation.
- Unless instructed otherwise.

2-30. PUBLIC SAFETY

(a) Stipulates interference with traffic to be avoided.
- Streets and waterlines to be kept clear.

(b) Calls for bridging and barricading ditches.
(c) Other directives.

2-31. COMPLETION EXTENSIONS

(a) Gives reasons for which granted.

2-32. TERMINATION OF CONTRACTOR'S RESPONSIBILITY

(a) Status of the contract permitting termination.
(b) Guarantees and warranties, how long in effect.

2-33. TRESPASS

(a) Crossing private property.
- Restrictions against.
- Who gets permissions.
- Who gets easements.

2-34. SCHEDULE OF WAGES

(a) Stipulates legal rate in area not to be violated.

2-35. QUALITY OF MATERIALS

(a) Minimums specified.

2-36. SUBMISSION OF BIDS

(a) Place.
(b) Form of submission.
(c) Numbering of alternates.

2-37. INDEMNITIES

(a) Situations calling for indemnification of owner.

2-38. WORK STOPPAGE BY OWNER

(a) Stipulates reasons for.
- Without charge to owner.

2-39. TEST BORINGS, PILE BEARING DATA

(a) Makes limitation to, or disclaimer of, owner's responsibility for accuracy of available data.

2-40. PROTECTION OF ADJACENT PROPERTY

(a) Describes responsibility of contractor and to what extent.
- To avoid damage and injury.
- And to replace or repair as necessary.

2-41. TRAFFIC MAINTENANCE

(a) Requires public convenience to be maintained.
- Or maximum period of interruption stipulated.
- States maximum frequency of interruption.

DIVISION 3

Special Conditions

Special Conditions encompass those nontechnical clauses of a specification made necessary by particular aspects of the job in hand. For example, facilities available for the storage of tools, equipment, and materials, or the assignment of responsibility for access ladders, ramps, and hoists.

As with general conditions, many special conditions can be carried over with minor changes from job to job, especially if their purpose is to establish a policy for handling certain classes of work. To this extent, the special conditions may become part of preprinted standard specifications.

The term *Supplementary General Conditions,* used by many specification writers, is synonymous in meaning with special conditions. Whichever is used can be a personal choice.

By either name, the clauses in this division of the specifications become part of the contract documents.

3-1. SCOPE OF WORK

(a) Describes required accomplishment by contractor.
- As a complete system.
- Including designated auxiliary items.

3-2. PERFORMANCE BOND

(a) Stipulates amount, if required.
(b) Approved carriers required.

3-3. INVESTIGATION OF SITE

(a) Responsibility for, prior to bidding
- Noting area characteristics.
- Transportation and handling facilities and limitations.

(b) Special conditions indicated for attention that affect handling of materials or prosecution of work.

3-4. PRELIMINARY DRAWINGS

(a) To be submitted with bids, number and kind.
- Manufacturer's prints.

(b) For approval by architect, engineer, or other.

3-5. PHOTOGRAPHS

(a) Number, size, and kind required.
(b) When required.

3-6. PROGRESS REPORTS

(a) Frequency to be made.
(b) Form of.

3-7. APPLICATIONS FOR PAYMENTS

(a) When and how to be made.
(b) Percentage retained.
(c) Final payment.
- Job status, requirements for.
- If any retained at this time, how much and for how long?

(d) Bond in lieu of percentage retained; state if acceptable.

3-8. POLICY FOR AWARD TO SUCCESSFUL BIDDER

(a) States whether right is retained to reject any and all bids.
(b) Or if award to low bidder is mandatory.
(c) If alternates are submitted, states whether right is retained to parcel out the work to more than one bidder in order to get advantage of lowest prices.

3-9. POLICY TOWARD SUBCONTRACTORS

(a) States if subcontractors must be approved by name.
(b) States if prime contractor may use subcontractors other than those whose names are submitted with original bid of the prime contractor.

3-10. TIME OF COMPLETION

(a) If stipulated by date, gives date.
(b) Granting extension of completion.
- Acceptable reasons listed.

(c) Penalties for failure to complete on time.
- Charges per day.

(d) Bonus for early completion.
- Amount per day.

(e) Definition of completion.
- Debris and equipment all removed.
- Other conditions to be met. Give details.

3-11. COORDINATION OF WORK

(a) Schedule to be worked out with other contractors.
- Delays as result of poor scheduling; warning against.

(b) Makes contractor responsible for not holding up other trades.

3-12. COMMENCEMENT OF WORK

(a) Date stipulated by specification.
- Bidder's alternate, if submitted.

3-13. SUBMISSION OF SHOP DRAWINGS

(a) When required.
- Items needing shop drawings; listing of.

(b) Who approves.
- Time limit, if any, for approval; days after receipt.
- Time limit for corrections; days after receipt.

3-14. STORAGE OF TOOLS AND EQUIPMENT

(a) Facilities provided by
- Owner.
- Contractor.

(b) State who is responsible for safety of.

3-15. SUPERVISION OF WORK BY CONTRACTOR

(a) Manufacturer's erection superintendent.
- Supplied by contractor if specified.

(b) Contractor's superintendent.
- Qualifications required of.
- Approval by owner.

(c) Grounds for removal.
- Contractor made responsible for unacceptable work.

3-16. SUPERVISION BY OWNER

(a) Stipulates not to relieve contractor of responsibility.

(b) Informs contractor of unacceptable workmanship or divergence from the specifications.
- Requiring prompt correction.

3-17. ESTABLISHMENT OF GRADES, LINES, AND LEVELS

(a) Designates by whom.

3-18. PROVISION OF COVERS FOR FIELD STORAGE OF MATERIALS

(a) Responsibility of whom.
(b) Gives nature of covering.

3-19. TEMPORARY USE OF STRUCTURES, ENCLOSURES, AND OFFICE SPACE

(a) If permitted by owner, where and what kind.
- Free or for rental? Rental rate stated.

(b) If provided by contractor,
- Requirements for construction and materials.
- Removal of, when and by whom.

3-20. SERVICE COSTS; HEAT, LIGHT, POWER, WATER, TELEPHONE

(a) Provided by owner, contractor, other.
- At whose expense?

3-21. SANITARY FACILITIES

(a) Toilets, by whom provided.
(b) Washrooms, by whom.
(c) Stipulates minimum sanitary requirements.

3-22. PERIODIC DISPOSAL OF DEBRIS

(a) How often and where.
- Whose responsibility?

3-23. MATERIALS AND EQUIPMENT FURNISHED BY OWNER

(a) Without charge.

3-24. COPIES OF DRAWINGS AND SPECIFICATIONS

(a) Stipulates number to be furnished by owner to contractor without charge.
(b) Stipulates number of materials and supplied items furnished by contractor without charge to owner.

3-25. SAMPLES FOR TESTING

(a) Identifies kind and number.
- Provided without charge to owner.

3-26. MODELS

(a) Fixes ownership.

3-27. GUARANTEE

(a) Applicable to materials and workmanship.
- For time stipulated.

(b) Performance guarantee. Penalties for failure to realize.
(c) Stipulates who conducts performance tests.

3-28. WARRANTIES

(a) Expressed.
- To be submitted by contractor in written form or printed form.
- Applicable to materials and design.

DIVISION 4

Power Plants, Steam

General Notes. The term "steam generator" is commonly used to distinguish a complete steam-producing complex from a simple steam boiler.

The specification for a steam generator therefore not only deals with the steel tubes and drums where the heat exchange and evaporation of water to steam take place, but necessarily includes such generation assistants as the fuel-burning equipment, the air preheater, the economizer, the forced and induced draft fans, and the superheater.

All these frequently are handled on a single contract, with the generator builder providing the components and being responsible for their proper installation and satisfactory performance. The guarantee is his alone so far as the purchaser is concerned; if the fans, the firing machinery, or some other vital part fails to do its share, it is up to this one vendor to correct the trouble, despite the fact that the ailing part was manufactured by somebody else.

Steam generators are of both water-tube and fire-tube design, and appropriate ways to specify them are given in subsequent sections. Since each of the components is specified individually by name, the specification writer for a single industrial boiler can simply omit the components of a steam generator complex that are not needed by the job in hand.

The net effect is a mechanical specifications checklist for steam power plants of any size and character.

4-1. STEAM GENERATORS, WATER TUBE, LARGE

(a) ***Scope of Installation, Describe***
- Indoor.
- Outdoor if weatherproofed or canopied.

(b) ***Capacity,*** pounds per hour
- At pressure of ________ psig.
- Boiler water concentration, ppm.
- Steam temperature, final (superheater outlet if superheated) ________ F.

- Feedwater temperature ________ F, to generator or economizer as appropriate.

(c) ***Fuel, Data by Owner***

- Coal, analysis, proximate or ultimate.
- Oil, analysis, proximate or ultimate.
- Natural gas, analysis by volume.
- Other gases, as by-product coke oven, city gas.
- Btu per pound or per cu ft as appropriate.

(d) ***Performance Tables*** submitted by bidder. See Tables 4-1, 4-2.

- Indicate number of points of operation and data required.
  - Point of guaranteed performance, and efficiency.
  - Efficiency at each point.

(e) ***Superheater***

- Describe type wanted.
  - Interdeck, suspended, radiant, convection, etc.
  - Materials of construction.
  - Maximum allowable pressure drop through.

(f) ***Steam Separator***

- Type; cyclone, plate, other.
- Quality of steam at drum outlet nozzle.

(g) ***Construction of Generator***

- Design pressure.
- Maximum steam working pressure.
- Drums, steam.
  - Thickness of sheet and heads.
  - X ray of welds, if required.
  - Certification of conformance with ASME Power Boiler Code.
- Drum, mud, form.
  - Design pressure.
  - Thickness of sheet and heads.
  - X ray of welds, if required.
  - Certification of conformance with ASME Power Boiler Code.
- If drums are formed by extrusion over a mandrel,
  - Maximum and minimum temperatures for working.
- Furnace tubes.
  - Finned or smooth.
    - Gage.
    - Minimum penetration of tubes into headers or drums.
- Convection tubes.

- Minimum penetration into drums.
- Gage.

- Downcomer tubes, inside or outside of casing.
  - Gage.
  - Penetration into drums or headers.

Note: Penetration is not usually specified, but occasional casualties resulting from a tube pull-out indicate that specification attention to this hazard is desirable.

Table 4-1. Performance Data, Filled in by Bidder Large Steam Generators

Job ____________ Make ____________
Fuel ____________ Model ____________

Item	Operating points, 1,000 lb/hr			
	#1	#2	#3*	#4
Excess air leaving boiler	%	%	%	%
Flue gas leaving boiler	Lb/hr	Lb/hr	Lb/hr	Lb/hr
Steam pressure, superheater outlet	Psig	Psig	Psig	Psig
Pressure drop in superheater	Psig	Psig	Psig	Psig
Steam temperature superheater outlet	F	F	F	F
Air through preheater or boiler, as appropriate	Lb/hr	Lb/hr	Lb/hr	Lb/hr
Temperature of air entering preheater or boiler, as appropriate	F	F	F	F
Temperature of flue gas to preheater, economizer, or stack, as appropriate	F	F	F	F
Temperature of air leaving preheater	F	F	F	F
Draft in furnace	In. W.G.	In. W.G.	In. W.G.	In. W.G.
Draft loss, boiler and superheater	In. W.G.	In. W.G.	In. W.G.	In. W.G.
Draft loss in economizer (if any)	In. W.G.	In. W.G.	In. W.G.	In. W.G.
Draft loss in air preheater	In. W.G.	In. W.G.	In. W.G.	In. W.G.
Draft loss in flues and stack	In. W.G.	In. W.G.	In. W.G.	In. W.G.
Friction loss, air ducts, wind box, etc	In. W.G.	In. W.G.	In. W.G.	In. W.G.
Solids in steam leaving boiler	Ppm	Ppm	Ppm	Ppm

* Guarantee point.

Note: The solids in the steam leaving the boiler, as given by the contractor, will assume a concentration in the boiler of ______ ppm at the moment the steam sample is taken.

- Casing, material.
  - Gage.
  - Insulation; kind and placement.
  - Finish.
- Furnace refractories.
  - Kind, firebrick, grade.
    - Minimum spalling temperature.
    - If coal fuel, specify fusing temperature of ash.
    - Required American Standard Pyrometric Cone Equivalent (PCE).
    - Where plastic refractory may be used.

Table 4-2. Heat Balance, Filled in by Bidder

Job ________ Make ________

Fuel ________ Model ________

Balance item	Operating points, 1,000 lb/hr			
	#1	#2	#3*	#4
Heat loss in dry gas	%	%	%	%
Heat loss, moisture in fuel	%	%	%	%
Heat loss, moisture in air	%	%	%	%
Heat loss by radiation from hot surfaces	%	%	%	%
Heat loss, unaccounted for	1.5%	1.5%	1.5%	1.5%
Total losses	%	%	%	%
Overall efficiency, boiler + superheater + air preheater + economizer, 100%, − total losses	%	%	%	%

* Guarantee point.

- Air tightness.
  - Furnace must be sufficiently tight to permit maintenance of the specified overfire draft, measured at points designated by the contractor.
    - And accepted by owner.

(h) *Tolerances Permitted*

- Steam outlet temperature, plus or minus ________ F.
- Superheater pressure drop, plus ________ per cent.
- Draft loss, plus ________ per cent.
- Efficiency, none at guarantee point.

(i) *Dimensions*

- Tubes, headers, drums, as shown on drawings.

- Boiler heating surface may be specified if desired.
  - Ask bidder to furnish figures.

(j) *Appurtenances, Drum Nozzles*

- Safety, number, size, opening and closing pressure.
- Feedwater nozzles, number.
- Blowoff nozzles.
  - Indicate if continuous blowdown is required.
- Water column connections.
- Vent nozzle.
- Auxiliary steam outlet.
- Feedwater treatment nozzle.

Note: Number and size of each category may be indicated in specification, or on drawing, or both.

(k) *Appurtenances, Water Column*

- Indicate type. As made by ______________ or approved alternate.
- With low-water alarm.
- And automatic feed.

(l) *Appurtenances, Soot Blowers (Boiler, Economizer, Preheater)*

- Kind. As made by ______________ or approved alternate.
- Steam, pounds of steam consumed.
- Air, pressure and volume required.
- Method of operation stipulated.
- Specifies automatic operation of the soot blowers.
  - Start when conditions indicate cleaning needed.
  - Length of blow controlled.

(m) *Appurtenances, Valves*

- Safety, make and model, or approved alternate.
- Size and number, as per ASME Power Boiler Code, for maximum actual rate of evaporation, pounds per hour.
- For blowoff, drains, vents, water column.
  - Blowoff valves arranged in tandem.
    - As made by ______________ or approved alternate
  - Makes and models of other valves ______________ or approved alternates.

(n) *Appurtenances, Gages and Thermometers*

- Main pressure gage, make, model ______________ or approved alternate.
  - Calibration of.

- Draft gage, make and model ______________ or approved alternate.
  - Calibration of.
- Thermometers, make, models ______________ or approved alternates.
  - Calibration of.

(o) ***Platforms, Ladders, Walkways***

- As shown on drawings.

(p) ***Supporting Steel***

- As shown on drawings.
- Designed to withstand a wind pressure (outdoor units) of _______ psf, or a wind velocity of _______ mph.

(q) ***Economizer***

- Type, as continuous tube, headers, etc.
- Minimum heating surface required, sq ft.
- Design working pressure, psig.
- Materials of construction.
- Wall thickness of tubes.
- Valves.
  - Safety, make and model. Capacity as required by ASME Power Boiler Code at full-load operation.
  - Blowoff valves. Number, make, and model.
  - Stop check valves, number, model, and make.
- Casing, gage of steel plate used.
- Inspection doors, number.
  - All as shown on drawings.

(r) ***Air Preheater***

- Type, plate, tubular, rotating exchange element.
  - Make and model.
  - Heat exchange surface, sq ft.
- Preheater supported as shown on drawings.
- Casing construction, plate gage.
  - Insulating materials, kind and thickness.
- Air ducts, preheater to boiler.
  - Construction as shown on drawings.
- Insulation required. Type and thickness.

(s) ***Firing Equipment, Stokers, Spreader, Underfeed, Water-cooled, etc.***

- Kind and model. As manufactured by ______________ or approved alternate.
  - Capacity, pounds of steam an hour under specified conditions.

- Inspection and access doors, number and size.
  - With safety latches.
- Smokelessness. Required performance, Ringelmann Chart.
- Smoke scanner, make and model. Mounted on flue as shown on drawing.
  - With audible alarm to warn of smoke.
- Motor drive. Electrical characteristics in accord with NEMA. Specify voltage available, phases and cycles.

(t) *Firing Equipment, Pulverized Coal*

- For details on the pulverizing equipment, see Sec. 4-7(h), Pulverizer Mills.
- Driving motor to meet NEMA standards.
  - Bidder to provide data on horsepower.
  - Temperature rise over a given ambient, stated by specification.
- Burners for pulverized coal.
  - Specify type, as horizontal, vertical.
  - Placement, for tangential firing, or cyclone furnace.
  - Specify capacity of each burner, _______ pounds per hour.
  - Specify the temperature, _______ F, and pressure, _______ in. W.G., of the primary air supplied to the burners.
  - Specify temperature and pressure of secondary air.
  - Give temperature of ambient air, _______ F maximum, _______ F minimum.

(u) *Firing Equipment, Oil Burners*

- Type, number, make, and model. Or approved alternate.
- Maximum capacity of each burner, gph.
- Temperature oil to burners, _______ F.
  - Kind of oil preheater, if any.
- Flameout safety device.
  - Kind and model.
  - Method of ignition.
    - Voltage for electrical spark.
- Inspection ports in furnace.
  - Number, size, locations.
- Burners to operate with zero carbon monoxide, at low ratings, guarantee point and below.
- Oil pressure as delivered to burners, _______ psig.
- Smokelessness and smoke scanner same as for stokers.

(v) *Firing Equipment, Gas Burners*

- Make and model ______________ or approved alternate.
- Gas pressure, as delivered to burners, _______ psig.
- To operate with zero CO up to and through the guaranteed output in pounds of steam per hour.
- Smokelessness, equal to or better than No. _______ Ringelmann.

(w) *Induced Draft Fan*

- Kind of fan. As made by ______________ or approved equal.
  - Housing of _______ gage plate steel.
  - Number of inlets.
  - Driven by _______ hp motor through a flexible coupling. Coupling by ______________ manufacturer.
    - Motor to NEMA standards.
- Bidder to provide operating data requested by Table 4-3.

Table 4-3. Induced Draft Fan Performance, Filled in by Bidder

Job __________ Make __________
Number of fans __________ Model __________

Performance item	Operating points, 1,000 lb/hr			
	#1	#2	#3*	#4
Gas handled per hour..........	Lb	Lb	Lb	Lb
Gas temperature..............	F	F	F	F
Fan suction..................	In. W.G.	In. W.G.	In. W.G.	In. W.G.
Cubic feet of gas per minute...	Cfm	Cfm	Cfm	Cfm
Fan speed....................	Rpm	Rpm	Rpm	Rpm
Fan horsepower...............	Hp	Hp	Hp	Hp

* Boiler output as guarantee point.

- Starting and control equipment.
  - Specify if furnished by owner or contractor.
  - Specify temperature rise above a _______ F ambient for the motor.
  - Specify bearings and how cooled.

(x) ***Forced Draft Fan***

- Make, model, or approved alternate.
  - Housing ________ gage steel plate.
  - Number of inlets.
  - Inlets to be screened.
- Driven by ________ hp motor, through a flexible coupling.
  - Motor to meet NEMA standards.
- Bidder to provide operating data requested by Table 4-4.

Table 4-4. Forced Draft Fan Performance, Filled in by Bidder

Job ____________ Make ____________
Number of fans ____________ Model ____________

Performance item	Operating points, 1,000 lb/hr			
	#1	#2	#3*	#4
Air handled per hour..........	Lb	Lb	Lb	Lb
Air temperature...............	F	F	F	F
Static pressure................	In. W.G.	In. W.G.	In. W.G.	In. W.G.
Air volume....................	Cfm	Cfm	Cfm	Cfm
Fan speed.....................	Rpm	Rpm	Rpm	Rpm
Fan horsepower...............	Hp	Hp	Hp	Hp

* Boiler output as guarantee point.

- Starting and control equipment.
  - Specify if furnished by contractor or owner.
- Specify temperature rise of motor above the given ambient.
- Specify kind of bearings and how cooled.

(y) ***Forced Circulation Boiler Circulating Pumps***

- Kind of pump, centrifugal, single stage, multistage, etc.
  - Temperature of water pumped, ________ F.
  - Materials in pump construction.
  - Capacity, gpm at ________ ft head (or psi head).
  - Motor data, NEMA standard. Horsepower and efficiency.
  - Pump circuit as shown on drawings.
- If the steam generator design is such that the manufacturer furnishes the pump, interconnecting piping, and

all appurtenances, the specifier may simplify his own work by stating under this heading that the supplier shall be wholly responsible for the design and performance of the forced circulation system, and shall submit data on the kind of pump—as centrifugal, multistage, single stage—and on the materials of construction, capacity, ________ gpm, horsepower, efficiency, etc.

(z) ***Supercharged Steam Generators***

Note: Supercharging, which means maintaining combustion under a positive pressure in the boiler furnace, is not yet a common practice. The checklist given here is therefore subject to possible changes with the improvements certain to come. This should be kept in mind by the specification writer, and the latest information should be assembled before specification is begun. See Fig. 4-1(a).

- The gas turbine has made supercharging attractive in some instances.
- A marine form of supercharged steam generator is shown in Fig. 4-1(b).
- Specify pressure in furnace at different rates of firing, in. W.G., or psig.
- Specify type of air compressor desired.
  - Volume delivered to furnace, pounds per hour.
  - Temperature, F.
  - Pressure of air leaving compressor, psig.
- GAS-TURBINE DATA
  - Horsepower generated.
  - Pounds of gas per hour.
  - Temperature of boiler flue gas entering gas turbine.
  - Temperature gas-turbine exhaust to stack.
- STEAM-TURBINE DATA
  - Kind of turbine.
    - Stages.
    - Steam at throttle, psig.
    - Steam temperature F.
    - Steam exhaust temperature, F.
    - Steam exhaust pressure, psig.
- Specify the operating control sequence of boiler, gas turbine, and steam turbine.
  - Muffler for gas turbine, intake and exhaust.
- General arrangement of system as shown on the drawings.

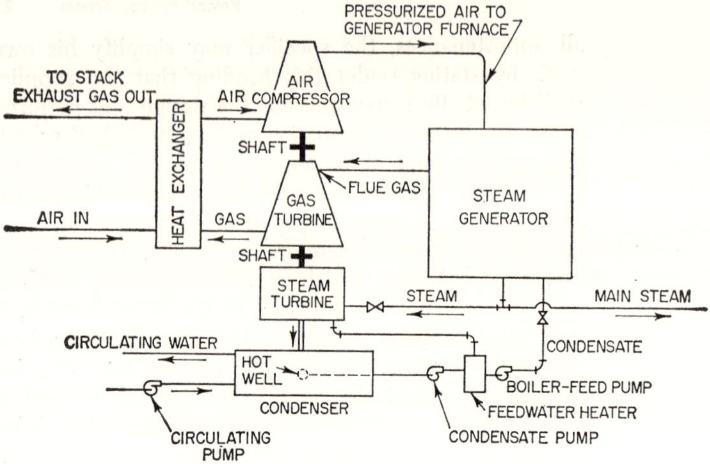

(a)

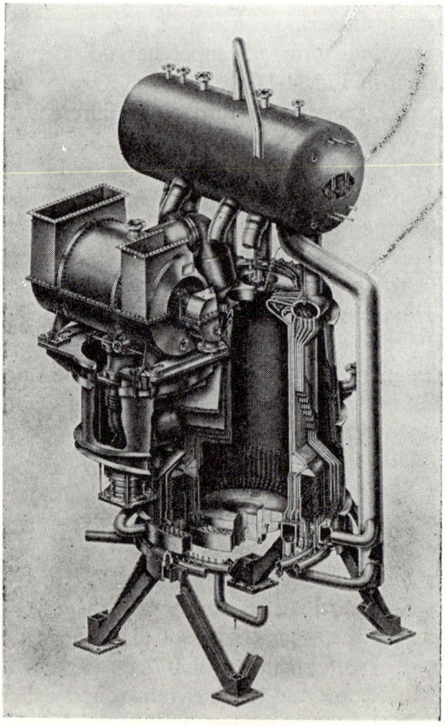

(b)

FIG. 4-1. (*a*) Supercharged flow diagram. (*b*) Supercharged marine boiler. (*Courtesy Foster-Wheeler Corporation.*)

- Specify all the other requirements for the steam generator that are not a part of the supercharging system. Special attention should be given to the duct system, since the air pressure to the boiler will be substantially higher than is the case with an air preheater.
- Require operating data at no less than 2 points in addition to the guarantee point.
- See Fig. 4-1(a) for an illustrative flow diagram. This is one of several possible arrangements, not necessarily the best for the project being specified. The designer should explore other suitable arrangements of the various components before deciding firmly on the specification.

(aa) *Starting Up*

- Starting-up procedures to be as directed by an engineer provided by the contractor, or steam generator builder, experienced in starting up equipment of this character.
  - After careful inspection inside and outside of generator for the discovery and removal of foreign materials and objects. Particular attention is directed to superheater header plugs, and the replacement of any temporary plugs by design material.
  - After boiling out, all water column glasses are to be replaced by new clean glasses.

(bb) ***Auxiliary Fuels***

- Specify if oil or gas auxiliary burners are to be provided on coal-fired installations.
- Specify if dual-fuel burners shall be provided for generators and indicate which fuel is to be considered as basic.

(cc) ***Desuperheaters***

- Specify location with regard to steam generator, as between stages of the superheater.
- Specify purpose of desuperheater, as to provide superheat final temperature control over the indicated range of generator operation, either with or without the assistance of superheater gas flow dampers. If desuperheated steam is required for process or other purpose, state this.
- Specify kind of desuperheater, as direct-contact water spray injection unit; or a coil submerged in the boiler steam drum; or the shell and tube external type.
- Specify the desuperheater temperature range, as

_______ F inlet and _______ F outlet at the operating pressure of _______ psig. If spray type, state temperature, _______ F, of the spray water and give its source, as boiler feed.

- Specify construction of desuperheater, materials of shell, nozzles, headers, tubes, tube sheets (for shell and tube), etc.

Table 4-5. Desuperheater Performance Data Filled in by Bidders

Job __________ Manufactured by __________

Kind of desuperheater __________ Bidder __________

Item	Boiler output, lb/hr or % rating		
	*	*	*
Desuperheater output	Lb/hr	Lb/hr	Lb/hr
Total temperature, steam in	F	F	F
Total temperature, steam out	F	F	F
Superheat, if any, remaining	%	%	%
Desuperheating water	Lb/hr	Lb/hr	Lb/hr
Desuperheating water, temperature in	F	F	F
Desuperheating water, temperature out	F	F	F
Steam pressure in (superheated)	Psig	Psig	Psig
Steam pressure out (saturated)	Psig	Psig	Psig
Pressure loss in desuperheater	Psig	Psig	Psig
Water pressure loss (noncontact)	Psig	Psig	Psig

* These points to be stated in the specification to bidders.

- Specify capacity, pounds of desuperheated steam per hour, stipulating conditions under which this capacity is to be attained, as temperatures of steam and water. In noncontact desuperheaters, stipulate desired temperature rise in the desuperheating water supplied to the unit.
- Require bidders to furnish drawings showing the construction and dimensions of the proposed desuperheater.
- Require bidders to submit complete performance data as called for in Table 4-5. On this form, include any other items of information that may seem desirable.
- Require a description of the control system being offered.

(dd) *Reheaters*

Note: Present favored practice is to return a volume of steam taken from some selected stage of the steam turbine to a coil, usually between the stages of the superheater. In construction, this coil is comparable to the superheater. Its purpose is to add heat energy to the steam without the expenditure of additional heat of evaporation. Separately fired reheaters may be encountered, or for some particular reason, desired. This is a design problem. The separately fired reheater usually has an anatomy similar to that of a separately fired superheater. For the steam-turbine connection, see Sec. 4-15(a), which refers to a boiler coil reheater. To develop a separately fired reheater or superheater specification, refer to the specification guide for a steam generator or boiler, and build up the specially desired specification accordingly. Separately fired installations generally are less economical, hence less used.

- Specify kind and location of reheater. Furnish bidders with drawing or sketch showing relative positions of reheater and turbine.
- Specify steam condition entering reheater, giving temperature, F, and pressure, psig.
- Specify condition of steam leaving reheater, giving desired temperature, F, and desired pressure, psig. This is to be the guarantee point.
- Specify conditions of the flue gas passing over the reheater, pounds per hour, temperature, F, and give analysis of the gas with particular reference to corrosive agents such as sulfur and water vapor.
- Specify maximum acceptable pressure drop of steam passing through the reheater.
- Specify the required range of reheating, from a minimum of ________ F to a maximum of ________ F.
- Require bidders to furnish sketches and drawings as necessary to make clear the reheater design, weights, and dimensions.
- Require bidders to describe temperature control system they plan to supply with their installation.
- Specify the electric service available in volts, phases, and cycles that may be used to power the controls.
- Require bidders to submit performance figures at three

magnitudes of steam volumes per hour passing through the reheater. These three points are to be stipulated in the specification.

4-2. STEAM GENERATORS, WATER TUBE, PACKAGED

(a) Specify items included in the generator package.
- Scope of installation for this contractor.

(b) *Capacity, Pounds per Hour*
- At pressure of ________ psig.
- With boiler water concentration of ________ ppm.
- Solids in steam not to exceed ________ ppm.
- Steam temperature, final (at superheater outlet if superheated), ________ F.
- Feedwater temperature ________ F.

(c) *Fuel Analysis, Specified*
- Oil.
- Gas.

(d) *Performance Data Submitted by Bidder*
- See data form, Table 4-6.

Table 4-6. Performance Data, Filled in by Bidder Package Steam Generators

Job ____________ Make ____________

Fuel ____________ Model ____________

Item of performance	Operating points, 1,000 lb/hr		
	#1	#2	#3*
Flue gas	Lb/hr	Lb/hr	Lb/hr
Excess air	%	%	%
Steam pressure, boiler or superheater outlet as appropriate	Psig	Psig	Psig
Steam temperature, final	F	F	F
Temperature, air to furnace	F	F	F
Temperature, flue gas	F	F	F
CO_2 in flue gas	%	%	%
Fuel to burners, oil	Gph	Gph	Gph
Fuel to burners, gas	Cfm	Cfm	Cfm
Pressure, gas to burners	Psig	Psig	Psig
Furnace heat release	Btu/cu ft	Btu/cu ft	Btu/cu ft
Overall efficiency	%	%	%

* Guarantee point.

(e) *Steam Separation*
- Plate scrubber in steam drum.
- Cyclone-type separators.
- Other forms of scrubber.

(f) *Design Pressure, psig*
- Maximum working pressure, psig.

(g) *Construction*
- Boiler tubes, material, welded or seamless.
- Drums, welded, X rayed, ASME Boiler Code.
- Supporting steel as shown on drawings.
- Steel casing, _______ gage.
  - How made gastight.
  - Refractory, kind, for _______ F maximum temperature.
- Insulation, kind, where, thickness.

(h) *Appurtenances, Valves*
- See Sec. 4-1(m).

(i) *Appurtenances, Water Column*
- See Sec. 4-1(k).

(j) *Appurtenances, Gages and Thermometers*
- See Sec. 4-1(n).

(k) *Appurtenances, Scot Blowers*
- See Sec. 4-1(l).

(l) *Firing Equipment, Oil Burners*
- See Sec. 4-1(u).

(m) *Firing Equipment, Gas Burners*
- See Sec. 4-1(v).

(n) *Forced or Induced Draft Fan*

Note: Package generators use either of these fans, but not both.
- Fan as made by _____________ or approved alternate.
  - Design capacity of fan, pounds per hour of flue gas or air.
    - At a temperature of _______ F.
    - With fan operating at _______ rpm.
    - Requiring _______ fan hp.
- Driven by electric motor through belt (coupling) as appropriate.
  - Indicate locked rotor current of motor.
  - Motor temperature rise, _______ F over the specified ambient.
- Motor to conform to NEMA standards.
- Require bidders to furnish details as indicated in Tables 4-3 and 4-4.

- Specify maximum volumetric capacity of fan, cfm, at _______ F.
- Specify suction at inlet of induced draft fan, _______ in. w.g.
- Specify discharge pressure, induced or forced draft fan, _______ in. w.g.

(o) *Dual-fuel Changeover*

- Specify method.

(p) *Shipping Data*

- Contractor responsible for equipment safe delivery to site.
- And selection of carrier and route, unless otherwise directed.

(q) *Starting up*

- See Sec. 4-1(aa).

(r) *Instrument and Controls Panel*

- Specify if desired.
- Listing items to be mounted on panel.
- Describing material and design of panel.

(s) *Electric Wiring*

- All materials and methods in conformance with National Electrical Code.
- Must be installed by licensed electricians.

(t) *Guarantees and Warranties*

- Required.

(u) *Automatic Controls, Complete or Partial*

- Designate desired control sequence of cause and effect.
- Describe system wanted, as electric, electronic, or pneumatic.

(v) *Placement at Site*

- Specification to note if unit must be handled by skids, cranes, or other lifting hooks.
- Indicate any access limitations present.

4-3. STEAM GENERATORS, FIRE TUBE, PACKAGED

(a) Specify scope of installation for which this contractor is responsible.

(b) Indicate general form and type, as a unit manufactured by _______________ or approved alternate.

(c) *Capacity, Pounds of Steam per Hour*

- At a pressure of _______ psig.
- With feedwater at _______ F.

(d) *Fuel Analysis as Specified*

- Oil, number _______.

- Calorific value of _______ Btu per pound.
- Preheated to _______ F.

- Gas, kind.
  - Calorific value, _______ Btu per cu ft.
  - Pressure delivered to burner _______ psig.
  - Pressure regulator furnished by _____________.

(e) *Steam Separator*

- Kind, dry pipe, drum, scrubber, or other.
- Quality of steam, per cent, at guaranteed rating.

(f) *Superheater*

- If required, specify type.
  - Final temperature of steam _______ F.
  - Pressure drop through superheater _______ psig.

(g) *Generator Construction*

- Number of passes _______; tubes, number _______; gage _______.
- Heating surface _______ sq ft per rated hp. (Designate minimum acceptable.)
- Boiler shell to satisfy ASME Boiler and Pressure Vessel Code, for stated pressure.
- Boiler shells for pressures below 15 psig steam or 30 psig water, to satisfy ASME Low Pressure Boiler Code.
- Inspection certificate by ASME inspector or National Board of Boiler and Pressure Vessel Inspectors to be furnished to owner.
- Specify number of manholes and lifting eyes.

(h) *Burners, Oil*

- See Sec. 4-1(u).
- Oil atomization.
  - To be steam or pressure.
    - State oil pressure, _______ psig.

(i) *Burners, Gas*

- See Sec. 4-1(v).

(j) *Dual Fuel Changeover*

- Specify method required.

(k) *Induced or Forced Draft Fan*

- For details applicable to the forced or induced draft fan, refer to Sec. 4-2(n).

(l) *Burner Ignition*

- Specify kind required.
- Specify kind of flameout protection desired.

(m) *Appurtenances, Valves, Water Column, Gages, Thermometers*

- See Sec 4-1(m), (k), and (n).

- Injector, make and type.
  - Capacity, _______ gpm.
  - With steam pressure of _______ psig.

(n) *Controls, Automatic*

- Specify control sequences from cause to effect.
  - Including steam pressure controls and stack switch.

(o) *Control Panel*

- Specify where to be mounted.
- List items to appear on panel, instruments, relays, switches, etc.

(p) *Electric Wiring*

- All wiring and equipment to conform to requirements of National Electrical Code.
- Installed by licensed electricians.

(q) *Shipment to Site*

- Contractor responsible for safety of unit.
- Selects carrier and route unless otherwise directed.

(r) *Guarantees and Warranties*

- Describe what is required.

(s) *Boiler Jacket and Insulation*

- Jacket sheet metal, _______ gage.
  - Finish painting by contractor.
- Insulation, kind and thickness.

(t) *Handling at Site*

- By crane or other lift hook.
- By skids.
- Access limitations, if any.

(u) *Start-up*

- Provided by field engineer of manufacturer.
  - With instructions in operation to owner's employees.

(v) *Drawings*

- Owner to be provided with complete dimensional drawings, also cutaway views, intended to make clear the design, construction, and operation of the generator.

(w) *Performance Data*

- Bidders may be asked to provide data indicated in Tables 4-1 and 4-2 as applicable.

4-4. STEAM GENERATORS, ONCE-THROUGH, SUPERCRITICAL

(a) *Discussion, Once-through Design*

The once-through generator may be somewhat less costly to build than other designs, since it eliminates both steam and mud drums. Without these drums, less space is required, another item of advantage. Since space saving is particularly desirable for ships, the once-through generator at subcritical or conventional pressures has been installed on shipboard, but its operation has had only limited success.

Sources of trouble are the return bends in the boiler tubes, where impurities precipitated out of the feedwater tend to adhere to the metal and may eventually cause overheating and burnouts. Another form of trouble is in the combustion controls. There is no appreciable water storage in this design, the boiler feedwater first entering a water-heating area of the boiler, then passing into the evaporating section, and finally to the superheating tubes in succession, so that all water is vaporized in the course of the journey.

The absence of reserve water requires almost instantaneous interruption of the firing cycle to prevent a disastrous boiler burnout, should the water supply fall below a specified minimum. This minimum is so close to the steam production that extremely sensitive controls are needed. In normal performance, the steam output in pounds per hour is exactly equal to the feedwater supply, pounds per hour. Any reduction of supply water is reflected in an immediate rise of steam temperature and degrees of superheat. This temperature change may be used to actuate the combustion controls.

(b) ***Discussion, Supercritical Design***

A critical temperature, as it applies to a gas, is that temperature at which the gas refuses to liquefy under compression alone. For steam, this is 705 F.

A critical pressure is that pressure, psi, at which a liquid vaporizes without the help of latent heat, and remains identical in volume with its liquid form. For water, the critical pressure is 3,206 psi, at the critical temperature of 705 F. The specific volume of the water under these conditions is 0.050 cu ft per pound, and the specific volume of the vapor is likewise 0.050 cu ft per pound.

Steam generators designed to operate at supercritical pressures, meaning above 3,206 psi, in combination with turbine-generators, water-treatment systems, pumps, controls, and other components also designed for these severe pressure and temperature conditions, produce power at comparatively high efficiencies. For this reason, supercritical plants are considered most seriously by public utility companies, since their profit depends on obtaining the best possible efficiencies.

With a supercritical plant, the expansion of the steam to a lower pressure produces a substantial drop in temperature, with the possibility of creating wet steam. To counteract this, reheat is essential after the first flow through a steam turbine's nozzles.

At present the initial range of temperatures for a supercritical design is from 1100 F to 1200 F with 1050 F on reheat. These temperatures are high enough to introduce problems in metallurgy, since many of the familiar alloys at such temperatures are unable to meet the pressure requirements.

The specification writer undertaking to prepare a specification for a supercritical steam generator, condenser, pumps, turbines, controls, and other components, will require assistance from designing specialists, such as metallurgical engineers, chemists and chemical engineers, and controls experts. To the extent available, he will also get advice from engineers versed in operating practices at high temperatures and high pressures. Nevertheless, the specifying engineer is referred to Sec. 4-1 with all those subheadings which may be appropriate to the supercritical project, such as the fuel-handling and ash-handling systems, to make sure that all components are being considered. Once identified as a part of the installation, the special features needed to serve the supercritical function can be supplied by the experts. Pulling all these together into a coherent specification is the responsibility of the specification writer.

(c) ***The Manufacturer's Function***

Close collaboration is necessary between the designing engineers, the specifying engineer, and the manufacturer's technical representatives, prior to the formulation of a suitable specification.

This requirement recognizes two facts: First, the once-through units, as presently available, are well covered by patents and are proprietary designs. To get what is wanted involves naming somebody's design in the specification and either promoting or rejecting certain specific features, which may or may not eliminate competition.

Second, prototypes of supercritical generators are presently in short supply. Each represents a design developed to meet one or several conditions of a particular project. Consequently, a supercritical plant is shaped by more original thinking and creative exploration than goes into the conventional subcritical designs. The important aspect here is, can the manufacturer build it? And does he want to? A specification that has no bidders is worthless. Preliminary discussions with the manufacturers' sales engineers therefore are essential. When the specification writer knows he can get what he calls for, he may proceed to prepare his specification accordingly. He can be sure it will be a highly specialized specification, and not founded on generalities.

All the guides given appropriate subheadings in Sec. 4-1 will require considerable augmentation in order to encompass the special features of these units.

4-5. STEAM GENERATORS, ELECTRIC

(a) *Scope of Installation,* items by contractor and items by owner.

- Indicate indoor or outdoor erection.

(b) *Capacity,* ________ pounds per hour.

- At ________ psig.
- Boiler water concentration, ________ ppm.
  - With an ohmic resistance of not less than ________ ohms per cu in. at 212 F.
- Total steam temperature leaving boiler, ________ F.
- Feedwater entering boiler at ________ F.
- Steam quality, ________ per cent dry.

(c) *Electric Rating of Generator* ________ kw

- Voltage.

(d) *General Construction*

- Thickness of shell, ________ in.
- Thickness of heads, ________ in.
- Number of electrodes, ________.
  - Material of electrodes.
- Size of bleed connection for controlling concentration of feedwater in boiler by means of continuous bleed, if continuous bleed is necessary to counteract high percentage of makeup.
- Size of blowdown connection.
- Thickness and construction of steam drum.
  - Shell.

- Heads.
- Indicate boiler and drums to satisfy ASME Boiler Code.

(e) *Controls*

- Describe; electric, pneumatic, electronic.

(f) *Transformers*

- Number _______.
- Type ________________.
- Provided by _______________.
- Electrical characteristics of _______________.
- NEMA Code.

(g) *Circuit Breakers*

- Number _______.
- Type ________________.
- Provided by _______________.
- Electrical characteristics of _______________.
- NEMA Code.

(h) *Wiring*

- National Electrical Code.
- Furnished and installed by _______________.

(i) *Performance Data*

- Contractor to provide curves, graphs, and tables as necessary to show relative economies of operation resulting from different combinations of boiler water concentration and water temperature.
- Indicating time lapse, _______ minutes, between energizing electrodes and production of steam at the nozzle at required pressure and temperature.

(j) *Boiler and Drum Insulation*

- Kind.
- Thickness, in. _______.
- Jacket, material and thickness.
  - How jointed.

(k) *Appurtenances*

- Refer to Sec. 4-1, all subheadings but soot blowers, which are not required.

(l) *Platforms,* ladders, walkways, supporting steel

- Refer to Sec. 4-1(o) and (p).

(m) *Economizer or Heat Exchanger*

- Refer to Sec. 4-1(q).

(n) *Diagrams*

- Contractor to provide wiring diagrams for information and guidance of owner after acceptance.

(o) *Start-up*
- Personnel to be furnished by contractor.
  - Including responsible superintendent.
- Personnel to be furnished by owner.
  - For observation and instruction.

(p) *Clearance*
- Contractor to clear installation and safety features with local authorities and codes.

4-6. BOILER BLOWDOWN SYSTEMS

(a) Describe system in general, such as continuous or intermittent, whether manual or automatic. Also state if blowdown is restricted to mud drum and possibly bottoms of waterwall headers, or includes a surface blow from the steam and water drums.
- State the required blowdown in pounds per hour from each blowdown valve.
- Giving pressure and temperature in the boiler.
- Describe type of blowdown valves to be furnished, as seatless, manner of opening and closing, etc.
- Specify separator or blowdown tank, and in either case, give size of connections and capacity, gph.
  - Require separator or tank to be built in accordance with the ASME Code for Unfired Pressure Vessels.
- Specify if heat exchanger (aftercooler) is to be provided.
- For addition of heat exchanger (aftercooler) and for sizing the separator, provide bidders with the size and number of boiler blowdown valves in the line to the separator. Also ft of drain piping horizontal and ditto, vertical; the pressure and temperature of the cold water supply; and the desired temperature of the drain discharge.
- Specify kind and type of temperature regulator.
- Require strainer and check valve on the cold water supply, as indicated on the blowdown system arrangement drawing.
- Bidders to furnish data on dimensions and weight of equipment furnished.
- Bidders to state size and height of vent pipe.
- For automatic control, require bidders to submit their recommendations. Alternatively, specify the control

methods and results which the bidders are expected to satisfy.

- Stipulate where the work of this contractor begins and ends.
- Specify all piping and valves in accordance with the pressure on the boiler, using appropriate schedule numbers.

4-7. COAL HANDLING AND STORAGE

(a) *Outdoor Storage*

- Describe scope of project, including enumeration of equipment to be furnished by contractor.

(b) Designate handling capacity required.

- Maximum per hour, tons.
- 24-hour rate, tons.

(c) *Drag Scraper*

- Refer to drawings for details of arrangement.
- Designate minimum cable requirements.
  - As plow steel with hemp center.
  - Bidder to provide diameter, make, and number of strands.
  - Design to meet all safety codes, including local and state codes.
  - Bidder to provide full data on scraper, as materials of construction, lubrication, etc.
  - Electric motors to meet NEMA requirements.
    - Specification to describe electric service available, as voltage, wiring, etc.
    - All wiring by contractor to satisfy National Electrical Code.
    - Acceptance test required.
    - Contractor to instruct owner's operators.
  - Controls; describe functions and kinds.

(d) ***Bulldozer*** (sometimes used instead of drag scraper)

- Indicate speed required.
- Coal to be moved, tons per hour.
  - Average distance traveled.
- Type of engine, diesel, gasoline.
- Contractor to furnish complete data on design, materials, including grade of steel, bearings, lubrication systems, engine and caterpillar sprockets, and treads.
- Contractor to identify safety features.
- Guarantees and warranties.

(e) *Coal-receiving Systems, Bucket Hoists*

- Designate scope of work for this contractor.
- Number of hoists.
- Capacity each, tons per hour.
- Specify size of buckets, tons (2,000 pounds).
- Refer to drawings for arrangement.
  - Giving height of lift, ft.
  - If unloading from barge, give extreme low-water elevation.
    - And elevation of bottom of barge at extreme low water.
    - Also general dimensions and capacity (in tons) of barges used to deliver the coal.
- Kind of coal to be handled, as run of mine, etc., on which capacity is to be based.
- Bearings requirements, as bronze bushed, roller bearing, etc.
  - Lubrication system.
- Cable (same as for drag scraper).
- Cable drum, main.
  - Specify material.
  - Specify remaining wraps of cable on drum when bucket is at lowest position.
- Speed of bucket hoist, fpm.
- Contractor to state motor characteristics.
  - Hp, open, closed, temperature rise.
  - Also full-load amps and amps with locked rotor.
  - Make and model.
- Brakes to be of a design and materials approved by the owner.
  - Must arrest the fall of the bucket in case of a failure of electric current.
- Bucket trolley arrangement to prevent derailment in event of overtravel of bucket.
  - Or fall, if axle breaks.
  - Bearings and lubrication.
- Data by contractor to include:
  - Drawings, detailed and general, as may be specified.
  - Full descriptions of all components, as bypass chutes, crushers, hoppers, picking tables, magnetic separators, control system, screens, gates, wearing plates, etc.
  - Owner's specification to include painting requirements.

- Kind of paint, number of coats, where and when applied.
- Belt conveyors, number, widths, lengths.
  - Specification to indicate friction standards by grade, minimum ply adhesions, pounds; minimum cover adhesions, pounds.
  - Specification to indicate belt cover standard, cover grade; type of service, abrasive, little abrasive; minimum tensile strength, psi, and minimum elongation, psi.
  - Belts to satisfy Code of the Rubber Manufacturers Association, all respects.
  - Specify belt idlers and sprockets, materials, lubrication, etc.
    - Or require contractor to submit such data for approval.
- Specify who provides steel for supports, towers, guards.
- Refer to owner's drawings if provided.

(f) *Coal-receiving Systems, Track Hoppers*

- Refer to drawings for layout and details.
- Specify if coal to enter hopper from gates or from car dumper.
  - Specify belts, sprockets, idlers, magnetic separators, picking tables, as previously checklisted for bucket hoists.
  - Specify car pullers, capstans, if required.
    - Noting speed, capacity.
    - Stipulating electrical service available.
      - Contractor to provide motor characteristics if not stated in specification as a requirement.
  - Coal-thawing equipment.
    - Infrared heaters.
      - Specify number, output in wattage or Btu.
    - Steam lances.
      - Specify steam pressure and quality, number.
- Car shakers or vibrators.
  - Specify method of operation.
  - Control methods.
- Car dumpers.
  - Contractor to provide data on motor horsepower, control of unit.
    - Time for a complete cycle.
    - Painting.

- Details of construction and dimensions.
- Electrical data to satisfy National Electrical Code and NEMA.
  - Also all local codes.

(g) ***Downspouts to Stoker Hoppers and Pulverizer Mills***

- Materials of spouts.
- Corrosion and erosion protection, glass liners, other means.
- Weighing devices on downfeed system.
- Flow-control gates.
- Weigh larries.
  - Kind, capacity, actuating methods.
    - Specify motor characteristics required.
- Nonsegregating devices, kind required.

(h) ***Pulverizer Mills***

- Specify make and type, or if wide-open specification, just type wanted.
- Specify capacity, tons per hour.
- Specify kind and sizes of coal entering mill.
  - Grindability, Hardgrove Scale.
- Specify fineness required, with percentage able to pass through a _______ mesh screen.
- Specify temperature of air to mill, cfm, and relative humidity.
- Indicate if driers are required.
- Specify transmittal system from mill to either burner or bin storage.
  - Describing vehicle carrying the pulverized coal, as pneumatic, belt, etc.
- Specify motor characteristics available and motor performance required.
- Specify starting, stopping, and other aspects of control system.
- Refer to owner's drawings as appropriate.
- Require compliance with appropriate codes.

(i) ***Silo Storage System***

- Refer to arrangement drawings.
- Specify bucket elevator characteristics.
  - Capacity, tons per hour.
  - Speed, fpm.
  - Require construction details showing bearings, sprockets, lubrication, etc.

  - Indicate source of coal to elevator, as track hopper, relaiming conveyor from field storage, etc.
- Specify capacity of silo, tons dead storage and tons live storage in live bin.
- Specify general construction of silo, as steel, masonry, or composite.
  - With minimum values of dimensions, strength, etc.
- Indicate power characteristics available.
- Require data on all electrical equipment.
- Require compliance with applicable codes and standards.

(j) ***Skip Hoists***

Note: Skip hoists are used for both coal and ash handling.

- Specify material to be handled.
- Specify principal parts this contractor is to furnish, as the hoist structure, skip bucket, hoisting engine and motors, automatic loader, skip guides, cables and sheaves, discharge chutes and gates, safety appliances, etc.
- Specify the capacity as _______ cu yd of (coal) (ashes) per hour, or alternatively _______ tons per hour. For cubic-yard basis, specify weight of coal or ash per cubic yard.
- Stipulate the source of the material, as a track hopper, pit, truck dump, etc.
- Specify characteristics of power available, as voltage, phases, cycles, etc.
- Describe type of automatic loader desired. Specify also that manual loading control by push-button station is to be provided.
- Describe normal cycle of operation, from loading to dumping to reloading.
- Specify skip hoist structure to be of steel, in accord with the applicable ASTM Specification No. _______. Tower to be self-supporting and must withstand wind, live and dead loads of _______ pounds plus _______ per cent to compensate for impacts.
- Specify capacity of skip bucket as _______ cu ft.
- Specify bucket construction as steel plate not less than _______ in. thick and stiffened with angles _______ × _______ × _______ in. thick.
- Specify kind of wheels, as cast iron, steel, etc. Also method of lubrication.
- Specify kind of bearings.
- Specify required details of hoisting engine, giving diam-

eter of drum, presence of lip on each end, kind of brake, etc. Also specify bearings and method of lubricating.

- Refering to the description of the operating cycle, specify the nature and kind of controls desired, as push-button stations, with automatic overload and no-voltage release, also a drum reverse. Limit switches as deemed necessary.
- Specify central control station, including mounting board, relays, main cutout switches, all other equipment suitable for mounting on the board. Stipulate where board is to be located.
- Specify the skip guides, as _______ pound rails, with bolted section to permit removal of skip if desired. All secured to the tower.
- Specify the hoisting cables to have _______________ center and to be made up of _______ strands of wire rope, each strand to incorporate _______ wires. Cable to be not less than _______ in. diam.
- Specify attachment of sheaves to shafts, as keyed and setscrewed. Specify bushings and lubrication.
- Specify safety guards around gears, moving shafts, and over potentially dangerous areas. Guards to be fabricated of steel of _______ gage or heavier.
- Specify chutes and gates as appropriate, giving sizes and dimensions.
- Specify that all electrical equipment and wiring shall meet the standards of the NEMA and the National Electrical Code; also any applicable local codes.
- Require bidders to furnish detail drawings showing all dimensions, arrangements, and other data. Also weights, method of protecting motors from coal or ash dust, full-load horsepower, and maximum starting current under full load. Also time of complete cycle, minutes.
- Specify painting, kind of paint, color, number of coats.

(k) ***Fire Protection for Coal Systems***

- Field storage piles.
  - Buried thermostats.
  - Exclusion of moisture by coverings.
  - Limitations on depth of piles.
- Bins and bunkers, including silos.
  - Smothering systems, foam or CO_2.
  - Provide drawing showing layout of system desired, locating foam generators or CO_2 flasks as appropriate.

- Designate number and kind of thermostats.
  - And description of alarm system.
- Specify minimum readiness requirements of smothering medium.

- Protection for conveyors.
  - In transfer houses and housed inclined lifts.
    - Describe method desired.
    - With alarm system.
    - Specify minimum time of operation acceptable at a full fire-fighting rate.
- For all fire protection systems, indicate nearness of vulnerable areas to water supplies and give availability of hoses, nozzles, and other water equipment.
- For pulverized coal systems, refer to NFPA No. 60, "Standard for the Installation and Operation of Pulverized Fuel Systems," latest edition.

4-8. ASH HANDLING AND STORAGE

(a) *Ash Gates under Boiler Furnaces*

- Specify type required.
- With operating features, as pneumatic, steam, electric, manual.
- Note water quenching if desired.
- Stipulate minimum aperture, or dumping capacity in tons per hour.
- Indicate if ashes are dumped from gates to trucks, narrow-gage cars, pneumatic or hydraulic conveying system.

(b) *Ash Storage Systems*

- Pneumatic systems, capacity, tons per hour.
  - Silo, materials and construction, including lining.
  - If steam-jet exhauster is used, stipulate rated capacity, pounds of steam per ton of material moved. Give steam characteristics.
    - Kind of steam condenser, surface, jet.
    - Kind and operation of discharge gates to truck, barge, railroad car.
  - List parts of pneumatic system to be furnished by vendor, as:
    - Ash-intake fittings and hopper adapters, nuts, bolts, fly-ash valves, vent for storage bin or silo, piping for fly ash, piping for main ash, steam valves as necessary for steam exhauster systems, collectors with dump gates.

- Require all piping, valves, and fittings to be of a material suitable for abrasive service, as cast iron, iron alloys, etc.
- Require bidders to furnish drawings illustrating design and arrangement of the system, dimensions, auxiliaries normally supplied, etc.

- Hydraulic systems, capacity tons per hour of dry fly ash.
  - Tons per hour of ash from furnace dumping gates for wet storage.
  - Indicate whether jet pump or centrifugal ash pump is to be used.
    - Describing driver, as motor or turbine.
    - Stipulate piping materials.
  - Stipulate kind of discharge controls wanted for storage gates and dump chutes.
- For both pneumatic and hydraulic systems, stipulate what part of system is by system manufacturer, and what parts by others.

(c) ***Ash Skip Hoists***

- See Sec. 4-7(j). Same as for coal.

(d) ***Ash Belt Conveyors***

- See sections on conveyors for coal.

(e) ***Car Pullers and Track Hoppers***

- See Sec. 4-7(f) for items to specify, page 46.

(f) ***Portable Conveyors, Belt, Flight, Bucket***

- Stipulate capacity, tons per hour.
- State characteristics of power available if motor-driven.
- Stipulate type and performance required of motor.
- For belts, see Sec. 4-7(e).
- Pulleys, idlers, lubrication: state requirements or ask bidder to submit full details.
- If engine-driven, stipulate whether diesel or gasoline engine desired.
  - Describe special features desired.
- State method of moving conveyor, whether manual, towed, or self-propelled.
- Bids to include full details and dimension drawings.

4-9. FUEL-OIL HANDLING AND STORAGE

(a) ***Outdoor Storage Tanks, Vertical, Small***

- Stipulate capacity, bbl or gal.
- Must comply with the Underwriters Laboratories Specifications for Vertical Tanks.

    - Limit height to 35 ft and diameter to 11 ft.
- Minimum thickness of bottom.
- Minimum thickness of shell for tanks up to 25 ft high, then between 25 and 30 ft, and from 30 to 35 ft.
- Designate material, as open-hearth steel plate, etc.
- Designate welding methods and welding rod.
- For sections fabricated elsewhere, close all openings for pipe connections with wooden plugs for shipment.
- Stipulate number and size of vents.
- Number and size of manholes.
    - Indicating how cover is locked or bolted.
- Stipulate dimensions of ladder.
- Describe required preparation of site.
    - Kind of earth material
    - Thickness of sand bed on top of graded earth.
- For multiple tank installation, specify walkways and railings.

(b) *Outdoor Storage Tanks, American Petroleum Institute, Large*

- Fabricate and erect in accordance with API specifications and standards.
    - Indicate diking required, height, width, and full containment of tank capacity.
    - Dike to suit requirements of National Board of Fire Underwriters.
    - Refer to API Standards 12C and 12D for width of courses and thickness of plate.

(c) *Outdoor Storage Tanks, Horizontal, Aboveground*

- Stipulate to meet Underwriters' requirements.
- Specify accessories and supports, vents, etc.
    - Foundations by whom.
- Stipulate number, size, and closure of manholes.

Note: Standard horizontal tanks range up to 25,000 gal in capacity.

(d) *Outdoor Storage Tanks, Horizontal, Underground*

- Give capacity, gal.
- Construction to meet Underwriters' requirements.
    - Indicate who excavates.
    - Stipulate depth and kind of cover.
    - Specify surface protection, as red lead, bitumastic, or other.
        - Number of coats.
    - Specify safety precautions in excavating, as provision

of shoring against cave-ins, continuous removal of water, how backfilled.

- Specify anchors to prevent tank from floating or from otherwise shifting position.
  - Specify concrete slab as per drawings, if used as anchor or as cover.
- Specify minimum size and number of vent and fill pipes. Fill pipe to have lock cover.

(e) *Storage Tanks, Concrete*

- Acceptable for oil storage when the oil is heavier than 35 degrees, American Petroleum Institute.
- Specify that design, materials, mix, aggregates, and all aspects of the tank must meet "Standard for the Installation of Containers for Storing and Handling Flammable Liquids," NBFU No. 30.
- Describe tank as being aboveground, underground, or partially underground.
- Expand on the requirements of the Standard as necessary to meet owner's needs.
- State capacity, bbl or gal.

(f) *Oil Storage Tanks, Inside*

Note: These tanks are of very limited capacity under the code. Unless enclosed, 275 gal in one tank and a total of 550 per building.

- Specify that tanks must be in accordance with NBFU Standard No. 30, "Standard for the Installation of Containers for Storing and Handling Flammable Liquids."
- Elaborate in the specification as deemed advisable
  - Describing the enclosing structure if one is to be provided, as brick or concrete.
  - Giving particular attention to locations and sizes of fill and vent pipes.
    - With lock cover for fill pipe.

(g) *Safety and Fire Protection for Oil Tanks*

- Flame arrestors for tanks over 1,000 gal.
- Indicate type wanted; plate, tube, or screen.
  - State if arrester is to be steam-traced against winter crystallization.
  - State flash point of oil stored, F.
  - State fire point of oil stored, F.
- Foam extinguishing systems
  - Stipulate kind of foam, mechanical or chemical.

- Indicate if special characteristics needed, as non-freezing at −20 F.
- Describe generating method required.
- Refer to NFPA No. 11, "Standard for Foam Extinguishing Systems."
- Refer to drawings showing system layout, location of water piping, mixing areas, and nozzles.
- Stipulate output required from each nozzle.

- Water spray systems
  - Stipulate water pressure available, psig.
  - Stipulate capacity of each nozzle, gpm.
  - Refer to location and layout drawings.
- Carbon dioxide systems
  - Show locations and size of CO_2 storage flasks by drawing.
  - Specify pipe materials and schedule of pressures.
  - Stipulate if system is to be in accordance with the High Pressure Standard or the Low Pressure Standard.
  - Stipulate filling density of storage flasks, per cent.
  - For either system, require conformation with the "Carbon Dioxide Extinguishing System" Standard.
  - Indicate if system is to operate automatically or manually.
  - Stipulate capacity of nozzles, pounds per minute for each nozzle.
  - If hose is required, state length, construction, and bursting pressure, psig.
  - Make note of any special hazards or conditions that extinguishing procedure must contend with, such as operations in a closed room for an indoor tank.

(h) *Tank Heating Systems*

- If steam coil is to be supplied in bottom of tank, specify steam pressure and quality available.
- Specify kind of metal in the coil, as steel pipe, wrought iron, copper tubing, etc., and kind of joints.
  - For welding, stipulate gas or electric and kind of rod.
  - Require welder to be certified by National Certified Pipe Welding Bureau or other approved authority.
- For hot water, give temperature and flow, gpm.
  - If coil must withstand corrosive attack from liquids in tank, so indicate, giving name of corrosive liquid.
  - If this bidder is to furnish control valves, specify kind of valves, as gate, globe, pressure reducing, etc. With minimum acceptable working pressure, psig.

- Specify kind of pipe joints and gaskets.
- If condensate traps also are to be furnished by tank contractor, specify kind, capacity, and working pressure.

(i) *Tank Testing Procedures*

- Specify how test is to be applied, giving source of water.
  - Indicating consecutive hours tank is to be observed for possible leaks, tilting, or any other significant phenomena.
  - Stipulating test is to be witnessed by owner's deputies.
- Specify arrangements, if any, to be taken to protect surrounding areas in the event the tank collapses under test. *Note:* Dikes take care of this on large tanks.

(j) *Fuel-oil Pumping Systems between Tanks and Burners*

- Specify type of pumps wanted, centrifugal, reciprocating, rotary.
  - Giving capacity, gph at ______ ft head.
  - With suction head, or suction lift, of ______ ft.
  - If steam-driven, give steam psig pressure and quality of steam.
    - Indicate type of pump controller to be provided.
  - If motor-driven, give characteristics of power available, as voltage and cycles.
    - State if motor is to be of open or closed type, whether it will function indoors or is exposed to the weather. State temperature rise, 40 or 50 C, above an ambient of ______ C.
    - Describe desired control, type of starter, automatic features, for motor drive.
- Designate pump foundation by whom.
  - If contractor, specify kind and bolting-down method.

(k) *Tank Gaging System*

- Indicate if pneumatic, electrical, or manual.
  - Stipulating kind of tubing and length of run with reference to drawings, if pneumatic.
  - And wiring to satisfy National Electrical Code if electric.
  - Size of scale on outside of tank if level is indicated by outside traveler that is synchronized with movements of float inside tank.
  - For remote readings, state approximate distance.
  - Indicate if compressed air is available or if auxiliary compressor must be furnished under the contract.

(l) *Oil Piping, Tank to Burners*

- Provide bidders with drawing showing layout of piping, and sizes.
  - Specify kind of pipe and schedule number.
    - With description of joints or welding.
    - Welding by certified welders as by National Certified Pipe Welding Bureau, or other recognized authority.
- If pipe is underground, specify who opens and prepares the trench.
  - Noting safety precautions by shoring and pumping.
- Specify bed for the pipe, as sand or gravel, forbidding use of cinders.
  - And treatment of pipe against corrosion.
  - And possible grounding against electrolysis.
  - If steam-traced, give steam conditions.

(m) *Fuel-oil Heaters*

- Specify type, and number of passes. Also working pressure.
- Specify temperature rise, starting at _______ F as oil enters heater.
- Specify capacity, gph, at a density of _______ Baumé.
- If heating medium is steam, give steam pressure and quality.
- If electric, specify electrical characteristics.
  - Wiring and materials in accordance with the National Electrical Code.
  - And NEMA standards as appropriate.
- For hot water, state temperature of water entering and allowed temperature drop, _______ F, at leaving.
  - Stating flow in gpm.
- For any type of heater, require thermometers on oil entering and oil leaving.
- Also describe form of temperature control desired.
  - As immersion thermostat operating steam or water valves, or actuating a relay for electric heating.
- Heater construction in accordance with ASME Code for Unfired Pressure Vessels, if appropriate to form used.

4-10. GAS FUEL SYSTEMS

(a) *Natural Gas Systems*

- Stipulate where work of this contractor begins, also indicating general scope of installation.

- Give characteristics of the gas, its constituents and calorific value. Also, minimum pressure.
- Provide and refer to applicable drawings.
- Stipulate type of meter, as iron, aluminum, steel, and maximum per cent of manufacturer's test pressure that may be used as the operating pressure.
- If series pressure regulators are used, primary regulator must be located outdoors.
- If liquefied petroleum gas is connected into system as a standby, install a check valve downstream from the meter as shown, to prevent the LP gas from backing into the meter.
- Describe weather and insect excluder methods desired for regulator vents.
- Show depth of cover required over outside supply piping.
- Require all piping and pipeline components, as valves, fittings, joints, to satisfy the American Standard Code for Pressure Piping. Designate applicable ASA standard for each class of component.
- Describe type and size of pressure regulators required.
- Stipulate methods for corrosion protection of underground piping, as painting with protective coatings, cathodic protection, etc.
- Stipulate safety precaution to be taken during construction, including a check of the ambient atmosphere for explosive mixtures before any welding or cutting is undertaken.
- Specify system of purging to be exercised when line is made ready for service. Require purge discharge to be delivered outdoors.
- Designate test pressure, method of application, and duration of the test.
- Stipulate that installation in general shall meet requirements of the National Fire Prevention Association "Standard for the Installation of Gas Appliances and Gas Piping," No. 54.

(b) *Liquefied Petroleum Gas Systems*

- Give general scope of the installation as accomplished by this contractor.
- State gas kind, as propane, butane, or any other.
- Stipulate that design and construction of the containing storage tanks shall be in accordance with the specifications of the Interstate Commerce Commission.

- The system, consisting of the container, vaporizer, carbureters, relief valves, pressure regulators, and other major components, shall be as tested and approved by the Underwriters Laboratories, Inc., or shall carry the approval of any other duly recognized authority.

Note: For large installations employing storage tanks not within the scope of the Interstate Commerce Commission, tanks must be constructed and tested in accordance with the ASME Code for Unfired Pressure Vessels, or comply with the API-ASME Code.

- Specify filling density.
- List valves and accessories to be provided with the system.
- Specify number and capacity of safety relief valves.
- State if vaporizer is to be operated with steam, hot water, or any other heat medium. Give steam pressure, psig and quality. For hot water give temperature. For electrical heating describe kind and electrical characteristics.
- Require this contractor to furnish name of vaporizer manufacturer, size, and type.
- Specify make and type of transfer pump or compressor if used for refilling tanks.
- Specify kind of level gage.
- Describe foundation for storage containers and state who provides.
- If dikes are required, refer to drawings for dimensions and call for compliance with the standards of the National Board of Fire Underwriters, No. 30.
- Specify piping, size, materials, schedule number, how protected against physical injury, etc.
- Specify that liquefied petroleum gas installation shall satisfy the requirements of the National Fire Protection Association, "Liquefied Petroleum Gases," No. 58. This is a joint issue with the National Board of Fire Underwriters.

4-11. SPECIAL-FUEL FURNACES

(a) ***Bagasse***

- Provide data on the weight, calorific value, and moisture content of the bagasse as delivered to the furnace. Give ultimate analysis.

- Describe fuel feeding system, as automatic or manual. If automatic, specify control system.
- Specify kind of hearth or stoker, with maximum firing capacity required, expressed in pounds of dry bagasse per sq ft of grate per hour at ______ in. W.G. air pressure.
- Provide drawing showing arrangement of furnace, firing opening, baffles, etc.
- If furnace is fitted with auxiliary oil burners or gas burners, specify these auxiliary units in accordance with the checklists in Sec. 4-1(u) and (v).
  - Giving all pertinent data needed to obtain the burner performance needed.
- If this contractor is to furnish and install the forced draft fans, specify their characteristics as checklisted in Sec. 4-1(x), including the performance charts. Size forced draft air duct and give plate thickness and construction. Refer to ASHRAE Guide for duct fabrication requirements.
- Specify refractory in furnace and baffles.
  - Giving maximum temperature for operation at which softening or spalling will not occur, and must be warranted.
- Show on drawing and specify how ash is to be removed.

(b) *Hogged Fuel*

- Provide data on dimensions and weights of fuel, and range of woods to be burned, as oak, pine, maple, etc.
- Provide drawing showing manner of injecting fuel into furnace.
- Specify kind of grates and tuyeres desired.
- Give proximate analysis, with moisture content on an "as fired" basis.
- Specify if grate is to be flat or sloped, and give number of cells if more than one is projected.
  - Noting dimensions of each cell.
- Specify capacity of forced draft fan, cfm, at ______ in. W.G.
- If preheated air is used, specify kind of preheater, as plate, tubular, or rotary.
  - Giving cfm to be heated from ______ F to ______ F. See Sec. 4-1(r) for checklist.
- Ducts to be fabricated in accordance with ASHRAE Guide dimensions, weights, and joints.

- Insulate preheated air ducts with materials and thickness to suit preheated air temperature.
- Capacity per cell to be specified in "units." Stipulate the volume of the unit rather than weight, 200 cu ft of wood being the standard, the weight varying from 1,700 to 2,300 pounds per unit, depending on the kind of wood making up the unit and its percentage of moisture.

Note: Average wood mixtures are given an approximate capacity of 2 units per cell per hour.

(c) *Sawdust Burners*

- Specify capacity required in pounds of sawdust per sq ft of grate area.
- Give analysis of sawdust to be burned, kind of wood, moisture content, etc., "as fired."
- Give furnace draft, in. W.G., at furnace outlet.
- Specify method of firing, as vertical drop on grates, or horizontal injection with air.
- Provide drawing showing arrangement of fuel hopper if used, dimensions, and metal gages to be used in the fabrication.

Note: Engineering Experiment Station, Oregon State College, is a source of data for sawdust burning projects.

- For overhead storage bins for sawdust fuel, specify angle of repose as well as all necessary fabricating data.
- Show or specify the kind of discharge gate desired.

4-12. FEEDWATER SYSTEMS

(a) *Raw Water Pumps*

- Specify kind of pump, as centrifugal, reciprocating, rotary, etc.
- Capacity, gpm, under conditions of ________ ft suction lift or suction head, and with a total dynamic discharge head of ________ ft. Give water temperature range, from ________ F to ________ F.
- If centrifugal, specify single or double suction, wearing rings and stuffing box design and material, whether thrust is balanced or compensated by thrust bearing.
- Specify shaft material and construction; require both static and dynamic balance; describe shaft bearings desired.
- Specify kind and type of driver, as steam turbine or electric motor, and kind of coupling.

- If steam turbine, give pressure, psig, quality including total temperature and superheat. Also back pressure on turbine exhaust, psia or in. Hg.
- Specify type of turbine, reaction or velocity, number of stages, form of overspeed trip, and operating governor required. Also indicate rotation.
- Specify turbine casing, material, how split, and size of casing excess pressure relief valve.
- Stipulate direct or reduction gear drive.
- Specify bearings design, material, and lubrication. Indicate if oil rings are required, and flow sight glasses.
- Stipulate design and packing of shaft glands.
- Specify thrust bearing, form and lubrication.
- Specify controls, whether by hand throttle valve or automatic governor. Give details of performance wanted.

- If pump is motor-driven, give electric power characteristics available.
  - Specify motor type and acceptable temperature rise.
  - Specify starting methods, as across the line, step increments, to suit conditions.
- Specify if pump and driver are to be set on a common base or baseplate, and how attached.
- Performance data to be submitted by bidders are shown on forms in Fig. 4-2, Tables 4-7 and 4-8.

(b) *Reduction Gears*

- Describe kind of gears wanted, herringbone, single or double reduction, etc.
- Specify hp to be transmitted, and range of speeds.
- Specify kind of bearings, line and thrust, also lubrication system.
- If oil is water-cooled, require bidder to submit data on temperature range and gpm. Also friction loss in cooler.
- Require that gears be balanced at factory, statically and dynamically.
- Require factory alignment of gears and driver prior to shipment.
- If gear case is to be dehumidified, specify methods and give acceptable range of relative humidity.
- Specify material and jointing of gear casing.
- Require bidders to furnish complete drawings showing

design and arrangement of the gears, driving connection for the lubricating oil pump, water, oil, and breather connections, dimensions, weights, and all other useful and appropriate data.

CENTRIFUGAL PUMP CURVES BY BIDDER

Submitted by:____________________ Job:____________________

Make, type and model of pump:____________________ Service: raw water pump no. 1

Quantity:____________________

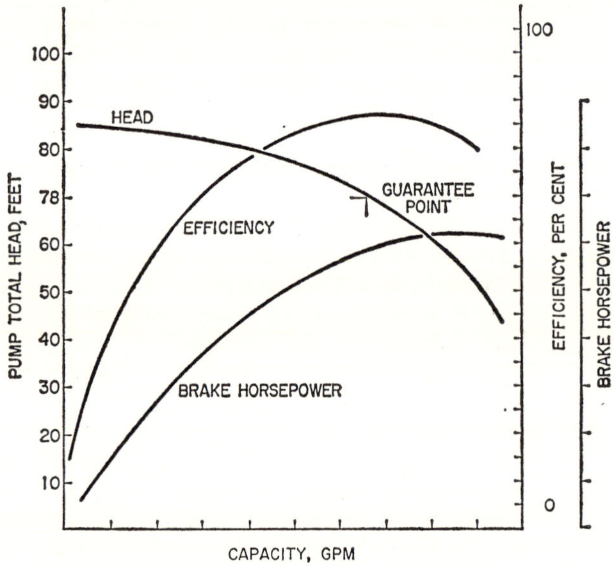

NOTE: BIDDERS TO SUBMIT SIMILAR DATA FOR RECIPROCATING AND ROTARY PUMPS WHEN OFFERED.

FIG. 4-2.

(c) *Filtration Equipment for Raw Water*

- Describe system, as for vertical or horizontal filtration, using gravity or pressure-type equipment.
  - Indicate if operation is to be automatic or manual.
  - Describe accessories to be furnished, as rate controllers, float valves, specialties.
- For gravity filters, stipulate the rate of flow, gpm, per ft of filter area. Set maximum required rate.

Table 4-7. Steam-turbine Performance Data, Filled in by Bidder

Submitted by ____________ Job ____________
Make, type, and model ____________ Service ____________
Quantity ____________ Drive: (direct or reduction gear)

Pump operation			Turbine operation		
Delivery, gpm	Speed, rpm	Bhp	Speed, rpm	Water rate, lb/hr	Hp developed

Note: When providing information on various gpm or head conditions, bidder must indicate which is the guarantee point for the steam turbine. The guarantee point for the pump is shown on the graph, Fig. 4-2.

Table 4-8. Direct-acting Steam Pump Data, Filled in by Bidder

Submitted by ____________ Job ____________
Make, type, and model ____________ Service ____________
Quantity ____________

Strokes per minute	Capacity piston displacement	Slip, %	Efficiency, %	Indicated steam hp

Note: If piston or pistons is driven from a crankshaft, as in duplex and triplex vertical pumps, bidder shall state crankshaft rpm, and if motor-driven, give motor horsepower and motor speed. In all cases, bidder must indicate the guarantee points.

- Require drawings from bidders showing the gradations of the sand, introduction of the raw water, and other details.
- Specify if filter tank is to be formed of steel, wood, concrete.
- Specify how filtered water is to be removed from filter—that is, by flume, pipe, pump—and if it is to be stored.
- Specify if coagulation and sedimentation are to be incorporated as such in the filter.

- Specify piping and valves, materials, schedule number of piping, whether valves are to be operated by extension stems and floor stands.
- Provide a drawing showing general arrangement of filter, or require bidders to furnish such a drawing illustrating their equipment.

- For pressure filters, provide and require information similar to that appropriate to the gravity filters.
  - Specify the working pressure of the filter shell and its material.
    - Tank to be fabricated in accordance with the ASME Code for Unfired Pressure Vessels.
    - Set maximum flow rate, gpm, per sq ft of filter bed. *Note:* This is generally less than for gravity filters, in the area of 3 gpm rather than 5 gpm of the latter.
- For all filters, specify filtering medium and methods of backwashing. For anthracite coal, give sizes.
- Specify range of temperatures in water passing through filter.
- Specify where this contractor's work begins and ends.
- Specify rate of backwashing, gpm per sq ft of filter bed.
  - Giving head of backwash water, psig or ft.
- Specify if bed is to be raked or washed with high-pressure sprays, either revolving or stationary.
- Disk filters, if used for extra-fine filtration, should be described by use or application and by type of filtering medium, as asbestos, terry cloth, fiberglass, paper, etc.
  - Specify performance in terms of solids removed, at ________ gpm, water at a temperature of ________ F.
  - Specify life of filtering medium, or frequency of renewal.
  - Bidders to provide drawings showing dimensions, sizes of connections, and all details of the parts arrangement.

(d) *Water-softening Processes: Lime-Soda, Cold*

- Cold-process softening is accomplished at the temperature of the raw water, usually from 40 to 90 F.
- For cold process, describe the equipment required, as for intermittent or batch handling, continuous sedimentation, or continuous slurry.
  - Specify construction of tanks, general form, plate thickness, and dimensional limitations if these exist.

- Note if system is to have one tank or two.
- Specify method of agitation, length of agitation in minutes, then length of time given for settlement.
- Specify point in the cycle at which lime is added, soda ash is added, or if both are put in at the same time.
- Specify capacity of system, soft water output, gph.
- Specify method of sludge removal, whether continuous or intermittent, automatic or manual.
- Either specify, or require bidders to provide complete information on methods of operating control.
- Specify piping, valves, and fittings, materials, kind, schedule numbers of piping, as appropriate.

(e) *Water-softening Processes: Lime-Soda, Hot*

- Hot-process softening is for water at high temperatures, up to and beyond the boiling point.
- Specify required capacity, gph.
- Describe equipment required, as tanks, float for controlling the input of raw water, the proportioner for feeding the lime, soda ash, and coagulant. Also pumps, if needed.
  - If deaerating is included, describe the trays, jets, or other devices to be used to deaerate the water.
- Stipulate the steam supply for heating the water to boiling or above, as ________ psig and ________ per cent saturated, or ________ F of superheat.
- Specify tank construction to comply with the ASME Code for Unfired Pressure Vessels.
- Specify all special valves, as pressure relief, vacuum breaker, water inlet control valve, and any others.
- Specify form and construction of the vent condenser, also form and materials in primary heater.
- Specify, or require full information on, the manner and control of backwash, including the wash-water pump, if used.
- Specify, or require full information on, the manner and control of sludge removal, whether periodic and automatic, manual, or a combination of both.
- For all electrical equipment as motors, relays, valve operators, etc., give characteristics of electrical power available.
  - And require compliance with the National Electrical

Code and the standards of the National Electrical Manufacturers Association.

- Require complete sectional and plan drawings from all bidders.

(f) ***Water-softening Processes: Zeolite***

- Describe system and list components to be furnished under this contract, as main softener tank, salt storage tank, number of units, etc.
- Specify construction of shell, design pressure of ______ psig and working pressure of ______ psig, complying with ASME Code for Unfired Pressure Vessels.
- Specify if downflow or upflow softener.
- Specify rate of flow, gpm per sq ft of bed area.
  - Set maximum friction head loss through softener at this maximum rate.
- Require all necessary controls, as rate of flow controller. Also meters for measuring flow of water through softener and flow of brine during reactivation.
- Specify if zeolite is to be natural or synthetic.
- Specify water to be handled, as raw water or filtered water, city water and sources, river, lake, artesian. Give temperature range, F.
- Specify total production of soft water in gallons, between reactivations.
- Require softening to a condition of zero hardness when supplied with water of ______ grains per gal hardness. Zero hardness shall be considered as ______ ppm or less.

  Note: One (1) grain of hardness per U.S. gallon in terms of calcium carbonate equals 17.1 parts per million.
- Specify loss of zeolite by attrition not to exceed ______ per cent over ______ years.
- Specify standards by which performance tests are to be made, as American Health Association.
- Specify brine tank parts, as strainers, telltale, float-operated valves, charging hatch, etc.
- Specify all piping valves and fittings to meet 125-pound Standard or Schedule 40, as appropriate.
- Specify painting as desired.
- Require full information from bidder, as cu ft of zeolite furnished for original charge; weight per cu ft; time required for reactivation, minutes; quantity of brine re-

quired per reactivation; dimension drawings of the tanks; and all other pertinent data.

(g) ***Deaerating Heaters***

- Specify construction, giving materials of shell and heads. Ditto trays, pans, heat transfer tubing. Stipulate ASTM specification covering this tubing. Stipulate material of tube sheets.
  - Describe methods of fitting tubes into tube sheets.
  - Stipulate ranges of pressures inside heater shell from a vacuum of ______ in. Hg to ______ psia. Stipulate maximum temperature of ______ F.
- Specify capacity in ______ pounds of water per hour, with entering temperature of ______ F and leaving temperature of ______ F.
- Specify maximum weight to be handled if necessary under emergency conditions, ______ pounds per hour. Bidders to provide data of performance under this maximum flow.
- Give pressure, temperature, and quality range of the steam supplied for heating the water.
- Specify oxygen content limitation on water leaving deaerator under the normal operating conditions, or require bidder to submit this information with his bid.
- Specify how much connecting piping, valves, and fittings are to be furnished by contractor.
- Specify form of water control to deaerator and form of steam regulating valve.
- Describe steam jet air ejector if required, giving steam pressure.
  - Bidder to provide capacity data.
- Specify pressure and vacuum tests to be made.
- Require bidder to provide complete plans and elevations of the equipment.
- And all necessary operating data over the full range of the deaerator capacity.

(h) ***Feedwater Heaters, Open Type***

- See Sec. 4-12(g) for deaerating heaters.
- Require addition of oil separating facilities in heater if steam for heating is exhaust steam and may contain oil.

(i) ***Feedwater Heaters, Closed Type***

- Describe unit in general terms, indicating if placement is vertical or horizontal.

- Specify construction, as material in shell, tubes, and heads.
- Specify working and design pressures for both shell and tubes.
- Give steam conditions to heater, psig, and total temperature, F. Note if live or exhaust steam.
- Give pressure and temperature of water entering heater, psig and F.
- Specify, or require bidder to supply, pressure and temperature of heated water leaving.
- Specify number of passes in heater.
- Require construction to comply with ASME Code for Unfired Pressure Vessels.
- Specify necessary openings, including vent and vacuum-breaking connections.
- Specify capacity in gpm of water passing through or pounds per hour when elevating the water temperature from ________ F to ________ F using steam as specified for heating.
- Describe control of both water and steam flow, indicating components to be furnished under the heater contract.

(j) *Evaporators*

- Describe kind of evaporator system wanted, giving number of effects.
- List components to be included in this contract, as the main evaporator shell, surface condenser, heat exchanger, recirculating pump, residual blow pump, drainers, motors, hot well, etc.
- Specify construction, as shell to be fabricated of steel plate, tubes formed of ________________ in accordance with ASTM Specification No. ________. Tube sheets to be of ________________ metal. Specify steam and water pressures the various evaporator components must withstand.
- List openings to be provided by contractor. Note that outlet for blowdown must be sized to accommodate a blowdown rate of ________ pounds per hour.
    - With effective prevention of the passage of scale from tube cracking, which otherwise would get into and choke the heat exchanger and residual blow pump.
- Specify construction of surface condenser. Tubes to be of ________ metal in accordance with ASTM Specifica-

tion No. _______ Tube sheets to be of _______ metal.

- Specify steam and water working pressures.
- Stipulate required hydrostatic test pressure.

- Specify construction and materials of the heat exchanger which is used on the blowdown line. Refer to ASTM for tube specifications and name shell material.
  - Specify weight of sludge or blowdown to be cooled per hour, and stipulate the range of cooling, from _______ F in to _______ F out.
- Pumps to be specified, including motors and couplings, bedplate, materials, glands, bearings, etc. See Sec. 4-12(a) for details. For boiler-feed pumps, see Sec. 4-12(b).
- Specify the capacity of the evaporator in _______ pounds per hour when supplied with filtered water at _______ F temperature, and containing _______ grains of solids per U.S. gal.
- Specify how scale cracking is to be accomplished and how it is to be removed.
- Provide bidders with adequate drawings showing arrangement of system. Minimum accompanying the specification should be a line diagram, similar to Fig. 4-3.
- Stipulate information that must accompany bids, as weight of the evaporator, empty and full; weight of the condenser, empty and full; weight and dimensions of heat exchanger, full; heating surfaces, square feet as appropriate. Also, plan and sectional drawings to illustrate in detail the equipment being offered.
- Specify how tests are to be conducted and by whom prior to acceptance of the system.
- Require compliance with National Electrical Code and the NEMA standards for all electrical devices and materials.
- Describe method of control, by float valves, meters, etc., all to be provided by this contractor. Give operating values to be maintained by the controls.
- When preparing the line diagram, Fig. 4-3, mark pressures, temperature, and weight of fluid flowing per hour on each line, so bidders may calculate their capacities.

(k) *Boiler-feed Pumps*

Note: For heat balance conditions, economy sometimes indicates that a battery of boiler-feed pumps is divided in pump drives, some being driven by steam turbines, some by electric motors.

- Specify number of pumps required, and their drives.
- Specify types, as horizontal, multistage, centrifugal, or other.
- Specify capacity of each pump in gpm against _______ ft head, when supplied with water ranging in temperature from _______ F to _______ F. Suction conditions

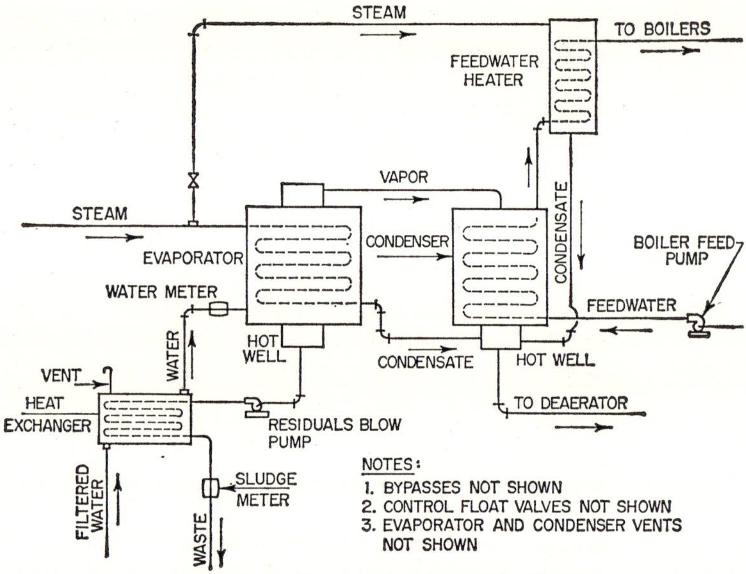

FIG. 4-3. Schematic diagram, evaporator system.

to be _______ ft positive head or lift, as applicable. Stipulate pump speed, rpm.

- If pump suction is taken from a deaerator or heater in which a partial vacuum is present, stipulate the vacuum _______ in. Hg column.
  - Require an equalizing line to be run from the vessel under partial vacuum to the pump casing.
- Describe pump construction desired, as casing material of cast iron, cast steel, bronze, or special alloy.
  - To withstand a working pressure of _______ psig.

- (Optional.) Stipulate the maximum tensile stress acceptable at the working pressure. This is to keep a check on later metal fatigue possibilities.
- Stipulate casing division, as horizontally split on center line.
  - With suction and discharge nozzles to be on the lower half of the casing, to permit lifting of the upper half while not disturbing the piping.
- Stipulate materials and design of impellers as all bronze, single suction, etc. Also wearing rings.
- Stipulate kind and method of lubricating the thrust bearing.
- Require bidders to furnish detailed sections of the pump, showing how leakage between stages is prevented.
- Require water passages to be smooth.
- Specify shaft material and balance both statically and dynamically.
  - Encase in bronze sleeves where shaft passes through stuffing boxes.
- Describe kind of shaft bearings wanted, as babbitt, split and capped, etc.
- Describe lubrication system for bearings desired, noting need for slings, spring caps, and cooling means.
- Specify material and design of baseplate, and if it is to provide a common base for both the pump and its driver.
- Require details on shaft sealing (very important for high-pressure pumps).
- Specify type of coupling. If pump is to operate at various speeds, and a hydraulic coupling is desired, specify type or manufacturer, giving model.
- Motor drives, specify kind of motor, temperature rise, degrees C above an ambient of ________ C.
  - To operate on ________ volt, ________ phase, ________ cycle current.
  - Require motor supplier to provide data on horsepower and speed, at stipulated pump discharges.
  - Motors to satisfy all applicable NEMA requirements.
- Steam-turbine drives, steam to be furnished live at ________ psig and ________ F total temperature.
  - Back pressure on exhaust not over ________ pounds absolute.
- Specify type of turbine, as single or multistage, and

give requirements in starting-up time from cold and no load, to full load, expressed in minutes. If noncondensing type, specify back pressure as in section above, psia.

- Describe kind of governor valve desired, and give range of control as from _______ per cent below normal speed to _______ per cent above.
  - Specify metal of valve, as cast-steel body, monel mountings, etc.
  - Specify kind of throttle valve, materials and general design.
- If turbine exhausts into a condenser, give condenser vacuum in in. Hg column.
- Require bidders to furnish detail drawings of the turbine, showing all important aspects of the design and construction.
- Turbine casing to be of (material) suitable for the temperature and pressure conditions stipulated, and split horizontally.
- Turbine wheels to be of (material) with buckets (milled in or shrouded). (Specify which.)
  - If balancing drum is necessary, so stipulate.
  - Stipulate data for thrust bearing and line bearings, and lubrication method and means.
    - Limit leakage through glands.
  - Require complete balancing at factory.
  - Stipulate limitation on number of hand valves acceptable, if this an item in the design.
- Performance of pump and drive to be filled in by bidders on the form shown in Table 4-9.
- If pump is to be of the barrel design, specify how rotor is to be inspected or lifted for repairs.
- Give all necessary data for working pressure, design pressure, temperature of water to be handled, whether suction lift or suction head, etc. This form of pump is centrifugal, with its application being mostly to very high discharge pressures, the lack of casing joints being an asset under these conditions.
- If pump is the steam-driven reciprocating type, refer to Table 4-8 for data to be provided by bidders.
  - Specification to give steam pressure, psig, quality, and back pressure in psia.
    - Also temperature of water _______ F.

Note: Some specifiers designate a delivery pressure above the steam pressure, to be maintained at all times. In these instances the steam pressure may be a variable owing to process demands, and the specification should stipulate the kind of discharge pressure control to be provided, either by the bidder or by the specifier.

Table 4-9. Boiler-feed Pump Performance

Gpm	Discharge, psig	Water,* F	Speed, rpm	Motor hp	Steam turbine†			
					Lb/hr	Hp	Rpm	Valves
Minimum...								
Normal.....								
Maximum...								

* If feedwater temperature is to be a variable, it is customary to obtain performance data at each of the specified water temperatures. This table can be repeated as often as made necessary by different ranges of temperature.

† Column for pounds per hour applies to pounds of steam per hour used by the steam turbine under the specified operating conditions. Valves pertains to the number of hand valves that must be open to provide the performance indicated.

Note: All blank spaces are to be filled in by the bidders, unless specifier decides to set the gpm rates to establish a common basis for the evaluation of the bids.

(1) *Condenser Condensate Pumps*

- Refer to drawing for information on position of the condensate pump or pumps, relative to the condenser hot well.
- Stipulate number and type of pumps, as centrifugal, reciprocating, etc.
- Stipulate capacity of each pump as ______ gpm when supplied with condensate at a temperature of ______ F, with a suction head (or lift) of ______ ft, and delivering the condensate against a total discharge head of ______ ft.
  - Pump to be operating at a speed of ______ rpm.
- Specify pump materials and construction as:
  - Casing (if cast iron, specify tensile strength as the

minimum permissible when operating at the specified head).

- Impellers of the (closed)(open) type, made of ________________ (material).
- Stuffing boxes method of sealing, and using ________ ________ packing.
- Shaft manufactured of ________________ (material). Fillets at changes in diameter to have a radius not less than ________ in.
- Glands to permit packing and inspection without any need to dismantle any other parts of the pump.

- Shaft sleeve (if provided) to be of ________________ (material) and held in position by ________________.
- Thrust bearing: describe type desired, as marine, ball bearing, roller bearing, etc.
- Lubrication for all bearings, line and thrust, to have large reservoirs with hinged and spring-loaded (or plain) cap.
  - Reservoirs to be provided with gage glasses and drain connections.
- Each pump driven by electric motor of the ________ type through a ___________ type of coupling.
  - Motor to meet all requirements of NEMA standards with a temperature rise of ________ C.
  - To operate on ________ volt, ________ cycle, ________ phase current.
  - For (describe starting method as automatic, manual, across the line) starting.
- Steam-driven pump to operate on live steam, ________ psig, and ________ F total temperature and ________ quality saturated or ________ degrees of superheat.
- For construction and materials of steam pump or turbine, see Sec. 4-12(1).
  - Bidders to fill in required data on condensate pump performance as shown by Table 4-8 for reciprocating steam pumps, and Fig. 4-2 for motor-driven or steam-turbine-driven centrifugal pumps.
- Subbase for motor-driven or steam-turbine-driven pumps to incorporate a catch basin from which the entire drip pan may be drained.

Note: If subbase and drip pan are to extend under both pump and driver, or if the drip pan is restricted to the pump end, the specification should so state.

- Describe control system in detail, Stipulate the codes and standards that must be satisfied by the components of the system. This includes safety devices.

(m) *Boiler Water Additive Equipment*

Note: In large plants the interrelations of filtering, evaporating, softening by lime-soda or zeolite, and deaerating, usually provide all the water treatment that is needed. In smaller units, process and manufacturing, the addition of chemicals is a prime form of feedwater treatment.

- Specify the form of water treatment required, as phosphate control, proprietary amines, etc.
    - Provide drawing showing location of chemical feeding connection to inlet side of feed pumps, directly into the feedwater heater, injection into the steam, etc.
    - Specify results to be accomplished in hardness, limitation on solids, the pH value; etc.
    - Describe any particular aspects of the system that are desired.
    - Specify the method and standards for determining the satisfactory performance of the treatment system.
- Tabulate the data that are to be supplied by bidders with their bids.
- Designate permissible frequency of treatment injections if these are intermittent rather than continuous.
    - Describe required form of injection control.
- For guidance of bidders provide information on the characteristics of the boiler feedwater as it enters the treatment station.

4-13. FLUE-GAS AND AIR SYSTEMS

(a) *Breeching from Boiler Outlet to Stack*

- Refer to drawing for illustration of general arrangement.
- Specify construction materials, thickness of steel plate, whether welded or riveted.
    - Specify number and positions of cleanout or access doors.
    - Specify pitch of breeching, in. per ft.
- Specify inside and outside finish, as bitumastic paint or anticorrosive proprietary paint.
    - Specify maximum flue-gas temperature and maximum velocity, fps.
    - Specify maximum weight of flue gas, pounds per hour.

- Specify normal analysis of flue gas, as ______ per cent CO; ______ per cent CO_2; ______ per cent water vapor; ______ per cent sulfur.
- Specify cleaning methods desired or required.
  - If soot blowers are included, specify if steam actuated or compressed air.
  - Give steam pressure and quality, or alternatively, compressed-air pressure, psig.
  - Specify method of operation.
- For induced draft fans, see Sec. 4-1(w).
- For air preheater, see Sec. 4-1(r).

(b) ***Fly-ash Removal***

Note: Fly-ash removal equipment is either mechanical or electrical, and may be mounted in several alternaative positions. It is important, therefore, to provide a location and connection drawing for the guidance of bidders. Cinder traps are considered to be in the fly ash category of equipment.

- For *mechanical separation,* describe the form of equipment required, as centrifugal, scrubber, multitube cyclone, etc.
- Specify the volume and weight of flue gas per hour that must be handled by the separator. Also the maximum and minimum temperatures.
- Provide bidders with analysis of flue gas and with information on burning equipment, as stoker, pulverized coal, oil, a wood-burning or bagasse furnace. Give maximum burning rate, pounds per hour, and per cent ash in fuel.
- Specify ______ per cent of separation that equipment must accomplish, under conditions given.
- Require bidders to submit dimensional drawings of the offered equipment, together with the gross weight.
- Require submission of performance data at no less than three points of boiler rating.
  - Giving efficiency at each point.
  - And per cent passed through a ______ mesh screen.
- For *electric precipitation,* describe required form, whether horizontal or vertically placed.
- Specify all characteristics of the fuel and flue gas, including an analysis. Also give firing method.
- Specify volume and weight of flue gas per hour that will pass through the precipitator.

- Specify the guarantee point of the precipitator relative to the load on the boiler, as _______ per cent separation at a boiler output of _______ pounds of steam per hour.
- Require bidders to submit complete data on design, arrangement, weights, and performance. Performance to be stated for at least three points of boiler rating.
  - Design to include a satisfactory method of cleaning adhering ash and dust from the collecting plates.
  - And a description of the collecting hopper and the gates or other means of ultimate removal of the fly ash.
- Specify that all electrical equipment must meet national codes, and as appropriate, the standards of NEMA.
- Give characteristics of the electrical service available at the site.
  - Bidders to state maximum amperage under full load.
- Bidders to submit data on automatic control of voltage and amperage fluctuations.
  - With description of high-voltage producer and method for quenching sparking, or preventing it.
- Note if rapping is sequential or otherwise.
- *Wet scrubbers* are seldom used for fly-ash and cinder removal, due to the acutely corrosive nature of the effluent. Wherever they are chosen:
  - Specify if the piping is to be of stainless steel or lead, or proprietary alloy.
  - Specify characteristics of the water, whether acid or alkaline, hard or soft, and the temperature, F, as it emerges from the nozzles. Also the pressure, psig.
  - Describe movement of gas through scrubber, as by an induced or forced draft fan, or by natural draft. Give draft value as _______ in. W.G.
  - Give full information on fuel, method of combustion, weight and volume of flue gas, as provided for other forms of separators.
  - Specify performance required, giving _______ per cent of elimination at at least three points of boiler operation.
  - Specify kind of control system.
  - Specify drainage system, with materials of collector and piping.
  - Designate on a drawing where this contractor's work begins and ends.

(c) *Steel Stacks*

- Classify desired stack as guyed or self-supporting.
- Provide drawing to show general construction, with plate thickness, width of courses, dimensions of the flare if self-supporting. For guyed stacks, show placement of guys, number, radius, etc.
- Show dimensions of foundations, number and size of anchor bolts.
- Specify painting, both inside and outside, giving kind of paint and number of coats.
- If stack is lined, specify lining material, as refractory, concrete, tile, etc.
- For welded stacks, specify that welders must be certified by a recognized certifying agency.
- Describe required tests for draft and tightness with particular reference to natural draft stacks.
- Require waterproof mortar for inside lining joints.
- Specify size of inspection door and its location.
- Specify ladder with safety cage (if necessary).
- Specify airplane warning lights if height of stack requires them.
- Provide location and dimensions of breeching connection.
- Specify form and construction materials of rain hood if provided.
- Specify drainage connections, locating piping, valves, and other items as required.

(d) *Masonry Chimneys*

- Specify if to be constructed of radial brick or concrete.
- Provide drawing showing height and all required dimensions, together with location of ladder, safety cage, gas sampling platform (if required), position and dimensions of lightning rods, conductors to ground. Also breeching connection.
  - Drawing to show all dimensions and weight of foundation, reinforcing rods, earth cover.
- Specify thickness of concrete walls, size and placement of reinforcing rods.
- Specify kind of lining or inside finish. If refractory material, give temperature that material must endure without spalling, cracking, or otherwise failing in service.
- Provide bidders with data on volume and weight of flue

gas to be handled per hour, and analysis of the gas with particular reference to sulfur compounds.

- Specify if draft is natural or mechanical, indicating whether fans are induced, forced, or both. In either case, give gas pressure at point of fan discharge into stack, ________ in. W.G. If natural draft, give required negative draft, in. W.G., measured at the base of the stack, or at point of breeching connection, as appropriate.
- Specify that stack thimble where stack passes through a combustible roof, must provide clearances in accordance with requirements of the National Fire Protection Association.
  - Minimum clearance for temperatures between 600 F and 1500 F, to be ________ in.
  - Over 1500 F, stack must be lined with firebrick, and minimum clearance to be ________ in.

Note: NFPA calls for 18 in. minimum clearance in both cases.

(e) ***Proprietary Stacks***

- Generally identified as the venturi or evasé form of stack, incorporating the induced draft fan in the base below the throat.
  - Describe the desired operation, as to whether the high-velocity jet is of combustion gas abstracted from the mainstream by the induced draft fan, or if fresh air is used.
- If combustion gas is jetted, give the analysis of the gas, its temperature range, and the volume and weight of the total gas output from the boiler per hour.
- Specify the maximum fan horsepower that will be acceptable under the guaranteed performance conditions.
- Specify all fan and motor chararteristics, requiring the bidders to furnish complete operating data. Refer to Sec. 4-1(w) and Table 4-3 for detailed requirements, including induced draft fan.
- Require bidder to furnish control data unless owner prefers to set up the sequence of controls. In the latter instance, furnish data to the bidders.
- Require bidder to submit data on emission of fly ash and cinders for at least three points of boiler operation. These three points to be specified to the bidders.
- Require bidders to submit horsepower used by fan at each of three points of boiler loading specified by the

owner. These three points may be the same as for fly-ash emission data.

- Bidders to supply all detail drawings as necessary to show dimensions and weights, including bleed of combustion air to the induced draft fan inlet, and arrangement of fresh air to fan as appropriate.
- Specify painting of ducts to fan inlet, both inside and outside.
- Specify combustion air connection insulation if required.
- For fresh air inlet, specify bird screen.

(f) *Spark Arresters*

Note: Generally not essential for gas or oil fuel; may be required for wood, sawdust, bagasse, and some coal, either to eliminate an atmospheric nuisance or to safeguard contiguous areas from ignition by sparks. Always needed for industrial incinerators and some domestic ones, especially apartment house installations.

- Specify that spark arrester must satisfy all local air pollution and fire prevention codes. Also must meet requirements of National Fire Protection Standard No. 213, as applicable to this installation.
- Specify materials of construction, as copper bearing steel, stainless steel, etc.
- Specify minimum and maximum acceptable mesh and give expected sizes of particles to be arrested.
- Specify that spark arrester and chimney combination must withstand the local wind velocities of ______ mph.
- Stipulate that chimney must be sufficiently strong to carry the spark arrester without undue strain.
    - Reinforcing the chimney or stack if necessary.
- Specify that arrester is to be assembled with welded joints rather than riveted, to ensure a longer life.

4-14. CONDENSERS FOR STEAM POWER PLANTS

(a) *Surface Condensers*

- Specify the source of the steam the condenser receives, as a steam-turbine, steam engine, industrial process, etc.
- Stipulate the steam pressure, if positive, in psig. If steam is in vapor form under some degree of vacuum, stipulate this in ______ in. Hg column against a

_______ in. barometer. For sea level, the barometric pressure is generally taken as 30 in.

- Specify capacity of condenser as _______ pounds of steam per hour condensed maximum, when supplied with _______ gpm of circulating water at a water temperature of _______ F.
- Specify the maximum acceptable drop in vacuum between the air pump connection and the steam inlet to the condenser as _______ in. Hg. Guaranteed drop is to be set by bidders.
- Specify the temperature of the condensate. This may be given as _______ F below the equivalent temperature of the vacuum, when this vacuum is measured at the neck of the condenser.
- Specify the construction of the condenser, giving material of shell, heads, baffles, baffle supports, etc. Itemize the important openings such as the inlet and outlet for the circulating water, steam inlet, condensate outlet or hot well, atmospheric exhaust, vacuum pump opening, air or overflow connections from other pieces of equipment. Make sure all openings needed are named to avoid later corrections of the oversight and added costs.
- Describe condenser supports, as spring supports, counterweighted supports, suspensions, etc.
- Describe type of access door required, including manholes in condenser heads. Give required materials as cast iron or steel, methods of making tight when closed.
- Specify paint and number of coats to be applied to both interior and exterior of the condenser.
- Specify materials and gage of condenser tubes and tube sheets, and designate which end of the tube is to be rolled into the sheet and which is to be packed.
  - Specify packing for tubes, giving materials.
- Specify packing elsewhere in the condenser, as between the steam source and the condenser joint, between the water box and water box cover, etc., giving materials, thickness, and placement methods.
- Specify number of gage glasses to be provided on hot well, also specify the strainer, size, material, and access.
- For circulating water, specify the various elevations involved, referred to a stipulated datum. Example: elevation of water at source (river, lake, etc.) at mean high

water, extreme high water, mean low water and extreme low water, given as plus or minus ft and in. to datum.

- Require bidders to submit operating data as indicated on the performance form, Table 4-10.
- Require bidders to submit condenser physical data as indicated on the data form, Table 4-11.

Table 4-10. Performance of Surface Condenser

Job ____________ Model ____________

Condenser manufacturer ________ Maximum capacity, lb/hr ________

Gpm	Temperature in, F	Temperature out, F	Absolute pressure in condenser, in. Hg	Temperature of condensate, F
Constant*...	40			
Constant*...	50	(To be filled in by bidders)		
Constant*...	60			
Constant*...	75			

* Require bidders to submit performance data as above, for at least three (3) different constant flows of circulating water. These may be classed as minimum, normal, and maximum gpm.

Specification should indicate at which gpm and which inlet temperature the performance is to be guaranteed.

In all cases, the inlet temperatures shall be selected to suit the conditions known to exist.

(b) *Condenser Auxiliaries*

1. AIR EJECTORS
   - Specify number, types, and arrangement of the air ejectors.
   - Specify condition of steam supplied to the ejectors, as the pressure, psig; quality; superheat, if any; total temperature.
   - If possibility is that ejectors may be obliged to work temporarily and occasionally on either higher or lower pressures and higher or lower total temperatures, specify the limits within which these temporary conditions will obtain.
   - Specify the construction and design of the ejectors, calling for inter and after condensers as necessary.
   - Specify number of stages and materials of parts.

- Specify capacity of ejectors in pounds of steam per hour; also give inlet temperature from condenser and absolute pressure to be maintained at ejector suction.
- Specify maximum back pressure against which ejector must discharge, in in. Hg.
- Specify maximum water temperature for the inter and after condensers.

Table 4-11. Surface Condenser Physical Data

Job ____________ Model ____________
Condenser manufacturer ________ Maximum capacity, lb/hr ________

Item	*Necessary information*
Total cooling surface, sq ft*	______________
Tube data:	
Material	______________
Number	______________
Length, ft and in.	______________
Outside diameter, in.	______________
Wall thickness, gage	______________
Thickness, in.:	
Shell	________ of ________ material
Heads	________ of ________ material
Supporting plates	________ of ________ material
Weight of condenser, empty, lb	______________
Weight of condenser, filled with water	______________

* Specification may stipulate a minimum of cooling surface acceptable, if considered desirable.

2. CONDENSER CIRCULATING PUMPS
   - Specify number and type of circulating pumps, as centrifugal, double suction, horizontal, etc.
   - Specify capacity of each pump in gpm when operating singly. If pumps may operate as a battery of two or more, specify capacity required of each in gpm under this condition.
   - Require pump to have sufficient discharge capacity to establish a flow through the condenser at start-up, and despite the absence of any syphon at the outlet of the condenser.
   - Specify the construction of the pump, as metals of casing and impeller.

- Including a minimum design strength of the casing to withstand a pressure of ________ psig. Some specifications limit the tensile stress created in the material by the maximum pressure. This is optional.

- Specify split of casing, as horizontal, to permit easy access to impeller.
- Specify a priming connection at the top of the casing.
- Specify the suction head (or lift) in ft, and give temperature of water. Also give range of water elevations as noted in Sec. 4-14(a) under the subhead referring to circulating water.
- Refer to Sec. 4-12(l) for additional construction characteristics of glands, lubrication, etc.
- Specify kind of coupling and stipulate if both pump and motor are to be mounted on a common base.
- Specify motors for circulating pumps, kind, type, etc. (induction, synchronous).
  - Giving electrical service available as voltage, phases, cycles, wires.
- Require bidders to provide the following information:
  Temperature rise, rotor ________ C
  Temperature rise, stator ________ C
  above an ambient of ________ C when carrying the specified load.
- All materials and construction to be in accordance with National Electrical Code and NEMA, as appropriate.
- Specify starting equipment and characteristics desired.
- Specify motor control safety features as transformers, relays, circuit breakers to disconnect motor in event of low voltage and to restart the motor on the return to normal voltage. These actions are to be taken automatically.
  - Giving voltages at which these actions are to occur.
- Specify mounting of circuit breakers and relays, giving form and material of panel.
- Require bidders to provide the information designated in Table 4-12 concerning motor performance.
- Bidders to state weight of motor assembled and weight of heaviest pieces.
- Bidders to provide detail drawings showing motor dimensions; also dimensions of starting equipment,

panel, and other components of the circulating water pumps system.

- Bidders to provide full information on the starting equipment, with catalogue identifications as appropriate.

3. ATMOSPHERIC EXHAUST VALVES

- Specify the size of the condenser in sq ft of heating surface.
- Specify the pounds of steam per hour to be vented to atmosphere through the valve at ______ psig, and give quality and total temperature, ______ F.
- Specify construction of valve in materials and general arrangement.

Table 4-12. Performance of Circulating Pump Motors

Gpm	Speed, rpm	Motor hp	Power factor, %	Efficiency, %	Starting amps*
Full load......					
¾ load........					
½ load........					

* At the given voltage.

Note: Bidders also to state

(*a*) Per cent slip at full load.

(*b*) Starting torque ______ per cent of full-load torque.

(*c*) Running torque ______ per cent of full-load rated torque.

- Specify location of valve with reference to condenser and operating floor. Provide a sketch or drawing showing location and vent to atmosphere.
- Specify the differential pressure range, psi, across the disk over which valve must be airtight.
- Specify size, type, and location of vacuum breaker on condenser.
- Require bidders to submit drawings showing all dimensions and giving weight of valve, including extension stem if needed.
- Have bidders also provide information on the pressure, psig, required to lift the disk from the seat, and on the pressure drop through the valve when it is venting the stipulated ______ pounds of steam per hour at ______ psig.

4. HOT WELL PUMPS
   - See Sec. 4-12(1) for condensate pumps. These are the same as hot well pumps and are included with the boiler feedwater system in this Handbook, rather than with the condenser auxiliaries.

(c) *Circulating Water Intake Systems*

1. TRASH RACKS
   - Provide drawing for bidders showing general features of construction, and giving dimensions and kinds of materials.
   - Specify maximum flow of circulating water in gpm.
   - State water elevations at both extreme and mean low and high water.
   - Stipulate desired methods of cleaning the racks, with data on manual or automatic cleaning.
     - If gate lifts, describe operating stands.
   - Stipulate where this contractor's work begins and ends.
   - Specify ice protection system. If compressed air, state pressure, ________ psig, and show size and placement of piping.
   - For concrete construction of forebay, specify the concrete mix, method for pouring under water, and days allowed for mix to reach ultimate strength.
     - Giving size and spacing of steel reinforcing rods.
2. TRAVELING SCREENS
   - For motor-operated screens, specify available voltage, cycle, phases, and wire.
   - Stipulate temperature rise, ________ C above an ambient of ________ C.
   - Motor to meet requirements of National Electrical Code and standards of NEMA.
   - Starting equipment, ditto.
   - Specify the number of screens required, to be furnished complete with trays, frames, etc.
   - Specify material and mesh of the screens, as galvanized wire cloth, ½-in. mesh, etc.
   - Stipulate construction of trays to avoid sagging and require close fits between adjacent trays to prevent trash from sliding through.
   - Specify construction of chains on which the trays ride. Set pitch.
   - Specify the painting.

- Specify automatic safety controls of motors for conditions of low voltage.
- Specify that guards for moving parts, around chain sprockets, etc., must meet indicated safety codes. These are frequently local as well as general.
- All wiring is to meet National Eectrical Code.
- Specify clearly the scope of this contractor's work.
- Require bidders to submit detail drawings of their equipment, showing dimensions, weights, and all pertinent data.
- Where gutters, refuse troughs, and the supporting construction of concrete are furnished by the owner or others, provide a drawing to bidders showing where and how the screen equipment must be placed.
- Specify manner of cleaning the screens. If by water nozzles, state water pressure and methods of operation.
- If motors and heads of screens are housed, so state.
- Specify how bearings and reduction gears (if any) are to be lubricated.
- Specify the speed of travel of the screen as ______ fpm.

3. CIRCULATING WATER CONDUITS, CONCRETE
   - Provide bidders with drawings showing location and dimensions of the conduits, pitch, and details of the circulating pump suction well.
     - Also show thickness of concrete, reinforcing steel, supporting means, etc.
   - Specify concrete mix and stipulate time for set or curing.
   - Specify kind of forms to be used, wood or metal, giving dimensions, thickness, etc.
   - Specify safety precautions during construction, as timbering and sheathing of excavation, dewatering, reference to local codes.
   - Specify kind and methods of tests, as slump test.
   - Specify how concrete is to be protected after removal of forms, if forms are removed early.
   - Refer to compliance with the standard specifications of the American Society for Testing and Materials as these apply to aggregates, and otherwise as appropriate.
   - Specify finish if desired; also waterproofing.

4. CIRCULATING WATER CONDUITS, METAL

- Provide bidders with drawings showing location, dimensions, pitch, and details of the circulating pump suction well.
- Specify material of conduit, gage thickness (or inches of thickness), kind of joints, and method for making them watertight.
- Specify safety precautions, the timbering and sheathing of the excavation, dewatering, and refer to applicable local codes.
- Specify method of supporting conduit, and give information on method of compensating for expansion and contraction.
- Stipulate the range of water temperatures, F. Also range of hardness and the presence of pollution, if any.
- Specify methods to be provided by this contractor for combating algae, if appropriate.
- Specify painting of conduit, both inside and out, noting materials and numbers of coats.
- Specify that all welding is to be accomplished by welders qualified by a recognized qualifying agency, such as the National Certified Pipe Welding Bureau.

(d) ***Barometric Condensers***

- Specify scope of work, listing components of the condensing system to be furnished and installed by this contractor.
- Name source of steam, as steam turbine, steam engine, process, etc.
- Specify steam conditions entering the condenser, as pressure, psia, total temperature.
- Specify temperature range of condensing water, with particular reference to the highest temperature normally experienced.
- Specify pounds of steam per hour to be condensed.
- Specify height above the level of the water in the hot well, to the water inlet opening in the head.
- At maximum guaranteed capacity, specify the vacuum desired in ________ in. Hg column.
- If dry air pump is to be included, specify kind and capacity.
- Specify if condenser is to be designed for parallel flow or counterflow.

- Specify materials of construction for head, baffles, water or splash trays; tail pipe.
- If tail pipe is not to be installed, specify the condensing water pump, giving its kind, form of drive, capacity in gpm.
- For motor drives, specify service characteristics, as voltage, phase, cycles, wires.
- Details of pump specification requirements may be taken from Sec. 4-12(1), Condensate Pumps.
- Motors and all electrical components are to meet requirements of National Electrical Code and standards of NEMA.
- Require bidders to submit information as follows:
  - A detail drawing showing all dimensions, all components, with weights.
  - Performance data including temperature of condensing water into condenser and temperature entering tail pipe, F, at three points of operation, including the guarantee point.
  - Vacuum established for each point to be given.
  - Also motor hp of the pump (if any) and the temperature rise, ________ C above an ambient of ________ C.
- Stipulate if this contractor is to furnish all erection steel needed for the installation.
  - If so, specificaion must include a drawing showing arrangement, dimensions, and weights.
- Specify time and method of testing condenser against its guaranteed and specified capacity.

(e) *Jet Condensers*

- Similar to barometric condenser, but has very short tail pipe, or none.
- Absence of tail pipe requires condensing water pump. Specifications for this pump are the same as for the barometric condenser pump described in the preceding section.
- Specify the vacuum breaker, giving materials of float and level of operation.
- Specify form and type of air ejector.
- Require this contractor to furnish all physical and operating data as designated for the barometric condenser.
- All data concerning water characteristics, temperatures, hardness, etc., must be furnished to bidders.
- Specify capacity of condenser in pounds of steam an

hour, under the specified conditions of water temperatures, vacuum, and gpm of condensing water.

Note: Jet condensers are loosely termed barometric condensers by some specifiers. There is a difference, although slight; and a properly written specification will recognize the difference and be worded accordingly.

- Both condensers are found in small plants.
- Their efficiency generally is lower than for surface condensers. Another important drawback: the condensate is mixed with the condensing water, and in most instances, the mixture requires treatment before it may be supplied to boilers. Treatment may include softening and deaeration. For details of the former as a specifying guide, see Sec. 4-12(d)–(f). Deaeration is covered in Sec. 4-12(g).

(f) ***Condensing Water-cooling Towers, Natural Draft, Wet***

Note: Cooling towers are classed as "wet" towers if designed for direct contact between the water and the cooling air; "dry" towers are those in which the water to be cooled flows through tubes (usually finned) and the air passes over the tubes so there is no direct contact. With the "wet" tower, cooling is accomplished mainly by evaporation, so that a continuous makeup must be provided. This makeup amounts to 5 per cent, more or less, of the total input to the tower. Specifications tend to limit this loss to acceptable values. Natural draft, induced draft and forced draft are applicable to both wet and dry towers.

- Specify kind and number of cells.
- Specify construction, as:

1. HYPERBOLIC TOWERS
    - Stipulate windage (usually 100 mph, or more in hurricane areas) that tower must withstand.
    - Set minimum thickness of concrete, both bottom and top.
    - Specify the kind of piping and water-distributing troughs, as asbestos cement, steel, brass, etc.
    - Specify kind of tower fill. Also kind or type of sprinkler heads.
    - Specify capacity of tower for delivering ________ gpm of water cooled from ________ F to ________ F, with an ambient wet bulb of ________ F and operating solely by natural draft.

- Specify maximum acceptable loss of water by evaporation in terms of per cent at the point of guaranteed performance.
- Provide information to bidders on bearing ability of ground where tower is to be located. State if this contractor is to furnish the foundation or to erect tower on foundation by others.
- Provide bidders with information on the character of the water to be cooled, its hardness as ______ ppm in terms of calcium carbonate, presence of sulfur or any other corrosive materials.
- For collecting basin, specify a float valve control for admitting the makeup water.
- Specify control of cooled water pump, float, temperature, etc.
- Require bidders to furnish detail drawings, giving dimensions and weights. Also to submit performance data with ambient wet bulb temperatures of ______ F, ______ F, and ______ F, the inlet water temperature in each instance to be ______ F. Performance data at these three points (one the guarantee point) shall include the percentage loss by evaporation.

2. WOOD LOUVER TOWERS
   - Specify capacity of tower as cooling ______ gpm from ______ F to ______ F with an ambient wet bulb of ______ F, operating solely on natural draft.
   - Specify construction as California redwood or cement asbestos siding. Specify a spray filled, or redwood filled tower, with spray nozzles of ______ ______ (material), distribution piping of ______ ______ (material) (galvanized).
   - Specify kind of hardware, bolts, nuts, screens, etc. (bronze, brass, galvanized, etc.).
   - Specify float control and valve, using ______ ______ material for each.
   - Specify if foundations or supports are to be furnished by this contractor or other.
   - Specify how much and what kind of piping, fittings, and valves are to be furnished by this contractor.
   - Specify windage tower must withstand in mph.
   - Provide bidders with information on characteristics

of water pumped to tower; also pressure at tower inlet in psig.

- Specify how flow of water to and through the tower is to be controlled, and what control equipment, if any, this contractor is to furnish.
- Require all bidders to furnish detail drawings showing dimensions and weights. Also bidders are to submit performance data with three (3) separate ambient wet bulbs with inlet water temperature of ________ F, and a windage or evaporation loss of ________ per cent. One of these points is specified as the guarantee point. Bidders may submit performance curves with the guarantee point marked, if they so desire.

(g) *Condensing Water-cooling Towers, Mechanical Draft, Wet*

Note: This form of tower is generally found to be most applicable to steam power plant requirements, although the use of dry towers is increasing. Whether forced or induced draft is selected is a matter for engineering study, in any given job.

- Specify number of cells and their locations.
- Specify cooling capacity of the installation in total and by cell, in terms of gpm cooled from ________ F to ________ F with an ambient wet bulb of ________ F and with an air flow through the tower of ________ cfm, this air to have a dry bulb temperature of ——— F.
- Specify windage tower must withstand, in mph.
- Specify construction of casing, as California redwood or steel. If redwood, stipulate framing required and placement of the siding, materials of nuts, bolts, nails, etc. If steel casing, stipulate the gage material, manner of assembly, as welding, bolting, etc. If welded, require welders to be qualified by a recognized qualifying agency. Also stipulate protection of steel against corrosion, as bonderizing, galvanizing, or painting with a material resistant to hot moist atmospheres, oxygen attack, etc. Specify kind of filling, means taken to prevent sagging.
- Specify kind and action of the basin float control valve, including materials.
- Specify piping, valves, and fittings of tower supply piping that are to be furnished and installed by this contractor.

- Specify who furnishes foundations and supports. If this contractor does, furnish data in the specification on the bearing capacity of the soil where tower is to be set.
- Provide bidders with data on water to be cooled, the degree of hardness, and the inclusion, if any, of corrosive agents such as sulfur, carbon compounds, boiler water treatment chemicals, etc.
- Specify whether induced draft or forced draft fans are to be provided.
- Specify current characteristics for fan motor operation, as voltage, phases, cycles, wires. All motors and electrical components to satisfy requirements of the National Electrical Code and the standards of NEMA.
- Specify how spray from tower is to be controlled, and limits acceptable. Include maximum percentage of makeup water that is acceptable for this job.
- Specify maximum noise level of the tower when operating at the guarantee point.
- Specify bearings of fan and motor, whether babbitt, ball bearing, roller bearing; and means of lubrication.
- Specify fan protection against corrosion from the atmosphere, including moist hot air in the placement of induced draft fans.
- Require bidders to furnish detail drawings showing all pertinent dimensions and weights.
- Require bidders to fill in and submit the performance data called for in the form in Table 4-13, plus such other data that might be desirable.

(h) *Condensing Water-cooling Towers, Mechanical Draft, Dry*

- Specify number of dry cells and their locations.
- Specify cooling capacity per cell, in terms of gpm cooled from _______ F to _______ F with an ambient dry bulb of _______ F.
- Specify wind pressure in mph that tower must withstand.
- Specify construction of casing, gage of steel used, method of making airtight joints. If welded, require welders to be qualified by a recognized qualifying agency.
- Specify protection of casing against atmospheric corrosion, as by bonderizing, galvanizing, painting, etc.
- Specify material and shape of tubes, as aluminum and elliptical.
- Specify size and number of fins per foot of tubing, also method of bonding fins to the tubes.

- Specify shape and working pressure of headers, whether box or cylindrical, cover plate type, forged, etc.
- Specify arrangement of tube bundle with access provided to tubes and headers.
- Specify fan (with variable pitch optional) and type of coupling that connects fan to driving motor or reduction gear.

Table 4-13. Performance of Mechanical Draft Cooling Tower

Job ____________ Kind of tower ____________ Bidder ____________

Item	Data to be filled in		
	Column A	Column B*	Column C
Ambient wet bulb†	F	F	F
Inlet water temperature	F	F	F
Outlet water temperature	F	F	F
Ambient dry bulb	F	F	F
Water-cooled	Gpm	Gpm	Gpm
Head loss in tower	In. W.G.	In. W.G.	In. W.G.
Evaporation	%	%	%
Air through tower	Cfm	Cfm	Cfm
Fan discharge pressure	In. W.G.	In. W.G.	In. W.G.
Friction loss in tower	In. W.G.	In. W.G.	In. W.G.
Motor horsepower	Hp	Hp	Hp
Sound level	Decibels	Decibels	Decibels
Wind loss	%	%	%
Fan speed	Rpm	Rpm	Rpm

* This column to be guaranteed by bidders.

† All three ambient temperatures to be specified. Bidders required to fill in the other spaces in these three columns.

- Specify lubrication system, with particular reference to the reduction gears, if these are used.
- Specify that all electrical components be in compliance with the National Electrical Code and with the standards of NEMA.
- Specify characteristics of electrical service available as voltage, phases, cycles, and wires.
- Specify manner of control, whether by automatic adjustment of the variable-pitch fan blades, by variable-

speed drive, by adjustable shutters, or any combination of these methods.

- Specify thermostatic arrangement that, actuated by the temperature of the condenser water leaving the tower, signals the need for the control sequence to adjust the volume of air being handled by the tower fan.
- Require bidders to furnish detail drawings of their equipment, showing dimensions and weights. Weight is particularly important if tower is to be roof mounted.
- Require bidders to fill in and submit with their proposals the performance data called for on the form in Table 4-13, with the exception of the water losses from evaporation and windage. These do not apply to towers in which the water is completely segregated from any contact with the atmosphere. In addition to the data requested on this form, other information may be requested as applicable to the job.

(i) *Condensing Water-cooling Towers, Natural Draft, Dry*

Note: This tower is recognized simply because it is theoretically capable of cooling water. Because of the relatively slow movement of natural draft over the tube bundle, the exchange of heat is slow, thus requiring a total tube surface that is too great to be economically justifiable. For this reason, specification data are omitted.

(j) *Condenser Water Spray Ponds*

Note: Spray ponds find considerable use where the land area is available, and where they contribute to the beauty of a setting by creating a lake and fountains. As a drawback, on windy days their blowing spray can be annoying.

- As with towers, specify capacity of the pond in gpm from ________ F to ________ F, with the wet bulb temperature at ________ F.
- Provide bidders with a drawing showing arrangement of spray nozzles and their piping, together with the dimensions of the pond, the thickness of the concrete, the method of supporting the nozzles and their piping, locations and sizes of drains and overflows.
- Specify kind of spray fence or other means for preventing blowing spray from becoming a nuisance.
- Stipulate size and kind of makeup valve, with the method of operation, as a float, thermostatic bulb, etc.

- Specify the concrete mix, size of aggregate, and set time of curing.
- Specify manner of waterproofing the concrete.
- Size and space the reinforcing steel.
- Specify the water pressure at the nozzles, ______ psig.
- Specify the materials of the nozzles and piping.
- Define clearly where this contractor's work begins and ends.
- Require bidders to furnish the following information at three conditions of wet bulb as enumerated in the specification, and at the nozzle water temperature also specified:
  1. Average final temperature of the pond.
  2. Percentage of loss by evaporation.
  3. Percentage of loss by windage.
- If pond-cleaning facilities are to be provided, specify what is wanted and by whom furnished.

  Note: Refer to swimming pool auxiliary cleaning catalogues for ideas on this. See Sec. 22-4.
- If design of pond is delegated to the contractor who will furnish the material and install it, specify the surface area to be provided in sq ft, for pounds of water sprayed per hour. For small plants, a rule of thumb provides 1 sq ft for each 150 pounds of warm water sprayed per hour. For the medium to large plants, the weight of water sprayed per sq ft may be increased by approximately 50 per cent.
  - Also specify the desired depth.

4-15. STEAM TURBINES

Note: Steam turbines functioning as auxiliary drives are specified with the auxiliaries driven. For example, steam turbines driving boiler-feed pumps are covered under the heading Boiler-feed Pumps, Sec. 4-12(k).

Turbines used for generating power are generally condensing, since back pressure even at the atmospheric level increases steam consumption and consequently reduces economy of operation. For large-capacity installations, 10,000 kw and above, the ASME has developed a preferred standard on the basis of 3,600-rpm, 3-phase, and 60-cycle power output. This standard should be consulted by all turbine-generator specifiers and carefully evaluated prior to preparing a specification in the indicated size range, since adherence to the standard, to the extent the job allows, is a way to conserve on time and investment.

For the smaller turbines, the specification writer's attention is directed to the American Petroleum Institute's turbine specification No. 615, together with the appended data sheets and checklist. Although this specification has been prepared for the general-purpose and special-purpose turbines used in refinery operations, it has valuable suggestive ideas for all turbines.

Because turbine-generator installations incorporate so many aspects which are peculiar to the job, a general specification is impractical. In using the guides given in this section, this must be remembered, and changes, additions, and deletions made as appropriate to the job.

See Table 4-14 for steam-turbine data by bidders.

(a) *General Turbine Characteristics*

- Specify type, as impulse, reaction, condensing (with _______ points of extraction if applicable), whether simple or compound, or reheat, exhausting into a (surface) (barometric) condenser, and driving through (gears) or (coupling) a _______ kw, _______ volt, _______ cycle, _______ phase electric generator.
- Turbine to operate at a speed of _______ rpm when supplied with steam of _______ psig and _______ F total temperature, and exhaust against an absolute pressure in the condenser of _______ in. Hg column and a _______ in. barometer.
- Steam extraction points (if applicable) shall be provided at the _______ and _______ stages, and sized to bleed _______ and _______ pounds of steam an hour, respectively, at the respective pressures of _______ psia and _______ psia.
  - Reheat bleed (if applicable) shall be taken off the turbine immediately after the _______ stage and pass _______ pounds of steam per hour at _______ psia and a temperature of _______ F for return to boiler reheat coil.
  - Reheat steam (if applicable) shall be introduced to the turbine low-pressure section as specified, and at a pressure of _______ psia and a total temperature of _______ F.
- Specify steam pressure range at throttle over which turbine must operate without danger or damage, as between

Table 4-14. Steam Turbine-Generator Data, Filled in by Bidders

Job ____________ Make, turbine ____________
Turbine, type ____________ Make, generator ____________
Generator, type ____________ Bidder ____________

Item	Steam pressure at throttle		
	Normal pressure	Normal plus ____ %	Normal less ____ %
Steam rate*	Lb/hr	Lb/hr	Lb/hr
Back pressure			
Condensing	In. Hg	In. Hg	In. Hg
Noncondensing	Psi	Psi	Psi
Steam bled for reheating	Lb/hr	Lb/hr	Lb/hr
Temperature			
Steam to reheater	F	F	F
Steam from reheater	F	F	F
Speed	Rpm	Rpm	Rpm
Cooling water			
Turbine lubricating oil	Gpm	Gpm	Gpm
Generator cooler	Gpm	Gpm	Gpm
Cooling-water temperature	F	F	F
Warm-up time before start	Hr	Hr	Hr
Safe temperature range, generator			
In rotor	____ F to ____ F	____ F to ____ F	____ F to ____ F
In stator	____ F to ____ F	____ F to ____ F	____ F to ____ F
Hydrogen cooling, pounds of charge	Lb of charge	Lb of charge	Lb of charge
Horsepower of turning gear motor ____ amps at ____ volts	Hp	Hp	Hp
Maximum axial shaft movement	In.	In.	In.
Clearance of wheel shrouds	In.	In.	In.

* Guarantee point, to be indicated.

a minimum pressure of ______ psig and ______ F total temperature, and a maximum of ______ psig and ______ F.

- Specify the speed variation that is acceptable when the turbine is running on its governor, as not more than ______ per cent above or more than ______ per cent below the normal operating speed.

- Specify if hand nozzle valves are to be provided for close control and for carrying the load at a dropped steam pressure.
- Specify the kind of running governor and governor valve that are to be furnished, as a spring-loaded, flyball, or other form of governor, and incorporating an overspeed trip that is adjustable, but is generally expected to shut the steam valve when the turbine speed exceeds the normal speed by ________ per cent. The steam valve to be of ________________ steel and suitable for a maximum pressure of ________ psig and total temperature of ________ F.
- Specify materials of buckets (or blades), and stipulate manner of attachment to the rotor wheel, or whether milled out of the solid metal.
- Specify material of stage diaphragms and manner of sealing into the casing to prevent pressure losses between stages.
- Specify turbine shaft material, stating minimum radius of fillets, and require both static and dynamic balancing.
- Specify if end thrust is to be compensated for by a thrust bearing, or other means.
- If an axial position micrometer is needed in permanent position, so specify.
- Specify material and arrangement of casing, as cast iron, cast molybdenum steel, horizontally split and bolted.
  - To withstand pressures of ________ psig and total temperatures of ________ F.

Note: Because of the wide difference of pressures between the high-pressure inlet end of a high-pressure turbine, and the greatly lowered pressures at the exhaust outlet and reheat sections, it is common practice to use two different kinds of casing material. Investment costs are lowered in this way. The specification should designate the two materials in these instances.

- Specify the gland sealing system.
- Specify that casing joints are to be machined smooth and made tight without gaskets.
- For turbines of heavy rotors and those which are shut down for more than 24 consecutive hours, specify a motor-driven turning gear to prevent the rotor and shaft from taking a permanent set.
- Specify the kind of coupling used to connect the turbine

shaft with the generator shaft, if these have not been manufactured as a single unit.

- Specify space available.
- Specify the source of the foundations and the kind, as reinforced concrete with dimensions and drawings provided, formed and poured by others. If structural steel, specify as appropriate.
- Describe crane and hoisting facilities at the site which are available for lifting and placing the individual elements or the unit as a whole.
- Specify the turbine lubrication system, including, as appropriate: (1) a lubricating oil reservoir heater; (2) the lubricating oil pump; (3) lubricating oil filtering system, or alternatively, a centrifuge for purification; (4) the lubricating oil cooler; (5) range of temperatures for oil entering the bearings, _______ F, and the temperature leaving, _______ F. (6) the lubricating oil pressure; (7) a dependable safety alarm to signal the failure of pressure, an extreme and dangerous rise in temperature, and interruptions of flow.
- Specify the electric generator's required characteristics as alternating form, _______ volts, _______ phase, _______ cycles, _______ kva rating at _______ per cent power factor, and a constant speed of _______ rpm.
  - Using an exciter mounted on the extended generator shaft. Or alternatively, excitation current to be provided by a house generator.
  - Excitation current to provide _______ kw at _______ volts.
  - Specify numbers and location of thermocouples for the automatic signaling of dangerous temperature conditions inside the generator.
  - Specify (if felt necessary) an automatic fire extinguishing system in the generator, using carbon dioxide gas.
  - Specify the kind of generator cooling system, as totally closed hydrogen gas or totally closed air. System to be capable of removing _______ Btu per hour, with the circulation of the gas or air maintained by the rotation of the generator rotor in suitable volumes and at suitable pressures.
    - Specify the heat exchanger, as finned tube type, with cooling water passing through the tubes at a rate of _______ gpm. Exchanger to be capable of trans-

ferring ______ Btu per hour from the gas or air to the water, when the water reaches a maximum temperature of ______ F. Give temperature range of circulating water from ______ F minimum.

- Require bidders to submit dimensional drawings.

(b) ***Turbine-Generators for Outdoor Placement***

- If equipment is to be exposed to the weather, specify a form of antirusting, anticorrosion treatment for the turbine and generator housing. Example: bonderized or parkerized steel; painting with anticorrosive paint.
- Alternatively, a canopy form of roof may be specified to prevent direct impact of snow and rain. All sides may be left open under such a canopy, unless the one facing the prevailing wind is enclosed as additional protection.
- Require provisions for heating the lubricating oil prior to the start of the unit, as an electric resistance heater in the lubricating oil reservoir. Also for holding the generator cooling water at temperatures above freezing. Require an adequate number of drains so that accumulations of water may be avoided in subfreezing weather. This includes steam and condensate lines.
- Specify some form of area heating, as by infrared overhead heaters, so that operating personnel can attend the installation without undue suffering from cold.

(c) ***Turbine-Generators for Topping Service***

Note: This is a high-pressure steam turbine exhausting into the steam supply header for one or more low-pressure steam turbines, and consequently it functions as a back-pressure turbine instead of condensing. Condensation takes place in the condensers already serving the low-pressure turbines.

Topping turbines are generally adopted to increase the output and efficiency of steam-electric generating plants that are becoming economically obsolete. The scheme permits the replacement of low-pressure steam boilers by high-efficiency, high-pressure units, without requiring new or additional plant space. As an alternate adaptation, instead of exhausting into a low-pressure steam header that serves other turbines, the exhaust steam is diverted to heating and/or process steam functions. In these instances, the topping turbine becomes an electric-energy-producing, pressure-reducing valve.

- Specify the type of turbine, as impulse, reaction, non-condensing, back-pressure unit, exhausting against a back pressure of _______ psig and at a total exhaust temperature of _______ F and driving a _______ kw, _______ volt, _______ phase, and _______ cycle electric generator.
- Specify turbine speed as _______ rpm, and supplied with steam at _______ psig and _______ F total temperature.
- Specify steam pressure range at the throttle, over which the turbine must operate without danger or damage.
- Specify the speed variations that are acceptable when the turbine is running on its governor, as not more than _______ per cent above or more than _______ per cent below the normal operating speed.
- Specify the kind and type of both running and emergency governors, as stipulated under Sec. 4-15(a), General Turbine Characteristics.
- Specify required details of bucket (or blades) material, the method of fastening to the rotor if not milled out of a solid forging. Also specify materials of stage diaphragms and turbine shaft, and details of placement and design features.
- Specify manner of compensating for axial thrust.
- Specify material of casing and general construction, as cast steel, horizontally split and bolted.
  - Indicating pressures, _______ psig, and temperatures, _______ F, casing must withstand.
- Specify the gland sealing system.
- Require casing flanges to be machined smooth, so that no gasket is necessary.
- If rotor is large, or subject to shutdowns lasting more than 24 hours continuously, specify some form of shaft turning gear, to prevent cooling distortion of the shaft.
- Specify kind of coupling between turbine and generator.
- Provide a drawing showing location of the topping turbine and its generator relative to the low-pressure steam header into which it discharges, and indicating what this contractor must provide and do in setting up the unit. Indicate handling facilities.
- Specify turbine lubricating system, as detailed in Sec. 4-15(a).
- Specify the electric generator, similarly to the guiding

items detailed in Section 4-15(a) for electric generators that are turbine-driven. This includes fire protection and generator cooling, excitation, etc.

- If turbine-generator is to be placed outdoors, see Sec. 4-15(b) for items of weather protection.
- Require bidders to submit complete drawings of their equipment, showing all important dimensions and weights. Also to fill in the form in Table 4-14 as appropriate.

4-16. STEAM ENGINES

Note: Because a steam engine is basically a slow-speed machine, increases in horsepower involve an increase in physical size. This characteristic limits its current fields of application. It is an important item in marine engineering, where ship speed is relatively low.

- Specify the location and nature of the unit the steam engine is to drive.
- Specify the kind of engine wanted, as horizontal, vertical, number of cylinders, type of valves, as slide or poppet; also, if uniflow arrangement, the number of expansions.
- Specify the desired indicated horsepower per cylinder at _______ rpm, with inlet steam at _______ psig and _______ _______ F total temperature, exhausting into the exhaust manifold at a back pressure of _______ in. Hg column.
- Specify the oil-cooler water temperature minimum, _______ F, and maximum, _______ F.
- Specify the engine materials for the cylinders, pistons, cylinder heads, bolting. If cylinder liner is provided and removable, specify material hardness and surface finish. Require ground joints.
- Specify kind and number of piston rings, with hardness and tensile strength of the material.
- Specify kinds of bearings, as babbitt sleeve, roller, ball bearing. Specify materials, as steel rollers and a babbitt of, say, 90 per cent tin, 8 per cent antimony, and 2 per cent copper. Bronze bearings may be specified in some applications.
- Specify kind of packing and method of sealing. Require lubricating valve connections to the various rings for the possible counteracting of excessive frictional heat in the rods. Require an arrangement that permits repacking with a minimum of disassembly boxes and cases.

- Specify oil cleaning, as by centrifuge or by strainers. Stipulate kind and materials of strainers, as a monel metal or stainless steel basket. Also fineness of mesh.
- Specify kind and materials of gaskets for flange joints wherever these occur.
- Specify type and materials of cylinder drain valves.

Table 4-15. Steam Engine Operating Data, Filled in by Bidders

Job ________	Number of expansions ________
Engine type ________	Engine builder ________
Horsepower ________	Bidder ________

Performance item	*Necessary information*
Steam rate	
Lb/hr at full load, indicated horsepower	________
Steam temperature at throttle	________
Steam pressure, throttle	________
Back pressure	
In. Hg	________
Psia	________
Superheat at throttle, F	________
Full load, rpm	________
Full load mean effective pressure, psi	________
Oil pressure to bearings, psig	________
Oil temperature to bearings, F	________
Oil temperature to oil cooler, F	________
Oil temperature leaving cooler, F	________
Cooling water at maximum water temperature, gpm	________
Cooling water at minimum water temperature, gpm	________
Water pressure drop in cooler, psig	________

Note: Instead of requiring bidders to submit some of this information, the specification may stipulate the various values and require the bidders to offer engines to suit.

- Specify type and material of throttle valve if a particular unit is desired; otherwise require bidders to submit complete data on the throttle valve they plan to furnish.
- Specify the engine governing system, as centrifugal, inertial, etc.
- Specify the tools that are to be furnished with the engine by the bidder.
- Require detail drawings, showing all dimensions, clearances, parts names, etc.

- Require bidders to submit physical and performance data as suggested on the forms in Tables 4-15 and 4-16.
- Specify space and weight limitations at the site of installation, if these exist.
- If exhaust steam from engine is to be used in processes, or condensed into boiler feed, removal of the entrained oil is essential. For a specifying guide to oil separators, see Sec. 4-18.

Table 4-16. Steam Engine Physical Data, Filled in by Bidders

Job ____________	Number of expansions ____________
Engine type ____________	Engine builder ____________
Horsepower ____________	Bidder ____________

Engine physical item	*Necessary information*
Number of cylinders	
Nominal cylinder bore, in.	
Nominal cylinder stroke, in.	
Piston rod diameter, in.	
Number and kind of piston rings	
Clearance	
Initial, piston, and cylinder, in.	
Permissible, piston, and cylinder, in.	
Clearance, top of piston and cylinder head, in.	
Number of lubricating oil sight feeds	
Number of revolutions per drop of oil in each sight feed	
Grade, kind, and viscosity of lubricating oil	
Flash point of lubricating oil, F	
Weight of heaviest part of engine, lb	
Total weight of assembled engine, lb	
Overall height, ft	
Overall width and length, ft	

Note: Other desired items may be included in this form.

- Specify where heat insulation is to be applied to the engine, as on the exhaust manifold, engine cylinders, etc. Specify the kind of insulation and its minimum thickness.
- Specify if this contractor or others is to furnish the engine foundations.
- Specify who is to furnish the lubricating oil cooler and its circulating water pumps.
- Describe availability of lifting gear for installation of the engine, including the maximum load it can handle.

- Require bidders to submit a list of spare parts and their individual costs, as recommended for on-the-job storage.

4-17. POWER PLANT MISCELLANEOUS PUMPS

(a) *Checklist of Miscellaneous Pumps*

1. Heater drain pumps.
2. Service water pumps.
3. Heating system pumps.
4. Water-treatment pumps.
5. Lubricating oil pumps.
6. Auxiliary pumps.
7. Oil transfer pumps.
8. Generator cooling pumps.
9. Refrigerant condensing pumps.
10. Evaporator pumps.
11. Cooling tower pumps.
12. Compressed air cooling inter- and aftercoolers.
13. Turning gear oil pump.
14. Auxiliary fuel pumps.

Note: These pumps may be centrifugal, rotary, reciprocating, screw type. See following sections.

(b) *Centrifugal Pumps*

- State service and kind of driver, whether motor, turbine, etc.
- Specify capacity, ________ gpm, against a total dynamic head of ________ ft, and including a suction lift (or head) of ________ ft, when handling (name of fluid) at a temperature of ________ F.
- Specify the nature of the fluid handled, as viscosity, density in pounds per gallon at the pumping temperature, whether corrosive or noncorrosive, hardness in parts per million, pH value, etc.
- Specify construction and construction materials, as iron, steel, bronze; number of stages, whether single or double suction; design pressure and factor of safety of casing; kind of glands; means to maintain balance or counteract thrust; whether casing is to be horizontally split or of the barrel type; design pressure of the casing; material of shaft sleeves; etc.
- Specify kind of coupling connecting pump and driver.
- Specify mounting arrangement, as pump and driver on a common bedplate of cast iron. Plate to be cast with raised lip and drain hole in a suitable area.

- Require bidders to submit detail drawings showing internal arrangement of pump, together with dimensions, method of assembly.
- Require bidders to submit performance curves of head, capacity in gpm, and hp. Also to state pump speed in rpm.
- For a specification guide for the steam turbine or electric motor driver, see Sec. 4-12(k) Boiler-feed Pumps.
- Specify desired manner of control.

(c) ***Reciprocating Steam Pumps***

- See Sec. 4-12(k) under heading Steam Pumps for a specification guide.

(d) ***Rotary Pumps***

Note: Included in this group are the gear rotaries, vane rotaries, and screw rotaries, all of positive displacement type. In general, their application is observed most frequently on oil pumping duty, as well as for other viscous fluids.

- Specify the liquid to be handled by the pump, giving complete data on viscosity, Saybolt; temperature; whether corrosive or noncorrosive; and pH value.
- Specify capacity as ________ gpm against a total dynamic head of ________ ft, including ________ ft of suction lift (or head), and handling the fluid at ________ F temperature.
- Specify materials and design pressure of pump casing, together with kind and position of casing joints.
- Specify materials of gears, vanes, or screws, and if these parts are machined as integral to the shafts, or are keyed and shrunk on.
- Specify type and kind of bearings, and their mode of required lubrication.
- If driven through reduction gears, specify the construction of these units, including the kind of gear teeth, method of attachment to the shaft, and if lubricating pump is driven by the gear or is to be provided as an external unit.
- Specify materials of reduction gear casing.
- If driven through a belt and pulleys, specify the kind of belt, as leather, rubber, composition; specify the shape as V, square, or flat; specify the kind of belt joint desired.
- Require an adequate belt and pulley guard as a safety feature.

- Specify kind of couplings to be provided.
- Specify electric motor service available, volts, phases, cycles, wires.
- Require motor to satisfy the NEMA standards and to meet all local codes.
- Require bidders to furnish detail sectional drawings of the pump and (if applicable) the reduction gear, showing arrangements and dimensions.
- Require bidders to furnish data concerning the speed, _______ rpm; the horsepower, and a family of curves showing the relationships between the pump speed, horsepower, and discharge head.
- If nature of pump control is not given by the specifier, require bidders to furnish information on the method of pump control that is proposed, subject to the owner's approval.
- Specify if pump and driver are to be mounted on common baseplate.

(e) *Duplex and Triplex Pumps, Motor-driven*

- Specify duplex or triplex; vertical or horizontal; driven by belt or through reduction gear.
- Specify capacity, _______ gpm, against a total dynamic head of _______ ft, including _______ ft of suction lift (or suction head), fluid temperature to be _______ F.
- Specify nature of fluid to be handled, its viscosity, seconds Saybolt; whether corrosive or anticorrosive; acid or alkaline as shown by the pH scale. If carrying suspended or entrained solids, describe these, and show concentration in ppm.
- If driven through a reduction gear, specify as indicated for rotary pumps, Sec. 4-17(d).
- If driven by belts and pulleys, refer to Section 4-17(d) also for data to include in the specification.
- Specify the system of lubrication and the pressure of the oil to the bearings if applicable; otherwise note how bearings are to be oiled, as manually or by automatic drip lubricator.
- In view of the fluid to be handled, specify the materials required in the cylinders, the piston, piston rods, and other parts in contact with the pumped fluid. In some refinery operations, very special alloys may be needed, as well as stainless steel, monel, bronze, etc.
- Specify pump control details, or require bidders to propose a system for approval.

- Specify that motor must operate on ________ volt, ________ cycle, ________ phase, ________ wire service.
  - And satisfy NEMA standards of construction and local codes.
- Require bidders to submit complete sectional drawings, illustrating pump arrangement and giving all pertinent dimensions.
- Require bidders to submit curves showing relationships between capacity, heads, and horsepower.
- Specify base, and if pump and driver are to occupy a common base. This base should have a raised lip around the periphery and a drain hole.

(f) ***Vacuum Pumps***

Note: Basically these pumps are classified as being "wet" or "dry," depending on whether they remove gas and water simultaneously, or deliver gas alone. All the usual forms of pump may be used to create a vacuum —that is, centrifugal, rotary, screw, reciprocating, and the steam jet ejector. Generally wet pumps are able to deliver the highest vacuum, and are particularly adapted to the return side of steam heating systems, since they separate and eject the air while returning the water to the boiler. The specification for a vacuum pump follows the usual specification for a centrifugal, rotary, or whatever kind of pump the vacuum pump happens to be, but with the emphasis on the degree of vacuum it is expected to produce.

- Specify the kind of pump: wet, dry, positive displacement or centrifugal, etc.
- Specify the vacuum required: ________ in. Hg column referred to a 30-in. barometer.
- Specify the volumetric capacity, cfm, when handling (air) (gas) from (name source) at a temperature of ________ F and discharging against an external pressure of ________ psia.
- Specify the general construction, and design, as number of stages, number of cylinders, form of impellers, materials for the various parts.
- If a wet pump of the hurling water type, specify the source of the water and its maximum and minimum temperatures entering into the pump.
- Specify the electric power available if pump is motor-driven, giving voltage, phases, and cycles.
- Specify the foundation and mounting of the pump, to-

gether with its driving motor. State if both are to be mounted on a common bedplate.

- Require bidders to submit sectional detail drawings of the pump, clearly showing its internal arrangement and giving all pertinent dimensions.
- Require bidders to submit curves showing relationship between vacuum created, pump speed, and horsepower.
- Require adherence to NEMA standards for motor and observance of all local codes and the National Electrical Code in wiring up the motor.
- Specify how the pump is to be controlled at the desired vacuum, or request bidders to submit their own control systems for approval.
- If pump is driven through reduction gears, see Sec. 4-17(d). Refer to same section for specifying guidance on a pump that is belt-driven.

(g) ***Fire Pumps, Centrifugal, Horizontal***

Note: Present standard practice favors centrifugal fire pumps, although in many older installations both rotary and reciprocating steam pumps have been installed. Generally a fire pump is energized only when fire or explosion places an abnormal demand for water on the primary water system, and it is not considered to be a replacement for a good public water system with adequate pressure. Centrifugal fire pumps are available in both horizontal and vertical forms; preferably they are located so that a suction head is provided for them, and this suction should enable the pump to operate at 150 per cent of its rated capacity.

- Specify the capacity of the fire pump as _______ gpm (including the 50 per cent above the basic load), when discharging against a total dynamic head of _______ ft and operating under a suction (lift) (head) of _______ ft.
- Specify the construction and materials of the pump, referring to the Fire Pump Standard of the National Fire Protection Association. This standard requires the fire pump, driver, bedplate, and all stipulated accessories and attachments to be furnished under a single contract.
- Specify the type of pump driver, as electric motor, steam turbine, diesel engine, gas turbine, or internal combustion engine, using gasoline fuel. Specify speed of pump as _______ rpm to suit driver.

- Specify the source of the water, as overhead tank, adjacent lake, river, or pond. If pump is to be placed in a pit to reduce or eliminate a suction lift, state elevation in relation to the supply, as ______ ft above or below.
- Specify kind and size of pressure relief if necessary to protect against excess pressure.
- Specify the number, locations, and sizes (usually 2½ in.) of the hose valves, mounted on the pump discharge or in an adjacent room to avoid water damage to the pump or personnel in the event of a hose burst.
- Specify the number, position, and size of test valves.
- Specify the number and size of the automatic air release valves mounted on the pump casing.
- Specify if pump is to be started manually, automatically, and if from remote control.
- Specify the number, size, and location of the circulation relief valves that function automatically to protect the pump from overheating.
- Require the pump manufacturer to furnish sectional detail drawings showing the internal arrangement of the pump and all pertinent dimensions.
- Require bidders to furnish a family of performance curves showing the varied relationships between the pump head, capacity in gpm, and hp.
- Specify the kind of electric power available, volts, phase, cycles, and state the source, as public utility company, private utility, a plant generator, etc. Indicate if this power is available 24 hours a day and every day in the year.
- Require all electrical equipment to meet the NEMA standards and the National Electrical Code, except as this code is modified by the Fire Pump Standard.

(h) *Fire Pumps, Vertical Turbine*

Note: This form of pump is generally considered when a horizontally set centrifugal pump would be obliged to function with a suction lift. If necessary, the turbine pump is submerged at the bottom of a well, thus allowing the water to flow by gravity into the impellers. If a well is used, it must first be tested and approved under the Fire Pump Standard for dependability of flow.

- Specify general design of the pump, as number of impellers, details of foot valve and strainer, couplings used by pump shaft, size of column pipe, etc. Also kind of driver, motor, engine, etc.

- Specify materials of shaft, impellers, foot valve and strainers.
- Specify well casing, size, material, and depth.
- Specify normal water level before start of pumping, maximum allowable drawdown, ft, in ________ minutes of pumping at rated capacity. Also time, ________ minutes, to reach maximum drawdown with pump operating at 150 per cent of rating.
- Specify required capacity rating as ________ gpm at a discharge head of ________ ft. If a suction lift must be accepted, state amount of lift, ft.
- Provide information on water characteristics, as hardness on the pH scale; whether corrosive or noncorrosive; amount of suspended materials, ppm, and their nature; presence, if any, of organic matter.
- Require strainer mesh to be selected by pump manufacturer on basis of the stipulated kind and quantity of suspended materials, both organic and nonorganic.
- Specify number and positions of hose valves, usually 2½ in. in size.
- Stipulate if pump is to start automatically or manually, and if the former, specify the kind and manner of the control.
- If pump takes suction from an open river, lake, or pond, provide a crib around the intake of the pump to prevent damage by the impact of floating or semisubmerged logs and other moving debris.
- For such open sources of suction, provide antifreeze arrangements, as compressed-air bubble systems or other water agitator. Although these protective devices will be placed by others, describe the method proposed in the fire pump specification so that pump bidder can approve the scheme as not interfering with the performance of his pump.
- Require the bidder to provide all fittings and attachments called for under the Fire Pump Standard.
- Require the installation of the pump to be made under supervision of the pump builder's representative.
- Specify characteristics of electrical service available for a motor driver, and require that all electrical equipment offered meet the NEMA standards, the National Electri-

cal Code except as modified by the Fire Pump Standard, and the local codes.

- Specify an antireverse ratchet on the motor drive to prevent reverse operation.
- If pump is driven by an engine or turbine, transmission or power is through a right-angle gear drive. So specify kind of gears, and require some form of device to prevent self-ignition in the case of internal combustion engines.
- Specify the gears as appropriate to Sec. 4-17(d), Rotary Pumps.
- Specify the steam turbine as suitable under the itemized list of Sec. 4-12(a), Raw Water Pumps.
  - For boiler steam pressures of 120 psig or less, specify that turbine must carry full load if pressure drops to 80 per cent of normal. If boiler pressure is higher than this, specify full load to be carried at 70 per cent of normal pressure. Specify that emergency governor on turbine shall stop turbine if overspeed reaches 120 per cent of normal.
- Specify the internal combustion engine to be in accordance with the recommendations of the National Fire Protection Association. This applies to diesel and gasoline fuel units.
- Specify that pump and engine drive must be assembled and test-operated at the plant of the pump manufacturer.
- Require engine to produce 120 per cent of the rated pumping horsepower, this extra horsepower being provided for taking care of auxiliaries and accessories and for the normal depreciation of power resulting from use.
- Specify number of engine cylinders, valve type, as slide, overhead, etc., also normal and maximum speed ______ rpm. Also arrangement of cylinders as in line, V, canted, etc.
- Specify type of bearings, crankshaft, connecting rods, thrust. Also methods of lubrication and engine-cooling systems.
- Specify starting equipment, as electric storage battery or compressed air, as suitable to the type of engine.
- For diesels, specify if 2 or 4 cycle, method of scavenging.
- Stipulate the ambient operating conditions, as the temperature is normally ______ F, and the minimum is

_______ F. If artificial heat is provided in fire pump room, provide 60 F for gasoline-fired engines and 70 F for diesels to facilitate starts. Alternative: Provide some concentrated form of engine heating, as infrared lamps, circulating hot water, etc.

- If bidders are to include the engine cooling system, specify the kind, as a shell and tube heat exchanger cooling tower, spray pond, etc. For specifying guides on towers and ponds, see Sec. 4-14(h)–(j).
- Specify that heat exchangers must have adequate exchange surface to maintain a cooling water temperature to the engine of _______ F when supplied with cooling water of _______ F maximum temperature.
- Specify construction of exchanger, as parallel, cross, or counterflow; include requirements for materials of shell, heads, baffles, and tubes.
- Specify maximum allowable pressure drop through the exchanger at maximum load operation.
- If exchanger is in the form of a radiator, specify location, providing data on maximum and minimum air temperatures to which the radiator will be subjected, also the direction and velocities of the prevailing wind. If air is contaminated with corrosive substance, as sulfur, so state in the specification and direct bidders to select radiator materials accordingly.
- Specify safety indicators, as gages to show serious drops in oil pressure, serious rises in engine temperature; also failure to start automatically, all to be signaled by a bell or horn.
- Require bidders to provide section drawings of the equipment being furnished, showing all details of the arrangement and all pertinent dimensions. Bidders also shall furnish performance curves showing relationships between pump discharge, gpm; discharge head, ft; and hp. Guarantee point to be marked on the curves.
- For steam engine drives, refer to Sec. 4-16, all paragraphs, and including data in Tables 4-15 and 4-16 as appropriate.
- Also stipulate ambient temperature conditions, minimum and maximum _______ F, indicating what precautions are provided to prevent freeze-ups in the engine and its connecting pipelines.
- Gas-turbine drives have been coupled to fire pumps in

several instances. In this service they have both advantages and disadvantages. They may be started up to full load in a matter of seconds—which is good. Since they are extremely high-speed machines, the noise, unless carefully muffled, could be intolerable—which is bad, because muffling adds to the cost.

- If space is limited, specify that the turbine must fit within given dimensions.
- Specify the kind of fuel, whether natural gas enters the combustor, or if a generator must be part of the installation.
- Depending on the fuel, require the turbine to satisfy National Fire Protection Standards Nos. 54 for natural gas, 58 for liquefied petroleum gases, and No. 37 for combustion engines and gas turbines.
- Specify that turbine and fire pump must be assembled and tested at the plant of the pump manufacturer before delivery to the service site.
- Specify the horsepower the turbine must deliver.
- Specify control sequences and the components of the system.
- Stipulate if turbine and pump are to be mounted on a common bedplate.
- Stipulate range of ambient air conditions of temperature, dry bulb and wet bulb, indicating at what point the guarantee is to be made.
- State if mufflers are to be fitted, either by this contractor or by others, to the air intake and the exhaust.
- Require bidders to submit complete sectional and detail views of the turbine-pump arrangement, showing all pertinent dimensions.
- Require bidders to submit complete performance data, showing fuel consumption at various horsepowers, speeds, and loads. Include the maximum allowed overload expressed as a percentage of normal full load.
- Bidders also shall indicate size and weight of heaviest piece.
- Specification should state if any attempt is to be made for heat recovery, involving the use of heat exchanger, or other described means. Also indicate if a simple cycle turbine, and approximate desired horsepower.
- If installation is to be made more than 1,000 ft above sea level, give elevation.

Note: Gas turbines may be used with either horizontal or vertical centrifugal pumps. Because of their relatively high speed, a reduction gear must be considered. For guidance on reduction gear specifications, see Sec. 4-17(d), Rotary Pumps.

(i) *Sewage Ejectors*

Note: General forms available include a horizontal centrifugal pump that is submerged in a pit or sump and is driven by an electric motor through a long vertical shaft. Another form incorporates a tank that is emptied by the action of compressed air. The pneumatic unit, like the centrifugal pump form, can be purchased as a package and is readily dropped into the prepared pit.

- Specify the kind of sewage ejection system desired, whether pneumatic or centrifugal, and stipulate the capacity in ________ gpm against a total dynamic head of ________ ft.
- Specify number of pumps required and method of automatically switching from one to the other.
- Specify construction of tank, as copper bearing steel, welded joints, etc.
- For pneumatic systems, specify kind of air compressor desired, as reciprocating or rotary.
  - Specify the compressor fittings, as strainers, mufflers, etc.
  - If water-cooled, give range of water temperatures from ________ F minimum to ________ F maximum.
  - Require bidders to furnish details of water flow control.
- For all ejectors, specify the motor requirements, as satisfying NEMA standards and the National Electrical Code. Also local codes.
- Give in the specification the power characteristics for the ejector in volts, cycles, phases, etc.
- Furnish sketch showing bidders how discharge is to be conducted, as to building sanitary drain, outside sewer, etc. In any danger of backflow from the drain, specify a suitable backflow valve and designate its placement.
- Stipulate who furnishes the ejector receiver pit or sump. State if concrete, tile, steel, etc.
- Require bidders to submit drawings illustrating the design and construction of the ejector, indicating materials and dimensions.

- Specify method of controlling ejector operation, as by float switch. Specify that an across-the-line starter is to be provided by this contractor.
- Specify the arrangement of ventilation for the sump or tank, indicating the size and terminus of the vent pipe. Also the means of fresh air entry.
- If a pneumatic system is chosen and is to operate from the plant compressed-air system, specify the available pressure, ________ psig.

4-18. SEPARATORS AND FILTERS, STEAM, AIR, AND GAS

Note: Fluids may be cleansed of their entrained and suspended undesirable materials by various forms of separation and filtering, as by centrifuging, baffling and impingement, electrostatically, and forcing through various kinds of filtering materials. Specifying guides for these forms of purification will be described under each of the identified fluids.

(a) *Steam Separators*

- Where separation is provided in the boiler drum, include in the specification for the boiler or steam generator. See Sec. 4-1(f).
- For steam separator mounted in the steam line, specify quality of steam entering the separator, ________ per cent, also the quantity flowing, ________ pounds per hour, the pressure, ________ psig, total temperature, ________ F.
- Specify the kind of separator, kind of baffling, and method for removal of condensate falling to well in base of the separator, whether automatic or manual.
- Specify materials of body and baffles.
- Specify the required quality of the steam leaving the separator, ________ per cent.
- Require bidders to furnish drawings showing all dimensions, weights, general arrangement of internals, etc.

(b) *Air Cleaners, Electrostatic*

- Specify the nature of the material to be removed, as dust, smoke, fly ash, or a mixture.
- Specify concentration of particles entering the precipitator, particles per cu ft of air or gas, and give sizes in microns or the fineness of the mesh through which these particles will pass.
- Specify the characteristics of the available electrical service, voltage, cycles, phases, etc.

- Provide a sketch showing chosen location for the precipitator.
- Specify the cleanliness of the air required, particles per cu ft, and also give the number of the mesh through which _______ per cent of the particles must not pass, and minimum micron size removed.
- Require electrical components to satisfy the standards of NEMA, the National Electrical Code, and local codes.
- Require bidders to describe in detail the methods proposed for collecting the removed particles and for their disposal.
- Stipulate the size and construction as well as the location of the collection receptacle into which the precipitator is to discharge.
- Specify instruments to be furnished by this contractor, including ammeters, indicating lights, the off-on switches, etc.
- Specify arrangement of precipitator as in modules or sections permitting the continued use of the device at a reduced rate in the event of failure of a given section.
- Specify how available a-c service is to be converted to d-c in the precipitator.
- Require adequate overload protection in event of short circuit.
- Specify the kind of test to be used in determining the efficiency of the unit, as the NBS, or National Bureau of Standards Spot Test; the DOP test with dioctyl phthalate smoke as developed by the U.S. Army Chemical Warfare Service; or any other selected test.
- If a particular weight of steel gage is desired, so state. Also specify kind and coats of painting if these differ from the manufacturer's practice.

(c) *Air Cleaners, Mechanical Filtering*

- As with the electrostatics, specify the capacity of the filter required in particles removed per cu ft of air passing through (or per 1,000 cu ft as appropriate). Stipulate the minimum size of particle that is to be permitted to pass through, as 0.3 microns, or larger.
- Specify efficiency of removal, _______ per cent, when air is introduced at a velocity of _______ fpm.
- Specify the effective filtering surface, sq ft.
- Specify the maximum allowable pressure drop through the filter, at the desired velocity of _______ fpm.

- Specify temperature range of air, ______ F.
- Specify filter medium, as glass fibers, glass-asbestos, or other inorganic material.
- Specify if filter medium is to be permanent and cleanable, or of throwaway type.
- Specify face velocity at guaranteed efficiency point.
- Specify number and arrangement of filter frames. Also stipulate if frames are to be fireproof or fire-resistant, or if combustible material is acceptable.
- If air or gas stream carries a large percentage of relatively big particles, a prefilter may be desirable to protect the continuing life of the final fine filter. If this is specified, stipulate, as above, the filter medium, whether washable or throwaway, and fire-susceptibility status of the frames.
- Specify efficiency test used for rating the filters, as NBS, DOP, Cottrell, or other.
- Require bidders to submit full data on their equipment with drawings that show the arrangement, dimensions, method of changing filters, weights, etc.
- Require bidders to submit operating data, as cfm of air passing through; pressure drop when new and when filter requires either cleaning or replacement, in. W.G.; efficiency of removal, per cent.
- Specify type of manometer to be furnished by this contractor for determining pressure drop through the filter.
- If filtering medium is on rollers and is passed continuously through a fluid cleaning bath, specify the kind of cleaning fluid used, and stipulate if bath is to be replaced or is to be cleaned by circulation through a wet-type filter. Require electric motors used in driving the main filter rollers or sprockets, and the fluid circulating pump, if any, to meet the standards of the NEMA, the National Electrical Code, and local codes.
- Require all starting and stopping equipment, whether manual or automatic, to meet electrical standards and code requirements as described in the preceding section.
- Require that such cleaning fluids be noncombustible.

(d) ***Gas Cleaners, Flue Gas***

- Refer to Sec. 4-18(b) for specifying guidance for electrostatic separation; also Sec. 4-13(b).
- For mechanical separation, see Sec. 4-13(b), Fly-ash Removal.

(e) *Gas Purifiers, Fuel Gas, Dry*

Note: These gases include coal gas, carbureted water gas, natural gas, producer gas, oil gas. Impurities are in the nature of tar and sulfur, both of which set up corrosion. Tar also chokes the fuel line to the burner and, if ignited, causes undesirable smoking.

Removal is by means of tar washing and chemical exchange for the sulfur. These processes are primarily the concern of the chemical engineer, and consequently are not specified in this Handbook. Reference to chemical handbooks and textbooks is recommended for specifying guidance in this field.

- For the filtration of oil gas, specify a dry type of filter, termed a scrubber, and mounted in the gas pipeline. Stipulate scrubber is to be of the centrifugal type.
- Specify the materials to be removed from the gas, as dust, pipe scale, distillates if any.
- Filter normally is placed vertically in the line, but if an appreciable amount of liquid is present, specify a horizontal placement with flushing out connections to a suitable storage tank.
- Specify the volume of gas passing through, ________ cfm.
- Specify gas pressure, ________ psig.
- Specify gas temperature, ________ F.
- Stipulate the construction of the centrifugal scrubber, including materials incorporated.
- Specify the kind of gas, with the analysis of its constituents.
- Require bidders to furnish arrangement and dimension drawings, together with weights.
- Require bidders to submit operating data, as pressure drop, psig, at no less than three (3) volumes of flow, one of which is the guarantee point.
- Specify if scrubber is to be mounted by this contractor or by others.

Note: This form of scrubber may be used on natural gas also.

(f) *Gas Purifiers, Fuel and Process Gas, Wet*

Note: In specifying wet gas scrubbers, care is necessary in the selection of materials, because of tendency to corrosion in some gas constituents when in water solution. This form of cleaning provides some control of humidity in the gas leaving the scrubber.

- Specify the nature of the gas entering the scrubber, giving its constituents by analysis. State the kinds of contaminants to be removed, as suspended or entrained solid particles, water-soluble vapors and gases. State if gas is acid or alkaline, and if this condition will be changed after passing through the scrubber.
- Specify the volume of gas passing through, cfm, and state velocity in fpm.
- State disposal of wash water, whether it is run to waste, reused, piped to a process.
- Specify if filtration and separation is by spray washing only, or is accomplished in combination with a bed of filtering material through which the water and gas pass.
- Specify materials of construction, piping, and spray heads in view of the nature of the gas being cleaned.
- Require bidders to submit detailed drawings of the scrubber arrangement, showing dimensions and weights.
- Require bidders to submit information on materials, water required, _______ gpm, and water pressure, _______ psig. Also water temperature range, seasonal or for other reasons, F, with maximum temperature acceptable indicated.
- Bidders also required to state pressure drop, water, between spray inlets and spent drawoff from bottom of unit. Loss of gas pressure, _______ in. W.G., also required at three (3) rates of operation, one of which is the guarantee point.
- Specifications should state who supplies and places the foundation and who makes the piping and duct connections to the scrubber.

(g) *Lubricating Oil Filters*

- Specify if filter system consists of a centrifuge, membrane filter, or a combination of both.
- Specify the oil duty, as to lubricate the bearings of a steam turbine, a diesel engine, other machinery.
- Specify the filtering procedure as continuous, bypass filtering, or batch filtering.
- Describe kind of oil, as straight run or additive.
- Specify kind of impurities to be removed, as water, air, sludge, oxidizing materials, etc.
- State temperature of oil entering the filter, _______ F, and give pressure, _______ psig.

- Specify filter body construction, as steel, monel, rubber, plastic lined, etc.
- If centrifuge is specified, consider special materials, as monel, stainless steel, etc.
- Specify capacity of filtering system in gpm or gph, as appropriate.
- Require all electrical components to satisfy the NEMA standards, the National Electrical Code, and local codes.
- Require bidders to furnish detail drawings showing arrangement of all components and giving dimensions and weights.
- Require data on speed, rpm, of the centrifuge if included. Also motor hp and temperature rise, C.
- If this contractor also furnishes the lubricating oil cooler, require data on the size and construction, including tube and shell materials, number of passes, and water flow, ________ gpm, when water enters at ________ F and leaves cooler at ________ F. Also bidder to guarantee reduction in oil temperature ________ degrees, when oil enters cooler at ________ F and passes through cooler at rate of ________ gpm, or gph, as applies.
- Bidders to state pressure drop through cooler of cooling water, ________ psig, and drop for the circulating lubricating oil, ________ psig.
- Specification to include a location sketch, showing relative positions of filters, coolers, and the machinery unit requiring the lubricating oil.
- If bag filters are to be used for removal of sludge, specify number, sizes, and materials.
- Specify desired system of control.

(h) *Water Filters, Pressure*

Note: Pressure filters are most familiar in industry, since they may be located as convenient, take up limited floor space, and can be bought off the shelf.

For community filtration, an extensive treatment is needed more often than otherwise, and involves sand filters rated as either "rapid" or "slow" in action. Slow filters may not require chemical coagulants, but are not suitable for muddy water supplies.

Since this is a power plant specification, it will be restricted to pressure filtration. However, filtration may be part of the boiler water-softening process, and in these cases, refer to Sec. 4-12(c). Data for specifying a simple pressure filter may be abstracted from this same section.

- Specify location of filter and required capacity in ________ gpm.
- Specify working pressure, ________ psig, and require construction in accordance with ASME Code for Unfired Pressure Vessels.
- Specify any limiting conditions of efficiency, weight, dimensions, etc.
- Require bidders to submit arrangement and design drawings of their equipment, and to furnish operating data as called for in Sec. 4-12(c).
- Specifications to include an adequate description of the water supplied to the filter, as its degree of turbidity, the nature of the suspended solids, dissolved materials, hardness range, and its pH value.
- Specify running period desired between backwashes. Also specify if operation is to be manual or automatic.

4-19. WATER STORAGE TANKS AND STANDPIPES

(a) ***Pneumatic Storage Tanks***

- State general features of tank, as location, horizontal, vertical, elevated, or on ground.
- Specify capacity in gallons.
- Specify maximum operating pressure, ________ psig, and minimum pressure, ________ psig, at which recharging by air becomes necessary.
- Specify if air recharging is to be under automatic or manual control, and explain source of air.
  - If auxiliary air compressor is required, specify kind, as reciprocating, rotary, centrifugal; give capacity as ________ cfm at a delivery pressure of ________ psig. State maximum free air temperature, ________ F, which is to be compressed and stipulate the free air temperature, ________ F, at which the capacity guarantees of the compressor are to be proved by test.
- Specify the electrical power characteristics available at the site.
- Require electric motor driving compressor to satisfy applicable aspects of the National Electrical Code and the standards of NEMA, as well as any special local codes.

- State if compressor and motor are to be mounted on a common bedplate. Indicate who provides the foundations both for the compressor and motor and for the pneumatic tank, if set on the ground.
- Specify the kind of foundations or supports for the tank.
- Specify the number of gage glasses, number and size of manholes.
- Specify kind of painting and number of coats.
- Specify steel plate material, as mild, open-hearth steel, or flange quality.
- Require tank to meet all requirements of the ASME Code for Unfired Pressure Vessels.
- Specify construction of tank, how welded, and require welders to be qualified by a recognized qualifying agency.
- If materials and construction are to conform to usual commercial practice rather than ASME code, so state.
- Refer to the standard list of sizes and numbers of openings for the tank desired, and if any changes to the standard are to be made, so specify.
- Specify if water level is to be maintained by float that actuates a pump.
- Specify the filling rate, ________ gpm.
- Provide bidders with water data, as hardness, iron content, turbidity, etc.

(b) *Water Standpipes*

- Define a standpipe as a flat-bottomed cylindrical tank having a shell height greater than the diameter.
- Define a reservoir as a flat-bottomed cylindrical tank having a shell height equal to or less than the diameter.
- Require all standpipes to meet the standards of the American Water Works Association.
- Provide bidders with information concerning the water, as hardness, turbidity, iron content, presence of organic matter, sulfur, etc.
- Specify the corrosion allowance to be made in the thickness of metal in contact with the fluid, and for metals not in contact with the fluid.
  - For beams and channels, add corrosion allowance to webs only.
- Specify the kind of foundation and thickness, providing a sketch showing dimensions. Specify size and length of foundation bolts.

- Specify wind pressure standpipe must withstand, ______ mph.
- Specify standard shell height in 2-ft increments between 20 and 50 ft; 5-ft increments between 50 and 100 ft; and 10-ft increments above 100 ft.
- Specify diameters similarly.
- Specify the capacity of the tank, ______ gal.
- Specify the size and placement of the manhole, pipe connection, overflow, vent in roof, and the ladders, both fixed external and swivel.
- Specify paint and number of coats.
- Require bidders to furnish detail drawings showing all dimensions and the locations of the various appurtenances.
- Bidders to furnish information on gross weight of standpipe empty and filled.
- Specify that welders must be qualified by an approved qualifying agency.

(c) ***Gravity Water Tanks, Steel, Elevated***

- Stipulate general design and arrangement desired, as capacity, ______ gal, height of bottom of tank above the ground or above the highest water outlet in the plant. Stipulate if tank is to be mounted on a roof or on a tower structure.
  - If mounted on a tower, specify if tower design is to be tubular, lattice columnar, or angle.
  - Specify height of tower.
  - Specify foundations for tower, providing drawing showing dimensions, and giving data on weight-bearing power of the earth at the tower location.
  - If tank-supporting structure is to be erected on a roof, provide all necessary drawings showing details of the design and materials. Specify who furnishes what, and make clear where this contractor's work begins and ends.

Note: If this gravity tank is to supply fire hose streams with water, bottom of tank must be not less than 75 ft above the ground level, and preferably will be more than 100 ft. Also, minimum capacity of such a tank shall be 30,000 gal of water.

- Specify the steel to be used in the tank construction, as ASTM Specification A-7.
- Specify material of roof, as wrought iron or copper

bearing steel, giving ASTM specification number to be complied with.

- If tank duty includes fire protection, require entire tank, supporting structure, and accessories to comply with National Fire Protection Association Code No. 22.
- Specify if tank and structure are to be riveted or welded. If welded, require compliance with the Rules for Field Welding of Storage Tanks as promulgated by the American Welding Society.
- If tank location is subject to earthquake shocks, specify how resistance is to be built into the structure. (Obtain information from shock areas for specification guidance, if necessary.)
- Specify how tank is to be protected from icing. If by means of heating coils or frost-proof casing, provide details for guidance of bidders.
- Specify number, location, and sizes of ladders, both inside and outside. Also specify walkways and balconies, if needed. Specify safety cages.
- Specify overflow location, as inside or outside.
- Specify number and sizes of openings, vents, etc.
- If pump or valve pit is located under the tank, stipulate material and labor with respect to this to be supplied by the tank contractor, if any.
- Specify added thickness of metals for resistance to corrosion.
- Provide bidders with complete data on the water which tank is to store, as hardness, turbidity, iron content, sulfur, etc.
- Specify kind of paint and number of coats to be applied to both inside and outside of tank and to the supporting structure. Alternatively, require cathodic protection.
- Specify design wind pressure, as ________ mph.
- Require bidders to submit drawings showing details of their design, giving dimensions, materials, and weights, both empty and filled.
- If considered desirable, specify unit stresses per sq in., referring to the table of stresses No. 13-88 in the 12th Edition of the Fire Protection Handbook.
- Specify kind of water-level indicator desired.

(d) *Gravity Water Tanks, Wooden, Elevated*

- Stipulate if water tank is to be set on an independent structure or is to be placed on a roof.

- Specify capacity of tank in gallons of water.
- Specify snow load, wind load, other stresses, if any.
- Specify spacing of tension hoops, and limit unit stress in hoops to 12,500 psi, for iron.
- Iron of hoops is to meet requirements of ASTM Specifications A-84 or A-189 as applicable.
- Unit stress of steel hoops may be imposed up to 15,000 psi.
- Specify number and size of openings to be provided.
- If tank performs fire duty, require the entire structure and appurtenances to conform with National Fire Protection Association Standard No. 22.
- Specify how tank is to be protected against icing.
- Specify number and location of ladders, inside and outside. Specify safety cages if considered desirable.
- Specify number, size, and positions of walkways or balconies.
- If tank support is on independent structure, specify construction form, as tubular, lattice columnar, or angle.
- Specify height of bottom of tower above ground and above highest plant water faucet or outlet.
- Specify nature of foundation and make clear who provides same. Provide drawing of foundation with dimensions, and showing number and sizes of foundation bolts. Also soil load bearing value.
- Specify kind of wood to be used in tank.
- Specify material and position of overflow pipe, as brass, inside; galvanized iron outside, etc.
- If lightning protection is to be provided, specify number of terminals on top of structure and number of conductors to the ground.
    - Specify materials of both points and conductors and specify depth and length of the buried grounds.
    - Require adherence to the National Fire Protection Association Code No. 78, "Protection Against Lightning."
- As with the elevated steel tank, stipulate who furnishes the pump and valve pit under the supporting structure.
- Specify kind and position of water-level indicator.
- Specify how water level in tank is to be controlled.
- Specify kind of paint to be used on the tank and supporting structure, and stipulate the number of coats.
- If any special wood preservative is to be applied to the

tank materials, specify when and where, and the degree of penetration desired.

(e) *Pressure Tanks for Sprinkler Systems*

Note: The amount of water stored in a pressure tank serving sprinkler systems is relatively small, for example, a minimum of 2,000 gal for an area of light-hazard occupancy, and a minimum of 3,000 gal for ordinary-hazard occupancy, Groups 1 and 2. For this reason, every pressure tank system must obtain specific approval from the National Board of Fire Underwriters, the National Fire Protection Association, or the insurance carrier before it is certified ready for use.

- Specify the capacity of the tank, ________ gal, and describe its location relative to the sprinkler system.
- Require design and construction to conform to the standards of the National Fire Protection Association and the ASME Code for Unfired Pressure Vessels.
- Specify pressure that is to be maintained in the tank, ________ psig.
- Specify that in no case is the pressure to drop below 15 psig at the highest sprinkler on the system.
- Specify an added allowance of at least 1/16 in. to allow for corrosion. If the increased thickness is still below the ASME Code thickness for the conditions to be encountered, the ASME Code shall govern.
- For horizontal tanks, specify a support for each course. Specify that no end shall overhang the nearest support by more than 4 ft.
- Locate the tank in a heated room, preferably of noncombustible construction.
- Specify number of openings and their size, with the fill pipe to be not less than 1½ in. diam. The air-supply pipe shall be at least 1 in. diam.
- Specify the number and length of the water-level gage glasses.
- Specify a pressure gage to be mounted on the tank.
- Specify the form of the tank supports, as of steel or reinforced concrete.
- Specify paint and number of coats for both the tank and its supports, particularly steel.
- If bidders are to provide supports, furnish them with a drawing in detail, showing dimensions, shape, and materials.
- Specify that tank discharge pipe is to project into the

tank bottom at least 2 in., in order to create a settling area and keep sediment from traveling into the sprinkler system.

- Refer to Sec. 4-19(a) for further details of pneumatic storage tanks. Combine these with this section on pressure tanks as appropriate, to develop a complete specification on pressure tanks as used with sprinkler systems.

4-20. CRANES

(a) ***Bridge Cranes, Power-operated***

- Itemize the governing equipment required, as a _______ ton, _______ (number) motor, single trolley, overhead traveling crane. Crane track to be mounted on columns set against side walls.
- Specify speeds as _______ fpm for bridge travel; _______ fpm for trolley traverse; lifting and lowering of hook, _______ fpm. Hook must raise and lower without twisting.
- Provide a location drawing to bidders showing spans and clearances. In addition, specify the span from center line to center line of runway rails as _______ ft; height from floor to top of rail as _______ ft; distance between hook and floor when hook is in lowest position _______ ft (usually zero).
- Require bidders to state distance from floor to center of hook when hook is in the extreme upper position.
- Specify the factor of safety to be designed into the bridge, as not less than _______ (5 is common, but subject to consideration).
- Specify the provision of a footwalk the full length of the bridge. Require standard handrails and toeboards.
- Specify that bridge brakes must be able to instantly stop bridge travel when carrying the full load.
- Specify both hand and automatic brakes for the hoist. Either brake to sustain at least twice the full rated capacity of the hoist. The automatic brake should function instantly and automatically to arrest the hoist wherever it happens to be at the moment when electric power either fails or is cut off for some reason.
- Specify the materials of the gearing, as cut from solid cast or forged steel. Specify attachment of gears to shafts, as forced fit, keyed on, or both.

- Specify the gear housing or guards, requiring them, if desirable, to be oiltight and dustproof. Require adequate inspection openings.
- Specify the materials and form of all wheels, both bridge and trolley, as wheels to be of steel, chilled iron, or other suitably treated material, and having treads machined to a uniform diameter. Require all wheels to be protected by guards of a design that will prevent persons on the track from being inadvertently injured. Specify that no wheels are to be overhung.
- Specify the kind of bearings, as phosphor bronze, roller, babbitt, etc. Specify means for lubricating all bearings.
- Specify the dimensioning and arrangement of the hoisting drum, as provision of a flange on each end of the drum, and requiring the drum diameter to equal at least ________ rope diameters.
- Specify the retention of at least ________ full wraps of hoisting cable in the drum grooves when the hook occupies its lowest position.
- Specify the ropes in detail, as the number of strands, diameter of strands, number and gage of wires, etc.
- Require bidders to furnish data on the material, tests. and factor of safety built into the hook itself. If felt desirable, require hook to be X-rayed to uncover any hidden defects.
- Specify the operator's cab construction, as all materials to be fireproof, with the lower panels to be formed of wire mesh screen from the toeboard to a height of ________ ft. Specify that cab is to be suspended from the bridge in a position that permits the operator instant visibility of signals from the floor. Require that a foot- or hand-operated gong be furnished in a position convenient to the operator for use as a warning signal. Require sufficient roominess to allow operator to be seated when manipulating his controls. Require that access to the cab shall not involve stepping across any open space more than ________ in. wide. Require handrails wherever necessary for the safety of the operator.
- Specify that electric motors and equipment shall satisfy the standards of NEMA, the National Electrical Code, and any local codes applicable. Specify characteristics of service available.

- Specify exactly what kind of control is desired, and the location of the levers in the cab or cage. Require overload and no-voltage protection for each motor. Require a total-disconnect switch in the operator's cab, and elsewhere in a convenient location.
- Specify the kind and operation of a limit switch on the lifting travel of the hook.
- Specify that all wiring must comply with the standards of the National Electrical Code and the National Board of Fire Underwriters. Require that the circuit for each motor be carried in a separate conduit.
- Require bidders to submit drawings and data showing how the bare wiring and current collectors for both bridge and trolley are arranged. Also indicate how these wires are protected against personal contact by individuals working on the crane.
- Specify the following safety precautions:
  1. Provide an approved guard on power feeds to prevent contact from a swinging hook.
  2. Setscrews of submerged type.
  3. Provide tool racks in cab or cage.
  4. Mount wheel bumpers or bridge girders.
  5. Specify factor of safety for gears and hoist.
  6. Provide safety guards over sheaves.
- Specify that hook is to be of swivel type with either ball or roller bearings and must turn freely by hand when fully loaded.
- Specifications should include information for bidders on available lifting and hoisting facilities for use in placing the bridge and the trolley.
- Require bidders to submit detail drawings of their equipment and to state manufacturer's name and model of purchased components that are used in the crane.
- Require successful contractor to furnish owner with at least two sets of wiring diagrams and one set of operating instructions that may be posted in cab.
- Specify kind and number of paint coats to be provided.
- Require bidders to furnish data on weight of equipment, electric motors, and controllers, and the design of the automatic brakes.
- Specify ________ per cent overload at which acceptance tests are to be made.

Note 1: If an auxiliary hoist for fast light handling is desirable, specify the details of this auxiliary hoist, being guided by the checklist on the main hoist.

Note 2: Make sure that columns carrying the runway are sized adequately to handle the moving loads. Furnish structural data to crane manufacturer for his approval of the loading capacity.

- Specify that crane shall meet the provisions of American Standard B-30.2, "Cranes, Hoists and Derricks," as appropriate.

(b) ***Bridge Cranes, Hand-operated***

- State general arrangement of crane, whether trolley travels on one I beam or is suspended from two parallel I beams.
- State if crane is entirely hand-operated, or if hook trolley is moved by power.
- Specify span of the bridge, ________ ft.
- Provide bidders with a drawing showing all important clearances.
- Specify height of center of rail above floor, ________ ft.
- Specify weight of runway rails and note type, as T rail.
- Specify the kind of braking required and the operating station location.
- Specify materials of gearing, gear guards, and methods of lubrication.
- Specify capacity of the crane, tons.
- Specify wheel materials, as cast iron, cast steel, milled from a solid piece of steel, etc.
- Specify the dimensions and arrangement of the hoisting drum, diameter to be not less than ________ times the diameter of the hoisting rope.
- Specify number of cable wraps on drum when hook is at lowest point.
- Specify a factor of safety of ________ for the ropes.
- Specify the kind and dimensions of the hoisting ropes or cables, as diam ________ in., formed of ________ strands containing ________ steel wires and with a center of ________________ (material).
- Require bidders to submit detail and dimensioned drawings of their equipment and in addition to furnish the following data:
  1. Standard lift of hoist, ________ ft.
  2. Distance, top of rail to highest point of crane, ________ ft, ________ in.

3. Wheelbase, _______ in.
4. Maximum wheel load, _______ pounds.
5. Total weight of bridge, trolley, and chain block, _______ pounds.

Note: When runway rails are to be supported on building columns, obtain weights as soon as possible to make sure that building structure is strong enough to bear the load.

- Specify painting, color and number of coats, preparation of metal.
- Specify crane shall meet the safety provisions of American Standard B-30.2, "Cranes, Hoists and Derricks."

(c) ***Gantry Cranes***

Note: Generally similar to bridge cranes, but the wheels travel on ground rails, and the entire equipment, including the operator's cab, requires weather protection.

- Specify the capacity of the crane on the main hoist. If auxiliary hoist is to be furnished, specify capacity of that also. Express capacity in tons.
- Specify the span as _______ ft, _______ in.
- Provide bidders with a drawing showing the area of operations, the layout of the tracks, particularly indicating switches over which crane must travel to reach various parts of the area.
- Specify the type of warning signal to be provided, as a gong, whistle, etc., operated from the cab.
- Specify the kind of wheel brakes desired. Also specify some means for making sure the crane is not blown along the track by a wind or storm, when not under the control of its operator.
- Specify maximum speed of travel over the ground, _______ fpm.
- Specify maximum speed of trolley, _______ fpm.
- Specify the maximum clearance required between the ground and the hook when the hook is at the top of its lift.
- Specify that cab construction permits weather closing in by operator when conditions require his protection.
- Specify how cab is be heated in winter and the desired temperature, _______ F.
- Specify access to the cab and all safety guards, as a cage around the ladder and adequate footwalk with hand-

rails. Avoid need for operator to step over any unguarded open space more than _______ in. wide.

- Specify automatic safety switches and relays as needed to:
  1. Arrest the hoist in its position at the moment power happens to fail or is cut off. This brake to be designed for safe operation with at least 25 per cent overload at time of failure.
  2. Provide low-voltage and no-voltage protection for all motors.
  3. Specify total power disconnect switch in cab and elsewhere as desirable.
  4. Require adequate safety guards around all wiring and other electrical equipment so that other workmen cannot come into contact with a live component.
  5. Specify the number of cable turns to remain on the drum when the hook is at its lowest position (usually ground), for both main and auxiliary hoists.
- Specify the size and materials of the hoisting cables, as previously indicated for the cables on bridge cranes.
- Require bidders to submit complete detail drawings showing dimensions of the steel structure, weight of columns, angles, plates, beams, etc. Also showing the interior arrangement of the cab.
- Require additional data, as material of wheels (if these have not been specified), weight of entire structure fully loaded, maximum weight on each wheel, material of all bearings and how lubricated, kind of all gears, material, and how lubricated. Also range of speeds over the ground from minimum to maximum, with speed at each step in _______ fpm.
- Specify that all electrical components and wiring must meet the standards of the NEMA, the National Electrical Code, and all local codes.
- Specify painting, color, number of coats, preparation of surface.
- Specify that load must not swing when suspended from hook.
- State nature of material to be handled and expected maximum dimensions of the load.
- Stipulate test loadings under which final acceptance of crane will depend. Include per cent overload.

- Require crane to meet safety provisions of American Standard B-30.2, "Cranes, Hoists and Derricks."

(d) ***Jib Cranes***

Note: Jib cranes are available in several designs. For example, the top-supported jib crane incorporates a vertical leg, mast, or column to be braced to a wall or ceiling, or may be built with guy ropes or struts. The general appearance is that of an inverted right-angle triangle, with a trolley or traverse mounted on the inverted base, and manipulated by hand from the floor. In a variation of this design, the vertical leg is self-supporting, with the trolley traveling on a horizontal boom, and likewise controlled by hand from the floor. This form is sometimes called a "pillar jib crane." For this category of crane a guide specification includes the following.

- Specify the capacity of the crane, in tons of lift.
- Specify the mode of affixing the jib crane to the wall, and the kind of support being furnished for a pillar jib crane.
- Specify the kind of gearing and hand cranks, indicating materials and methods of lubrication.
- If crane swings in a horizontal arc, specify the arc in degrees, as 180, etc.
- Specify length of boom, ______ ft, ______ in.
- Specify kind of hoist brakes required, with emphasis on the automatic brake that functions if the load takes charge and commences to run away. Also brake that permits stoppage of the hoist at any point of its travel, and is able to hold there the design load plus at least 25 per cent.
- Specify factors of safety required for gears, hoist, boom, every stressed component.
- Specify painting, color, coats, and preparation of surface.
- Specify what preparatory work is to be accomplished by the owner, such as making ready a foundation for the pillar crane, or wall reinforcement for the simple jib cranc.
- Specify speed of trolley, ______ fpm.
- Require bidders to furnish detail drawings showing dimensions of all steelwork, kind and arrangement of gears, protective devices around gears and sheaves.

- Specify hoisting cable, diameter, number of strands, wires, center material, etc.
- If jib crane is arranged to run on a floor track, with the top of the vertical leg riding on a rail fixed to the wall, it may be called a "walking crane." In these cases specify the size of the rails and the length of the run, ft. Tell bidders who is furnishing and placing the rails.
- Require cranes to meet safety standards of the ASME, American Standard B-30.2, "Cranes, Hoists and Derricks," as applicable.

(e) ***Locomotive Cranes***

Note: Originally built as steam cranes, to carry a boiler and steam engine hoist, most of these cranes are now being constructed as diesel or gasoline engine hoisting units. Therefore the specifying guidance in this section will be restricted to cranes operating with internal combustion engines.

- Specify the maximum load, _______ tons, to be handled at a boom radius of _______ ft, when lifted at right angles to the track and without the use of outriggers.
- Specify the maximum load in these circumstances with outriggers in position.
- Specify number of trucks, wheels, and gage of track.
- Specify clear lift of hook block at a boom length of _______ ft.
- Specify clear lift of bucket (if bucket is used) at boom length of _______ ft.
- Specify the diameter and design of the hoisting drum, and the kind of gearing. Specify materials.
- Specify hoisting cable as previously described for bridge cranes.
- Specify size of bucket, _______ cu ft, if bucket will be used.
- Specify construction of bucket, thickness of steel, dimensions of reinforcement angles, etc.
- Specify speed of crane along track, _______ mph.
- Specify arc through which boom may be rotated.
- Require the mounting of a capacity chart in the cab, showing the load the boom is capable of lifting at various radii.
- Require bidders to submit drawings giving dimensions and arrangements of all components of the crane.

- Require bidders to furnish the following information:
  1. Wheelbase, _______ ft, _______ in.
  2. Clearance to swing cab, _______ ft, _______ in.
  3. Minimum radius track curve crane can negotiate, _______ ft, _______ in.
  4. Capacity at minimum boom radius, _______ pounds.
  5. Capacity at maximum boom radius, _______ pounds.
  6. Maximum length of boom, _______ ft, _______ in.
  7. Weight of crane, unloaded, _______ pounds.
  8. Kind and hp of hositing engine; make and model.
  9. Kind and horsepower of locomotive engine; make and model.
  10. Kind and lubrication of all bearings.
- As applicable, require crane to satisfy safety provisions of American Standard B-30.2, "Cranes, Hoists and Derricks."
- Stipulate kind of brakes on boom to prevent the accidental drop of the boom under load.
- Stipulate kind of brakes on the locomotive wheels.

Note: When these cranes are fitted with rubber-tired wheels for operation over highways, they are termed "*transit cranes.*" Specify similarly to the locomotive guide, but either specify the size and kind of rubber tires to be furnished, or require bidders to submit their standard practice or intentions in this regard.

- For CATERPILLAR TRACTOR TREADS:
  - Specify width of treads, thickness, and kind of material in treads.
  - Specify material and method of forming of caterpillar sprockets, as forged steel, machined to size.
  - Specify size, placement, and materials of idlers, if included in the design.
  - Specify kind of sprocket bearings and methods of lubrication.
  - Require drawings showing how the tread joints are made, and how broken or defective treads may be replaced with a minimum of dismantlement. Bidders to furnish weight of each segment.
- For all cranes in the self-propelled category, bidders shall state the number, dimensions, and arrangement of outriggers, and their effect in increasing the lifting capacity of the extended boom.
- For all cranes in which power is furnished by a diesel

or gasoline engine, bidders shall submit data on the engine cooling system, as radiator and fan and maximum operating temperature at ______ F ambient temperature.

4-21. POWER PLANT INSTRUMENTS

(a) *General Discussion*

- When specifying steam power plant instruments, the designer should keep in mind the purposes of the various kinds, and should relate these purposes to the cost of the instruments in place and to the likelihood of the plant's personnel being able to use the information thus provided, effectively. Small industrial plants, unable to afford a test or instrument engineer, might or might not be able to interpret to the plant's advantage what the instruments show. The point is, instruments are essential for safe operation, and beyond that basic safety, additional costs must be evaluated against the usable results. Instrument purposes may be classified as follows:
  1. To provide operating guidance for the watch engineers.
  2. To provide information on which the economy of the plant operation can be evaluated. For example, determination of the pounds of steam produced by 1 pound of coal or by 100 cu ft of gas, etc.
  3. To allow comparisons of the daily operation of the plant with what it has been designed to do.
  4. To establish information on which costs can be segregated into various accounts. For example, steam flowmeters, or condensate meters, may be used to determine what percentage of the steam plant's installed value is chargeable to the various departments using steam for heating, for process, perhaps for both.
- INDICATING INSTRUMENTS are those which show values of the moment. They may be mechanical in the way they function, as a mercury-column thermometer; electrical, as a thermocouple pyrometer; or a combination of both, as in certain models of flowmeters. Indicators provide operating guides.
- RECORDING INSTRUMENTS not only show values of the moment, but make a record of changes in those values as these occur. Recorders are essential for purposes 2, 3, and 4.

Table 4-17. Chart of Boiler House Instruments

Area of measurement	Kind of measurement	Kind of instrument
Main steam, saturated as it leaves boiler	Pressure	1—Indicating
		1—Recording
Main steam, superheated as it leaves superheater	Pressure	1—Indicating
		1—Recording
As above, saturated	Temperature	1—Indicating
As above, superheated	Temperature	1—Indicating
		1—Recording
Main steam, superheated	Flow, lb/hr	1—Recording
Combustion air, overfire	Draft, in. W.G.	1—Recording
		1—Indicating
Combustion air, underfire	Draft, in. W.G.	1—Indicating
Combustion air leaving air preheater	Temperature	1—Indicating
		1—Recording
Combustion air entering air preheater	Temperature	1—Recording
Flue gas leaving boiler	Temperature	1—Indicating
		1—Recording
Flue gas leaving economizer	Temperature	1—Indicating
		1—Recording
Flue gas leaving air preheater	Temperature	1—Indicating
		1—Recording
Flue gas, CO_2 meter	% CO_2	1—Indicating
		1—Recording
Flue gas, CO meter	% CO	1—Recording
Flue gas leaving economizer	Draft, in. W.G.	1—Indicating
Flue gas leaving air preheater	Draft, in. W.G.	1—Indicating
Coal-air from pulverizer to burners	Temperature	1—Indicating
		1—Recording
Flue gas, smoke present	% smoke	1—Recording
		1—Indicating
Feedwater entering economizer	Temperature	1—Indicating
		1—Recording
Feedwater entering boiler	Temperature	1—Indicating
		1—Recording
Fuel firing equipment	Pulverizer, rpm	1—Indicating
	Stoker, rpm	1—Indicating
Fuel oil to burners	Pressure	1—Indicating
	Temperature	1—Indicating
	Flow, gpm or lb/hr	1—Recording
Fuel gas to burners	Pressure	1—Indicating
	Flow, cfm or 100 cfm	1—Recording
Coal to stokers or pulverizer	Weight, lb/hr	1—Indicating counter
Boiler blowdown meter	Blowdown, lb/hr	1—Recording
Boiler room clock	Time	1—Indicating
Portable pyrometer	Temperature	1—Indicating

Table 4-18. Turbine and Engine Room Instruments

Area of measurement	Kind of measurement	Kind of instrument
Main steam to throttle	Pressure	1—Indicating
	Flow, lb/hr	1—Recording
Each bleeder nozzle	Pressure	1—Indicating
	Temperature	1—Indicating
Lubricating oil system	Pressure	1—Indicating
	Temperature	1—Indicating
Main condenser	Vacuum	1—Indicating 1—Recording
Main condenser hot well	Temperature	1—Indicating 1—Recording
Circulating water, in	Temperature	1—Indicating 1—Recording
Circulating water, out	Temperature	1—Indicating 1—Recording
Generator cooler, water in	Temperature	1—Indicating 1—Recording
Generator cooler, water out	Temperature	1—Indicating 1—Recording
Generator coils	Temperature	1—Indicating 1—Recording
Feedwater from final heater *Note:* Place indicating thermometers at each heater.	Temperature	1—Indicating 1—Recording
Large room clock	Time	1—Indicating
Feedwater, boiler-feed pump discharge	Pressure	1—Indicating 1—Recording
Feedwater, boiler-feed pump discharge	Flow, lb/hr	1—Recording
Compressed air system, various locations	Pressure	1—Recording
Compressed air, after cooler	Temperature	1—Indicating
Water for air compressor coolers, in	Temperature	1—Indicating
Water for air compressor coolers, out	Temperature	1—Indicating
Various liquid-level gages	Liquid level	1—Each, indicating
Evaporators, water in	Temperature	1—Indicating
	Flow, lb/hr	1—Recording
Steam to evaporators	Pressure	1—Indicating
	Temperature	1—Indicating
	flow, lb/hr	1—Recording
Condensate from evaporators	Flow, lb/hr	1—Recording
	Temperature	1—Indicating

Table 4-18. Turbine and Engine Room Instruments (Continued)

Area of measurement	Kind of measurement	Kind of instrument
Steam to various auxiliary turbines, pumps, etc.	Pressure	1—Each, indicating
Exhaust steam header	Pressure	1—Indicating 1—Recording
	Temperature	1—Indicating 1—Recording
Main turbine	Speed	1—Indicating 1—Recording
Main turbine bearings	Temperature	1—Indicating with alarm
Steam engine, each expansion	Steam temperature	1—Indicating
	Steam pressure	1—Indicating
Steam engine, final exhaust	Steam pressure	1—Indicating
	Steam temperature	1—Indicating
Regenerative steam, leaving turbine	Temperature and pressure	1—Indicating 1—Recording
Regenerative steam, leaving boiler	Temperature and pressure	1—Indicating 1—Recording
Barometric conditions	In., Hg	1—Recording

- As with indicating instruments, recorders are available that function mechanically, electrically, or both.
  - The specifier must decide not only on which classification of recorder he wants, but also on the use of strip or circular charts. Each has certain advantages and certain drawbacks, depending again on the purpose for which the record is being taken. For example, the strip chart usually will run longer without changing; but it may be that the records are to be filed daily, so that a 24-hour circular chart change is desirable in view of the filing system.

(b) ***Checklist of Steam Power Plant Instruments***

- For a chart of boiler house instruments, see Table 4-17.
- For a chart of instruments common to turbine or engine rooms as appropriate, see Table 4-18.

(c) ***Pressure Gages, Indicating***

- Specify piping, tank, machinery component, etc., on which gage is to be mounted.

- Specify range of pressures to be provided on gage dial.
- Specify if gage is be back-connected or bottom-connected.
- Adjustable mounting is available if desired, so the gage may be pivoted or tilted to make easier reading.
- Specify materials of cases, as stainless steel, other alloy; brass bezil, etc.
- Specify accuracy desired as within ________ per cent of the entire range.
- Specify actuating element, as Bourdon spring, bellows, or other.
- If this supplier is to furnish pigtail (if wanted) and segregating valve in piping, so specify.
- Specify diameter of dial, ________ in.
- Specify Bourdon tube as tube element, spiral or helix, and name material as phosphor bronze, steel, etc.

Note: An absolute pressure gage incorporates a barometric compensator. Specify this if wanted.

- Specify signal contactors to give warning light or sound signal, if such monitoring is necessary.

(d) ***Pressure Gages, Recording***

- Specify time period over which record is to be made, as 8 hours, 12 hours, 24 hours, 30 days, etc.
- Specify kind of chart, as strip or circular.
- If more than one value is to be recorded on the same chart, as pressure, temperature, and flow, identify each measurement by name and specify the color of ink to be used for each.
- Specify kind and materials of sensitive element, as with indicating gages.
- Specify kind of transmission system desired between sensitive element and recorder, as pneumatic, electric, or electronic.
- Specify method of chart movement, as clockwork, electrical, etc.
- Specify that electrical connections and service must comply with National Electrical Code and local codes.
- Specify voltage available, whether line or reduced.
- Specify kind of signaling desired, if any; such as a simple warning bell or whistle; a two-position signal of red light for warning, green for normal conditions; or a combination of lights and sounds.
- Specify values at which signals warn.

- Specify distance of transmission between gage element and the recording instrument.
- If capillary tubing is used for transmission, specify size and material.
- Stipulate where recorder is to be located, as on panel, machine control board, etc.
- If transmission is pneumatic, specify compressed air as available from plant compressed-air system, and give pressure, _______ psig. If air is not available, specify a compressor to be provided with the instruments.
- If recording pen is attached to the element without intermediate transmission, so specify.

(e) *Draft Gages, Indicating*

- Specify if gage is to be of the inclined type, or U-tube manometer.
- Specify position of mounting; on panel, on a duct, etc.
- Specify if indication is to be single, or multiple, usually accomplished by either a battery of gages or a tubing manifold and segregating valves.
- Specify graduations and range on the dial or dials.
- Specify if dial is to be back-lighted by bulb inside casing.
- Specify liquid in tube, as mercury, oil, or proprietary with gage manufacturer.
- Specify gas to be gaged, as flue gas, combustion air, furnace, etc.
- Specify degree of accuracy.

(f) *Draft Gages, Recording*

- As with recording pressure gages, specify means of transmission; pneumatic, electrical or electronic, or by direct mechanical connection to the pen.
- Specify chart as strip or circular, and time period.
- Refer to Sec. 4-21(d) for items to be specified.
- Require bidders to submit drawings and data showing details of transmission.

(g) *Vacuum Gages and Barometers, Indicating*

Note: Vacuum is sometimes shown on the same dial as positive pressure, termed a compound gage. Use the guide for pressure gages, indicating, plus the following:

1. Identify the gage as a compound gage.
2. Give the range from vacuum to pressure.
3. Stipulate required accuracy, within plus or minus _______ per cent.

4. If warning of a loss of vacuum is required, specify nature of signal; light, sound, or both.

- For a simple vacuum gage, without positive pressure, specify whether gage is to be actuated by a bellows or diaphragm, and compensated by a spring and bellows; or whether actuation is by means of a mercury column.
- Specify kind of well required for the instrument, if any, as stainless steel or brass.
- For vertical glass column indicators, specify kind or color of liquid in the tube, as mercury, red alcohol, etc., or as provided by the manufacturer.
- For barometers, specify whether manometer or aneroid.
- Specify required accuracy.
- State where instrument is to be mounted, as gage board, outside, etc.
- Require bidders to submit information on adjustments, as variable settings provided, setscrews, etc.

(h) *Vacuum Gages and Barometers, Recording*

- Refer to Sec. 4-21(d), Pressure Gages, Recording. Use guidance given in that section as appropriate to vacuum gages and barometers. Also:
  - Specify kind of barometer, whether aneroid or manometer type.
  - Specify accuracy desired, ________ per cent.
  - Require bidders to submit drawings showing details of design and operation.

(i) *Thermometers, Indicating*

- Expansion-type thermometers operating on an expansion of a liquid, such as mercury or alcohol, that is proportional to the temperature, may be specified as follows, glass column and either angle or straight:
  - Specify the liquid to be used.
  - State where thermometer is to be mounted, and specify if full or partial immersion is desired.
  - Specify temperature range and markings on the glass column.
  - Specify the scale wanted, as Fahrenheit, Centigrade, Kelvin, Réaumur.
  - Specify kind of case to be furnished, naming metal.
  - Specify accuracy required, ________ per cent.
- For expansion thermometers using a bimetallic element, specify the two metals making up the element.
- Specify temperature range.

- Specify accuracy required, plus or minus ______ per cent.
- Specify diameter of dial, and if calibrations extend through the entire 360 degrees or are to be restricted to an arc.
- Specify length of stem, ______ in., and material.
- Specify scale as Fahrenheit, Centigrade, etc.
- Specify if dial is to be bottom- or back-connected to the stem.
- Specify case details, as corrosion resistant, removable bezel.
- Specify internal or external adjustment.
- Require bidders to submit drawings, cuts, or photographs showing details of construction and arrangement.

(j) *Thermometers, Recording, Expansion*

- Refer to Sec. 4-21(d), Pressure Gages, Recording.
  - Write the thermometer specification similarly, specifying the same items, as appropriate. Also specify if transmission is to be by liquid or gas expansion, or is designed to function on a vapor.
  - Specify desired liquid, as mercury, acetone, ethyl alcohol, etc.
  - Specify desired gas, as nitrogen, carbon dioxide, etc.
  - Specify desired vapor, as ethyl chloride, ether, propane, etc.

Note: Before specifying the transmitting agent, the varying effects on the true temperature recording of such factors as ambient temperatures, head, and dip for individual agents should be considered. For example, changes in the ambient temperature around the transmitting capillary affect both mercury and gas transmission, but do not affect vapor transmission. Head affects vapor but does not affect either mercury or gas. Evaluation of these factors may strongly influence the choice of the transmitting agent.

(k) *Pyrometers, Thermocouple*

- Specify the application of the instrument, as to a boiler furnace.
- Specify temperature range over which pyrometer is to register.
- Specify required accuracy, plus or minus ______ per cent.
- Specify scale of temperatures, Fahrenheit, Centigrade, etc.

- Specify metals in thermocouple, as chromel-alumel; iron-constantan; chromel-constantan, etc.
- Specify the span, _______ F or C, for a developed emf of 10 mv.
- Specify the melting temperature of the couple, _______ F or C.
- Specify if couple is to be used bare or inserted in a well. If bare, state if atmosphere is corrosive, contaminating, or otherwise likely to be detrimental to the function of the couple.
- Specify gage wires to be used, as 8, 14 or 20 B & S wire gage.
- State if emf generated is to be measured on a millivoltmeter or potentiometer.
- Specify distance between thermocouple and the indicating instrument as _______ ft. Also specify if connecting lead wires are to be of same material as thermocouple, or something else.
- If couple is set in a thermal well, specify material of well most suitable to resist oxidation and other corrosion as bronze, monel, stainless steel, etc.
- Specify length of scale over which pointer is to travel on either millivoltmeter or potentiometer.
- If instrument is to make a record, specify if chart is to be strip or circular.
- Require bidders to furnish complete data on method of reference-junction compensation. This is to be automatic. Also to show how the battery emf, against which the thermocouple emf is compared, is standardized.

(1) *Pyrometers, Optical*

- Specify area of application of the instrument, giving nature of the target surface.
- Specify the range of temperatures over which the instrument is to function, as between 1200 and 2800 F, etc.
- Specify method of reading desired, as by comparison with the brightness of a screen illuminated by a tungsten lamp, the current to the lamp being adjustable and calibrated in terms of temperature.
- Specify accuracy, plus or minus _______ per cent.
- Require bidders to furnish complete data on operation and construction of the instrument, including weight.
- Specify if instrument is to be hand-held or mounted on a tripod.

Note: A new form of optical pyrometer called an "infrared thermometer" is now available and can be used in some models from 50 F below zero to anything desired but below 1000 F. Average range is from 100 F to 300 F. Increase of the effective range may be available momentarily. These will be helpful in many plant practices, since the usual pyrometer, both thermocouple and optical, is most effective from 1000 F up.

(m) ***Thermistor Thermometers***

Note: The thermistor is a heat-sensitive resistor, developed as the result of semiconductor findings. The complete arrangement includes the heat-sensitive resistance unit, or semiconductor, a source of electric current, and a device to measure current flow. In operation, the current introduced into the thermistor finds increasingly greater resistance as the temperature falls, and a lessening of resistance as the temperature rises. By measuring the flow of current, as permitted by the resistance, the temperature is determined. For example, at 32 F the resistance may be, say, 7,500 ohms, but at 140 F resistance has fallen to perhaps 530 ohms, while no more than 150 ohms is encountered at 212 F. By contrast, a copper wire as used in the older resistance thermometer offers a resistance of only 10 ohms at 32 F; nickel, 100 ohms to 300. Resistance rises slightly as temperature rises. The advantage of the thermistor is in its extreme sensitivity in the middle temperature range, reflecting comparatively wide ranges in resistance for very small temperature changes. Their best performance is obtainable at temperatures below 500 F. Their temperature coefficient is negative.

- Specify the application of the thermistor, as measurement of a surface temperature by contact of the thermistor with the surface; insertion of the thermistor in a material, etc. (Note: Standard thermistor "probes" cover 100 or more interchangeable units to operate with a single indicating instrument.)
- Specify temperature range for each condition for which a probe is needed.
- Specify source of necessary current, whether instrument is to be battery-powered or otherwise.
- If several temperature ranges are to be used with the single instrument, specify number and gradations of

scale over which the pointer travels. Note whether Fahrenheit or Centigrade.

- Require bidders to submit all necessary data on the equipment being offered, including the diameters of the probes, the range limits, kind of indicating instrument, weights, etc.

(n) *Steam Flowmeters*

Note: The indicators and recording instruments, together with the transmission components, are generally applicable to both water and gas as well as to steam. The differential means for evaluating flow may be different, as a thin plate orifice for steam and gases and a venturi for water. The following specifying guide distinguishes between the various ways of measuring flow.

- For a thin plate orifice installation, specify the steam pressure, psig, and temperature, F. If superheated, state degrees of superheat; if saturated, state quality, ________ per cent.
- Preferably, furnish bidders with drawing showing size and position of piping with reference to instrument location. In lieu of drawing, specify pipe size, elevation with reference to instrument, and total distance, ________ ft and ________ in., between orifice and the instrument connections.
- Specify material of orifice.
- Specify the transmission agent, as mercury-mechanical or mercury-electrical.
- Specify if transmission is to be directly proportional to flow, or if compensations for differences in steam density shall be provided.
- Require bidders to submit sectional drawings of their equipment, showing internal arrangements of bells, seals, floats, and other mechanisms. Also to include a list of recommended spare parts that should be stocked.
  - Drawings should show dimensions, pipe sizes, valve sizes, and details of orifice installation between flanges, including diameter of orifice.
- Specify required accuracy as plus or minus ________ per cent.
- Specify accessories to be furnished by this contractor, as mounting brackets, mercury, pipe and nipples, meter valves, etc.
- Specify chart, as strip or circular, or designate the unit

as indicating and integrating with a suitable counter to totalize the flow in pounds per hour.

- Alternatively, differential device may be an *adjustable orifice,* or a *flow nozzle.*
  - For the *adjustable orifice,* specify the material of the orifice plate, as monel; and call for a pointer on the orifice stem to indicate the status of the orifice. Also specify the working pressure of the orifice body to suit the steam passing through, as ______ psig and ______ F total temperature.
  - Require bidders to supply information on how the adjustment on the orifice, once made, is locked in position so it cannot creep or otherwise change its setting. Also bidders shall indicate the range of movement of the orifice in its different positions, in pounds of steam an hour.
- If a *flow nozzle* is to be used, specify the material of construction.
  - Require bidders to furnish full data on the nozzle, its dimensions and range of flow in pounds per hour, for which the required accuracy may be guaranteed.
- For electric meters, specify the characteristics of the available current, voltage, phases, cycles.
  - Specify if alternating current is to be used or direct current.
- For *shunt-type flowmeters,* specify the required capacity range for which an accuracy of ______ per cent is required.
  - Specify that rotating elements, counters, etc., may be removed from the instrument without disturbing the section bolted into the main steam line.
- Specify an electrically operated chart recorder, or totalizer, if remote readings are desired. Give distance between the in-the-line instrument and the recorder or totalizer.
- Require bidders to furnish a correction-factor curve to be used for periods in which steam pressure and temperature may vary from the conditions for which the meter is calibrated at the factory.
- Specify the steam pressure, ______ psig, and total temperature, ______ F, minimum and maximum values, at the intended operating conditions.
- Specify the desired accessories, as gage glasses (for use

with the rotor-damping water), demand contactors for measuring hourly demands, limiting orifices, etc.

- Require bidders to furnish sectional drawings or photographs illustrating the internal functioning of the meter parts. Also to furnish information of the metals to be used in the meter construction.

(o) *Steam Condensate Meters*

- Specify that the piping carrying the condensate to the meter will be of _______ in. diam, of (iron) (steel) (copper) material, and pitched to the meter.
- Specify that traps discharging to the meter will be of the continuous flow type, rather than cyclical.
- State the flow to the meter will be by gravity from an adequately vented pipe or receiver.
- State that a check valve will be installed in the meter discharge piping to prevent any backup from that direction.
- State that condensate entering the meter will not exceed a temperaure of _______ F.
- If gravity flow is to enter a receiver before draining to meter, state capacity of receiver and projected difference in elevation, _______ in., between the bottom of the receiver and the connection to the meter.
- State if condensate is pulled through the meter by a vacuum pump. If so, specify placement of a thermostatic air valve on the meter outlet piping.
- Specify amount of vacuum, _______ in. Hg.
- Require bidders to submit detail drawings showing general arrangement and dimensions of the meter and its immediate piping connections.
- Require bidders to enumerate spare parts recommended for stocking.
- Specify required accuracy of measurement, _______ per cent.

(p) *Flue-gas Analyzers, Multiple Data*

Note: These analyzers are available for both on-the-site and remote indicating and recording. Normal measurement is of the volume of oxygen and of combustibles in the flue gas.

- Specify the gases for which a determination is required.
- Specify the minimum range as zero to _______ per cent, and the maximum range as zero to _______ per cent.

- Specify if instrument is to be mounted at site, or is to indicate and/or record at a remote position.
- Specify relative positions and distances between the point of measurement and the indicating or recording instrument.
- Specify kind of chart, strip, circular.
- Specify desired accuracy, ______ per cent.
- If a simultaneous temperature record is to be made, so stipulate, giving temperature range over which the thermocouple is to function.
- If transmission of measurements to instrument is accomplished by electricity, state available service, voltage, cycles, phases, etc.
- If transmission is pneumatic, state air pressure available.
- Require bidders to furnish complete drawings and data, showing internal arrangements, dimensions, weights, etc. Also bidders shall state their requirements of power, or compressed air, and any other limitations or restrictions on performance.
- Alternatively, specify if analyzer is to be of portable design.

(q) ***Flue-gas Analyzers, CO_2***

Note: Two major types are available. One operates on the Orsat principle, using a chemical that is gas absorbent; the other functions on the change of resistance, or thermal conductivity, that follows a change in CO_2 percentage of the flue gas. For the chemical type, a guide specification includes the following:

- State if analyzer is to be portable, for periodic testing, or is to be permanently mounted.
- Specify location of sampling connection, and state if sample is to be taken from a single connection or a perforated tube that is inserted in the flue gas.
- Specify the frequency with which the chemical absorbent is to be renewed.
- Specify the accuracy of the measurements, ______ per cent.
- Specify if indication of the CO present is also to be provided by the same instrument. Specify the frequency with which the CO absorbing chemical is to be renewed.

Note: Application of the Orsat principle permits the measuring of CO_2, CO, and O_2, using different absorbing chemicals. This is a common procedure when tests are being made of boiler combustion efficiency, for acceptance tests, or for other particular reasons. Industrial practice is more likely to be limited to a simple determination of CO_2, since this is a direct indication of the combustion efficiency. The specifier should clarify the ultimate purpose of the analyzer before expanding the CO_2 requirement.

- For electric resistance analyzers, specify the following:
  - The expected range of flue-gas temperatures.
  - Stipulate the location of the point of sampling.
  - Specify if indicating and/or recording are to be accomplished at the site, or on a remote instrument.
  - Since this form of analyzer employs a standard cell of various dimensions, stipulate where the cell may be located.
  - Specify the electric service available, voltage, phases, cycles, etc.
  - Specify the accuracy required, ________ per cent.
  - Require bidders to furnish complete data on their equipment, giving dimensions and weights and showing internal arrangements.
  - Require a statement on the estimated life of the resistance element.

(r) ***Boiler Meters***

Note: As a guide to combustion control, boiler meters generally record on a single chart the simultaneous relationships of steam flow, thousand pounds per hour and air flow, also thousand pounds per hour, and frequently with a third pen showing the averaged flue-gas temperature. Each factor is recorded by ink of a particular color, as red for steam, blue for air, black for temperature.

The mechanism producing these results consists of flow orifices or nozzles for the steam and air and thermocouples for temperature. Specifications for boiler meters will include the following:

- Specify the range of boiler output to be measured, in thousand pounds of steam an hour.
- Specify the range of air flow to be metered, also in thousand pounds of steam an hour.

- Specify the range of air temperatures, corresponding to the output of steam at various boiler loadings.
- State if boiler meter is to be mounted on the boiler or on an adjacent panel.
- Specify average temperatures to be measured and recorded, stating from what areas of the boiler passes the thermocouples will report to the recorder.
- Specify ink colors desired.
- Specify accuracies required.
- Refer to Sec. 4-21(n), Steam Flowmeters, for detailed checklist of items to be specified for the flowmeter component of the meter.
- Require bidders to submit complete details of their equipment, showing operating features, dimensions, weights, etc.
- Specify characteristics of available electric power.

(s) *Smoke Scanner*

- Describe generally the installation of this unit, stating the location of the density transmitter, as in the boiler breeching, stack, boiler pass, etc.
- Specify if scanner is to indicate only, or make a chart record. Specify terms in which density is to be expressed, as Ringelmann, etc.
- Specify kind of chart for recording instrument, strip or circular.
- Specify characteristics of electric power available for use by the recorder.
- If excessive smoke is to actuate a signal, specify kind of signal and point of smoke density at which it is to function.
- Require bidders to furnish complete data, including installation drawings, dimensions, and electrical connections that must be provided.

(t) *Boiler Blowdown Meter*

Note: All steam plants do not need a blowdown meter, since the water lost in that way may be computed from the metered input to the boiler, compared with the metered steam output. Where used, it serves the purpose of recognizing and metering any undesired leakage through the blowdown system. Operating on the principle of the V notch, the metering tank is combined with the blowoff tank, and usually is fed by a gravity flow of blowdown that first passes through a heat exchanger, where it is cooled to a value acceptable to the city or other-owned sewer. If this meter is to be included, specify as follows:

- Specify the capacity of the tank in pounds per hour. Also state maximum allowable dimensions of the tank, in view of its projected location.
- Specify the tank construction, gage of steel, how braced, as with inside or outside angles, and stipulate painting in material, colors, and number of coats.
- Specify the range of temperatures, _______ F minimum through _______ F maximum, of the water entering the tank.
- Specify site and position of tank vent, and where this contractor's work stops.
- Specify the location of the indicating, recording, and integrating equipment, as top of tank or elsewhere.
- Specify the drive for the instruments, as mechanical clock, electrical, etc.
- In order to check on blowdown leakage, specify the provision of two pens, one to record maximum rate of flow, the other to show on an expanded scale the rate of leakage. Ratio of the maximum flow to leakage shall be (5) (10) to 1, as deemed desirable for easy reading of the chart.
- Require bidders to furnish drawing of the equipment proposed, showing dimensions, weights, and the locations of the instruments.

(u) *Tachometers*

- Specify use of tachometer; whether engine, turbine, etc. Also if portable or to be panel-mounted, or hand type.
- Specify speed range tachometer is to function over; rpm.
- Specify if to be indicating or recording. If latter, specify kind of chart.
- State distance between point of speed measurement and indicating or recording instrument, _______ ft.
- Specify mechanism, as vibrating reed, etc.
- State characteristics of electrical power available for electrically operated instruments.
- Require bidders to furnish complete data of design and arrangement of tachometer, including drawings or photographs, with dimensions and weights as appropriate.

4-22. POWER PLANT TEST CODES

(a) *ASME Test Codes*

- For a chart of the available test codes, refer to Table 4-19.

(b) *Boilers, Fire Tube*

- "Standard Test Procedure for Packaged Fire-tube Boilers," American Boiler Manufacturers Association.

(c) *Capillary Tubes*

- American Society of Heating, Refrigerating and Air-Conditioning Engineers, "Standard Method for Testing Capillary Tubes" (1957).

(d) *Coal for Stokers*

- "Tentative Standard Procedure for Testing and Evaluating Bituminous Stoker Coals, (1952)," Stoker Manufacturers Association and Bituminous Coal Research, Inc.

(e) *Mechanical Condensing Units (Refrigerant)*

- "ASHRAE Standard Methods of Testing for Rating Mechanical Condensing Units (1959)," ASHRAE Standard 14-59.

(f) *Coolers, Drinking Water, Self-contained*

- "Standard Methods of Rating and Testing Self-Contained Mechanically Refrigerated Drinking Water Coolers," ASHRAE Standard 18-56.

(g) *Cooling Towers, Mechanical Draft*

- "Accepted Test Procedure for Water Cooling Towers, Mechanical Draft, Industrial Type (1955)," Cooling Tower Institute Bulletin ATP 105.

(h) *Fans*

- "Standard Test Code for Air Moving Devices," Air Moving and Conditioning Association, Bulletin 210.
- "Testing and Rating Ventilating Fans, Axial and Propeller Types (1951)," Commodity Standards Division, U.S. Department of Commerce.

(i) *Air Cleaning Devices, Filters*

- "Code for Testing Air Cleaning Devices Used in General Ventilation, Unit or Panel Type Air Filtering Devices (1953)," Air Filter Institute.

(j) *Electric Motors, Single Phase*

- "Test Code for Single Phase Motors (1958)," American Institute of Electrical Engineers, Report 502.

(k) *Stokers*

Table 4-19. ASME Power Test Codes Applicable to Steam Plants

Area of application	PTC number	Provides
General	1	Rules and practice for making acceptance tests.
General	2	Definitions and nomenclature.
Fuels, solid	3.2	How to determine fuel data and values.
Fuels, solid	4.2	How to test coal pulverizers.
Fuels, gaseous	3.3	How to determine gas data and values.
Steam generators	4	Rules for various combinations of steam generating components.
Steam engines	5	Rules for testing reciprocating steam engines and steam jackets.
Steam turbines	6; 6.1	Instructions for all types.
Steam pumps	7; 7.1	Instructions for reciprocating pumps and pulsating pumps.
Centrifugal pumps	8.1	Testing of all types.
Rotary displacement units	9	Testing for compressors, vacuum pumps, and blowers.
Centrifugal units	10	Compressors and exhausters, including superchargers.
Fans	11	Tests for fluid density less than 7 per cent.
Feedwater heaters	12.1	Both open and closed heaters.
Steam condensers	12.2	Testing instructions
Deaerators	12.3	Testing instructions, both spray and tray types.
Evaporators	14	A guide to test practices.
Steam turbine-generators	20.1	Testing the speed governing system only.
Dust separation	21	Tests for all types.
Dust concentration, measuring	27	For use with airstreams in ducts.
Water-cooling towers	23	For natural draft, forced draft, and spray ponds.
Ejectors and boosters	24	Test measurements.
Safety and relief valves	25	Tests and checks instruction.
Instruments, pressure	19.2.5	Guide for testing.
Instruments, temperature	19.3	This is an entire book on temperature instruments.

Table 4-19. ASME Power Test Codes Applicable to Steam Plants (Continued)

Area of application	PTC number	Provides
Instruments, flow..........	19.54	Tests of orifices, nozzles, Venturis.
Horsepower, shaft..........	19.7	How to determine.
Horsepower, indicated......	19.8	How to determine.
Steam quality and purity...	19.11	Methods of measuring.
Rotary apparatus, speed....	19.13	How to determine speed.
Smoke density.............	19.2	How to measure.
Specific gravity............	19.16	Determination of.
Viscosity..................	19.17	Determination of.
Condenser leakage.........	19.21.1, 2	Determination of.

- "Code for Determination of Rated Capacities of Anthracite Underfeed Stokers (1940)," Stoker Manufacturers Association.
- "Code for Determination of Rated Capacities of Bituminous Underfeed Stokers (1944)," Stoker Manufacturers Association.

(l) *Wiring*

- "National Electrical Code (1959)," National Fire Protection Association, or National Board of Fire Underwriters, No. 70.

(m) *Testing Efficiency of Air Filters*

Note: There are three methods that may be specified. They are described as follows:

- TESTING BY WEIGHT

A selected weight of dust is put into an airstream that next passes through a filter. By comparing the weight of dust collected by the filter with the weight originally put into the air stream, filter efficiency is determined.

- TESTING BY DISCOLORATION

The ability of a filter to remove the staining ingredient of atmospheric dust is determined by measuring the degree of staining accomplished before and after air passage through the filter. Originated by the U.S. Bureau of Standards, it is sometimes referred to as the N.B.S. Discoloration Test Method for Atmospheric Dust.

- TESTING BY DUST COUNT

  Carried out by counting the particles in the airstream before and after the air passes through the filter. Counting particles by the dioctyl phthalate smoke test (DOP) was developed by the U.S. Army Chemical Warfare Service for use on the particulate filters in gas masks. Equipment involves a special DOP smoke generator and optical-electronic methods for measuring the amount of smoke collected by the filter.

- Specify the kind of filter test desired.
- Specify the percentage of filtration required, as 99.97 for 0.3-micron particles.
- Specify the filter surface area, ________ sq ft, that is to be used in the tests.
- Specify the time duration of the test, ________ hours, ________ minutes.

DIVISION 5

Power Plants, Diesel and Gas Engine

5-1. DIESEL ENGINES

(a) *Engine Classification and Details*

- State service in which engine is to perform, as driving an electric generator, rated at _______ volts, _______ kw output, _______ power factor, direct current or alternating current, etc.
- State if drive is direct or through a reduction gear.
- Specify type of engine, as _______ cylinders, (2) (4) cycle, (air) (mechanical) injection, supercharged, etc.
- Specify horsepower rating and speed in rpm.
- Give characteristics of the fuel to be supplied to the engine, as a proximate analysis. Also temperature and pressure as delivered to the injectors.
- Stipulate dimensions of space available, with particular limitations. Provide a location drawing if possible, showing general arrangement and construction of housing.
- Specify manner in which cylinder liners are to be pulled.
- Specify piston details, as cast iron, cast steel, nickel steel, etc., and note special design shapes if desired. State minimum clearance, in.
- Specify number and design of piston rings, also the materials used in them.
- Specify manner of piston cooling.
- Describe desired form of injection, as solid, airless, common rail, etc.
- Specify kind and material of bearings, connecting rod, and crankshaft.
- Describe bearing lubrication method desired, including means for lubrication oil cooling. State minimum and maximum temperatures of available cooling water, F. Give water analysis.

- Specify kind of engine cooling system required, giving temperature and analysis of cooling water available.
- Specify elevation above sea level, ________ ft, at the proposed location for the engine.
- Bidders to supply complete design, construction, and performance details of the equipment being offered.

(b) *Engine Starting System, Storage Battery*

- For electric storage battery starting, specify voltage to be provided.
- State if storage batteries are to be furnished by engine contractor, or by others.
- Specify that batteries must satisfy the National Electrical Code, as applicable.
- Specify the maximum ampere-hour discharge rate, and the voltage at temperatures of 80 F and 40 F.
- Specify kind of battery, as lead-acid, or alkali type, or nickel cadmium.
- Specify number and kind of meters and gages necessary to provide continuing information on the operating status of the battery.
- Require this contractor to furnish complete data on the batteries proposed, including dimensions, weights, charging equipment, materials in plates and case.
- State maximum and minimum temperatures to be expected in storage battery room, and describe kind of room ventilation being provided.

(c) *Engine Starting Equipment, Compressed Air*

- Require bidders to state pressure of compressed air necessary to start the engine, ________ psig.
- State what items of the starting system are to be furnished and installed by the engine contractor, as air compressor, air tanks or flasks, connecting piping, relief valves, stop valves, gages, and thermometers. If these are to be furnished by others, so state, and require the engine contractor to coordinate his work with that of the accessories supplier.
- Specify if air-tank charging is to be accomplished with a single compressor arranged for alternate driving by an electric motor and a gasoline engine. The engine is for the initial charging of the tank, after which power from the main generator becomes available and is used to keep the tanks up to pressure through the compressor motor. If local electric utility service is available, the

gasoline engine may be omitted, unless failure of the utility power supply might have serious consequences and must be guarded against.

- If two air compressors are to be used, specify that one is to be driven by an electric motor, the other by a gasoline engine.
- Specify the number and sizes of the air tanks, and capacity.
- Specify a design pressure for these tanks at least 50 per cent above the expected necessary starting pressure, _______ psig.
- Specify that tanks must conform to the requirements of the ASME "Code for Unfired Pressure Vessels."
- Specify if tank construction is to be riveted or welded.
- Specify how tanks are to be painted.
- Specify number, sizes, and locations of all tank openings.
- Describe projected location of tanks relative to the engine they serve, giving distance between, _______ ft.
- Specify the kind and type of air compressor to be used, as two-stage, reciprocating, rotary, etc.
- Specify capacity, _______ cfm, of free air per minute at an intake temperature of _______ F compressed to a pressure of _______ psig.
- If air cooler is required, specify kind, construction, circulating water pressure, range of temperatures summer and winter, F, and source. Also note analysis and possibilities of corrosion, materials deposit, etc.

(d) *Air-intake System, Ducts and Filters*

- State if air intake is to be located outside or inside the building. (*Note:* for large engines, an outside intake becomes imperative.)
- State how much of system the engine contractor is required to furnish, as piping or duct, rain hood, filters, silencer, etc.
- If possible, provide bidders with a drawing showing arrangement and location of intake, size of air duct, kind and gage of duct steel or aluminum, kind of joints, supports and suspensions of ducts, location and kind of expansion joints.
- Specify the range of air pollution concentrations, from a minimum of _______ grains per 1,000 cu ft to an average maximum of _______ per 1,000 cu ft.

- Specify kind of air pollution, as dust, soot, ashes, corrosive gases, etc.
- Specify the cleanliness value of the air going to the engine, determined by one of the tests described in Sec. 4-22(m). Samples of air to be taken _______ ft downstream from the air filter.
- Specify the kind of air filter to be provided, as dry, viscous-impingement, oil bath, or electrostatic precipitator.
- For *dry filters,* specify kind of filtering medium, as fabric, felt, spun glass, or other.
- For *viscous-impingement* units, specify if filter is to be of the continuous automatic type, or is to be cleaned periodically by hand. Limit velocity to _______ fpm. Stipulate maximum acceptable pressure drop through the filter, as expressed in in. W.G.
- For *oil bath* filters, specify the minimum time between cleanings that is acceptable. Also specify the velocity of air passing through, _______ fpm.
  - Since the oil in this filter is sprayed into the airstream, specify the total amount of oil in the filter, including both the reservoir and the spray.
  - Require bidders to submit data on oil pump, its kind, capacity in gpm, discharge head, psig; also data on the pump motor, horsepower, temperature rise, C; voltage, phases, cycles, etc.
  - Bidders to state if motor and pump are direct-connected, mounted on a common baseplate, and what kind of starting equipment is used.
  - Specify that motor, wiring, and starting equipment must satisfy the requirements of the National Electrical Code, the NEMA, and local codes.
- For *electrostatic precipitators,* sometimes used for cleaning air to diesel engines, provide bidders with information on degree of air pollution the precipitator must contend with, also the volume of air passing through, from a minimum of _______ cfm to a maximum of _______ cfm. Specify the degree of cleanliness required, using tests as described in Sec. 4-22(m). Also refer to Sec. 4-13(b).
- Require filter contractor of whatever kind, to submit dimensional drawings, weights, and information about the construction materials and arrangement of his equip-

ment. Suppliers of electrostatic precipitators and other electricity-using equipment shall submit also the needs of their equipment in power characteristics, as volts, cycles, phases, etc.

(e) ***Air-intake System, Mufflers***

Note: Designers tend to favor the creation of a large-volume plenum chamber on the air inlet, usually between the filters and the inlet to the engine. Alternatively, several forms of muffler are available, and their performance is very good.

- Specify the volume of air being admitted to the engine through the intake system, _______ cfm.
- Specify the range of temperature in this intake air, from a minimum of _______ F in winter to a maximum of _______ F in summer.
- Specify nature of air contaminants, with particular reference to corrosive gases.
- Specify the degree of silence required, expressed in decibels.
- Require bidders to submit drawings and data fully illustrating the design and materials used in the device.
- Bidders also are to state their guarantee conditions, of _______ decibels when handling _______ cfm of _______ F air, passing through at a velocity of _______ fpm.

Note: Some mufflers are combined with an incorporated filter arrangement. This may be specified if desired.

(f) ***Exhaust System***

- State how much of the exhaust system is furnished by the engine contractor, as ducts or piping, mufflers, supports, and heat exchangers.
- Specify the length and diameter of the exhaust pipe from the engine to the muffler.
- Specify the height to which the muffler tail pipe is to be carried. Also give diameter of pipe, materials of which constructed, and forms of support.
- Specify if exhaust is to be carried through a heat exchanger, and if so, whether it will be used to warm the engine intake air or to provide comfort heating, hot water, etc.
- Specify design and capacity of heat exchanger, as shell and tube, with water in tubes entering at _______ F and flowing through at a rate of _______ gpm and a final

temperature of _______ F when exhaust gas is at _______ F, and going through the exchanger at a rate of _______ pounds per hour. Also specify friction loss in exchanger of _______ psi for water and _______ in. W.G. for the exhaust gas. Specify materials of exchanger.

- Specify if heat exchanger is to be in the form of a waste-heat boiler, generating steam at a pressure of _______ psig saturated, and at a rate of _______ pounds per hour with feedwater at _______ F and entering exhaust gas at _______ F, the gas passing through the boiler at a rate of _______ pounds per hour.
- Specify type of muffler required, giving noise limit in decibels.
- Specify materials and construction of muffler and designate location.
- If possible, provide bidders with a drawing showing arrangement of exhaust system from engine to open end of tail pipe. Drawing to show where and how the various components of the exhaust system are supported or suspended. Also indicate where and how exhaust pipe is to be compensated for thermal expansion.
- Require bidders to furnish drawings and complete information on the parts of the system being furnished, with particular care being shown for the dimensions, weights, and arrangement of the muffler and heat exchanger.
- Require bidders to submit data concerning the operating range of the muffler and heat exchanger, from a minimum to a maximum, at various rates of exhaust gas flow expressed in pounds per hour. Bidders also should indicate the effect of changes in exhaust gas temperature on heat recovery in the exchanger, and the friction loss, in. W.G.

(g) *Speed Governors*

Note: The promptness with which a speed governor acts to keep an engine from either running away or stalling is a function of the flywheel. The heavier the wheel, the less urgent the action, since the flywheel effect serves to release energy or store it to meet the load fluctuations. Conversely, a small, light flywheel requires fast action by the governor.

Some definitions will be helpful in specifying governor performance, as follows:

An *isochronous governor* maintains engine speed between

no-load and full-load conditions within 0 to 3 per cent of the full-load speed, change of load to be gradual.

A *nonisochronous governor* maintains the speed from no-load to full-load conditions within 5 to 10 per cent, when the load change is gradual, either up or down.

Stabilization period is the time in seconds that elapses between the greatest amplitude of a momentary speed change and the establishment of a steady speed.

The *sensitiveness* of the governor refers to the per cent of speed change required to produce a corrective action in the fuel feed control equipment.

Stability is the term for the governor's ability to prevent hunting or overcorrection of the load effect.

- Specify the kind of governor desired, as isochronous or nonisochronous.
- Specify the desired stabilization period and the sensitiveness of the governor.
- Specify speed range over which governor is to act.
- Require bidders to furnish complete data on the form of governor being offered, together with cross-sectional drawings showing the arrangements of springs, weights, balls, etc.
- Specify form of overspeed trip to be supplied, and state per cent overspeed at which the trip shall function and stop the engine.
- Specify kind of lubricating oil pressure alarm designed to stop the engine when lubricating oil pressure drops to a dangerous level. Designate actuating pressure.
- Specify kind of cooling water high-temperature alarm designed to stop the engine if jacket cooling water temperature rises to a dangerous level. Designate the actuating temperature as ________ F.
- Specify the provision of a visual or audible danger signal, or both, to inform the personnel present of the development of dangerous pressure or temperature conditions.
- If engine is to drive an electric generator in parallel with one or more other generators, specify kind of synchronous control that is to function with the governor.

(h) *Engine Foundations*

- Provide bidders with complete data on kind of soil on which engine foundation is to be placed. Include load-bearing value per sq ft, ________ tons.

- Specify if engine contractor is to furnish and place the foundations, or if this will be done by others.
- Specify a minimum depth of foundation that will prevent settling from frost.
- If pilings are required, specify kind, as wood or reinforced concrete, and specify manner of driving, giving weight of hammer and the sinking, ________ in., of the last ________ blows (usually 5). Alternative: Drive to refusal.
- Specify the mix of the concrete and the pouring, as 1 part cement, 2 parts sand, 4 parts broken stone or gravel of ________ in. maximum size. Also, pouring is to be carried to completion once started.
- Specify that newly poured foundation shall be covered and wet down ________ times per day (usually 2) until forms are removed.
- Specify that engine shall not be set on the new foundation until at least (10) (15) days have elapsed.
- Specify that engine shall not be started after bolting to foundation, for at least (10) (15) days after being set on foundation.
- Specify that engine is to be leveled by wedging as necessary, then grouted in place. Wedges are to be removed after grouting.
- If there is a possibility of excessive vibration becoming a nuisance, support the engine by means of cork or spring vibration absorbers. Specify the kind and size of the vibration absorbers.

(i) *Diesel Engine Instruments*

- A list of those instruments usually needed by diesel plants appears on the chart, Table 5-1.
- As with the instruments required by steam plants, specify whether unit is to be indicating or recording, giving calibration of dial in each instance, the kind of chart, as strip or circular, pneumatic or electric operation, etc.
- Refer to the instrument chart, Table 4-18, and to the various subheads in Sec. 4-21 for guidance in preparing the specification for indicating and recording instruments. These specification guides may be applied, as appropriate, to the specifications for individual instruments as these appear in the chart, Table 5-1.
- Note that all the instruments listed in Table 5-1 may not be needed for every diesel plant, or for some reason may

Table 5-1. Instruments for Diesel Engine Power Plants

Area of measurement	Kind of measurement	Kind of instrument
Engine starting air	Pressure, psig	Indicating Recording
Injection air	Pressure, psig	Indicating Recording
Scavenging air	Pressure, psig	Indicating
Lubricating oil	Pressure, psig	Recording Indicating
Fuel oil	Pressure, psig	Indicating
Engine jacket cooling water, in	Pressure, psig	Indicating
Engine jacket cooling water, out	Pressure, psig	Indicating
Lubricating oil cooler, water in	Pressure, psig	Indicating
Lubricating oil cooler, water out	Pressure, psig	Indicating
Engine jacket, cooling water in	Temperature, F	Indicating
Engine jacket, cooling water out	Temperature, F	Indicating Recording
Lubricating oil cooler, water in	Temperature, F	Indicating
Lubricating oil cooler, water out	Temperature, F	Indicating Recording
Exhaust, each cylinder	Temperature, F	Indicating
Fuel oil to engine	Temperature, F	Indicating
Engine exhaust heat exchanger, in	Temperature, F	Recording
Engine exhaust heat exchanger, out	Temperature, F	Indicating
Waste-heat boiler, in	Temperature, F	Recording
Waste-heat boiler, out	Temperature, F	Indicating
Ambient air	Temperature, F	Indicating
Intake air to engine	Temperature, F	Indicating
Engine tachometer	Speed, rpm	Recording
Water, heat exchanger, in	Temperature, F	Indicating
Water, heat exchanger, out	Temperature, F	Indicating
Generator coils	Temperature, F	Indicating
Waste-heat boiler, steam output	Pressure, psig Temperature, F Flow, lb/hr	Recording Indicating Recording
Barometer	Atmospheric pressure In. Hg	Indicating
Air compressor coolers, water in	Temperature, F	Indicating
Air compressor coolers, water out	Temperature, F	Indicating
Starting air tanks	Temperature, F Pressure, psig	Indicating Indicating
Fuel oil to engine	Flow, lb/hr	Recording

be changed by the designer, as from indicating to recording or vice versa.

- For temperatures below 300 F, the mercury type of thermometer is commonly used. Above this, thermocouples are the general practice, the materials of the couple depending on the temperature encountered. For example, up to 1600 F, iron-constantan may be selected; from this up to 2000 F, chromel-alumel will do. Still higher temperatures require other combinations.
- Bidders to submit list of instruments being offered under this specification, as makes, sizes, guaranteed accuracy.

(j) *Vapor Phase Cooling*

Note: In some instances engine manufacturers may offer vapor phase engine cooling as an improvement on circulating water cooling in the jacket, since the water always boils in the top of the jacket no matter how cool the incoming water happens to be. This results in uneven cooling of the cylinder liners and is a cause of considerable liner warpage. Although the vapor phase system means operating at engine temperatures 50 to 75 F warmer than with water, experience indicates that maintenance otherwise is no greater than with water and efficiency is higher.

- Specify water temperature, _______ F, to be introduced into engine jacket, and pressure, _______ psia. If cooling is by steam exclusively, stipulate temperature and pressure of steam available and entering the cooling jacket.
- For a cooling system with water entering at approximately 212 F, and later flashed into steam, describe kind of condenser to be provided by this contractor (or by others).
  - State entering and leaving temperatures of condensing water.
- Specify design and construction of vapor phase condenser, as shell and tube. Specify permissible pressure loss of condensing water in passing through condenser.
- Provide a drawing showing relative positions of condenser and engine, the size, kind, and arrangement of the interconnecting piping.
- Specify insulation on vapor phase piping, if any, by kind, thickness, and form of joints.
- Specify range of temperatures, _______ F, acceptable

between the bottom and top of the engine liners. State required average temperature, _______ F.

(k) *Dual-fuel Diesels*

- If diesel is to be accommodated to the burning of gas, as manufactured, natural, sewer gas, etc., in an emergency, so state.
  - Provide information on the kind of gas and provide a proximate analysis. State pressure, psig.
- Describe desired means and method of changeover from oil fuel to gas.
- Describe any projected change in the operation of the unit while burning the gas, resulting from the variations in exhaust temperatures, volumes of exhaust, pounds per hour and cfm, different efficiencies, etc.
- Specify required efficiency while operating on the gas fuel, giving fuel conditions on which efficiency guarantee is to be based.
- Specify probable percentage of total running time during which engine will burn gas rather than diesel oil.
- Specify Btu per bhp-hr when operating on oil and also when operating on gas.
- If small amounts of diesel oil are to be introduced into the cylinder along with the gas to make sure of ignition, specify amount of this pilot and the compression ratio.
- Specify kind and type of gas compressor, how driven, and characteristics of the driver if a separate motor is used.

(l) *Diesel Fuel-oil System*

- Provide a drawing showing piping to engine, indicating location of oil tank for storage plus a day tank if this is to be used.
- Specify kind of fuel-oil preheater, as steam heat exchanger, electrical, or other means, if used.
- State if heating coils are provided in the storage tank, and give temperature, _______ F, of oil as it enters the preheater (if any).
- Specify required temperature of the fuel oil as it enters the engine, _______ F.
- Specify the kind of *fuel pumps* to be provided, as rotary gear pumps or another form of positive displacement.
  - Specify capacity, gpm, at a suction pressure of _______ psia and temperature of _______ F and _______ rpm.
  - Specify driver, as electric motor, operating on _______

volt, ________ phase, ________ cycle, ________ wire service.

- Specify if direct drive or through reduction gears. Specify kind of coupling, as flexible, hydraulic, etc.
- Require electric motors and starting equipment to satisfy the National Electrical Code and the NEMA standards as appropriate.
- If pump and driver are to be mounted on a common bedplate, so specify. Require bedplate to have a raised lip and facilities for the drainage or retention of leaks as the installation indicates to be proper.

- If a day tank is to be used, state capacity in gallons, whether tank feeds engine by gravity or pump, and who is to provide the tank and its related piping and auxiliary equipment.
- Specify materials, thickness of plate, and general construction of day tank, as welded or riveted.
- Specify gage glass or glasses for indicating level of oil in the tank.
- Specify a float and low-level alarm to inform engine attendant if fuel-oil level falls to a value that requires action to prevent a shutdown of the engine from lack of fuel. Similarly, a high-level alarm is desired to call attention to overflow conditions with the supply pumps still running.
  - Stipulate if alarm is to be by bell, whistle, flashing light, or a combination.
- If day tank is to be mounted on an elevated structure, state who builds the supporting structure.

(m) *Fuel-oil Strainers and Filters*

- Specify kind and type of fuel-oil purifier, as filter screen, cloth bag, single or duplex arrangement. Latter arrangement permits switching the purifying action from one side to the other, thus avoiding interruption during cleaning.
- Specify degree of cleaning or separation required, as requiring the removal of foreign particles from the oil down to a size of ________ microns.
- Specify minimum acceptable times between the cleanings of the filter element, as ________ hours.
- Specify desired method of cleaning filter, as removal of filter element, opening of plug or handhole, etc.
- Specify pressure of oil entering filter, psi, and the acceptable friction drop through the filter, psi.

- Bidders to submit details of design, construction, and performance.

(n) *Oil Centrifuges*

Note: Although the specification for a centrifuge is given here under fuel headings, the same centrifuge may be used for lubricating oil. The switch from fuel oil to lubricating oil, or vice versa, is accomplished by opening and closing the proper valves in the respective fuel and lubricating systems.

- Specify the type of centrifuge, as a simple unit or a combination of centrifuge and filter.
- Specify the method of operation, as continuous or putting oil through in batches.
- Specify desired capacity in gph under each method of operation.

Note: Under continuous operation, 1 gal *lubricating oil* an hour for each 10-engine hp usually is satisfactory. Under batch operation, 1 gph for each 15-engine hp may be acceptable. For *fuel oil,* all the oil must go through the centrifuge if the unit is placed in the supply line to the engine. If the day tank is a large one, batch handling might be practical for maintaining the oil in the tank at a suitable purity.

- State temperature of oil entering centrifuge, ________ F.
- If oil is to be preheated, stipulate kind of heater, as electric, gas-oil heat exchanger, etc. See Sec. 5-1(l) regarding temperature of oil entering the diesel engine.
- State electric service available, as ________ volts, ________ phase, ________ cycle, ________ wire.
- Require bidders to furnish detail drawings, photographs, and cuts of the centrifuge being offered, so that a complete disclosure of its design and working methods is provided.
- Require bidders to state degree of cleanliness accomplished at the flow and temperature stipulated.
- Bidders to state motor hp, speed in rpm, weight of equipment operating.
- Bidders to describe kind and capacity of incorporated pumps, if provided by this contractor.

(o) *Diesel-oil Storage Tanks*

- Fuel-oil storage tanks, outside, inside, aboveground, belowground, horizontal, vertical, concrete, fire protection, etc., together with connecting piping and details of tank heating, are guide-specified in Sec. 4-9(a)–(m).

Table 5-2. Physical and Performance Data for Diesel Engines

Job ____________ Engine make ____________
Fuel (as dual) ____________ Model ____________
Number of units ____________ Driving ____________

Item	*Information by bidder*
Number of cylinders	____________
Arrangement (Horizontal)(In Line)(Vertical)(V Type)	____________
Supercharged	____________
Cycles (2)(4)	____________
Cylinder diameter, in.	____________
Piston stroke, in.	____________
Rpm at full load	____________
Piston travel, full load, fpm	____________
Weight of engine, operating, lb	____________
Overall dimensions, ft and in.	______ Height, ______ Width, ______ Length
Brake horsepower, full load	____________
Fuel injection pressure, psig	____________
Fuel injection temperature, F	____________
Pressure of air entering cylinders, psig	____________
Temperature of air entering cylinders, F	____________
Fuel consumption, oil, lb/hp-hr	____________
Fuel consumption,* gas, cu ft/hp-hr	____________
Pressure, lubricating oil system, psig	____________
Lubricating oil cooling water, gpm	____________
Water temperature entering cooler (range)	______ F to ______ F
Water temperature leaving cooler (range)	______ F to ______ F
Oil temperature drop in cooler, F	____________
Oil pressure drop in cooler, psi	____________
Water pressure drop in cooler, psi	____________
Water for engine cooling system, gpm	____________
Water temperature entering engine jacket, F	____________
Water temperature leaving engine jacket, F	____________
Water pressure required entering jacket, psig	____________
Circulating water through engine cooler, gpm	____________
Temperature, circulating water entering cooler	______ F to ______ F (max)
Temperature, circulating water leaving cooler	______ F to ______ F
Temperature, exhaust gas leaving engine, F	____________
Normal noise level of engine, full load, decibels	____________

* As a dual fuel engine, when operating on gas, the consumption rating is to be on the basis of fuel gas having a calorific value of 1000 Btu per cu ft, lower heat value. For a radiator installation using air for cooling the circulating water, see Table 5-3 on gas engines. See the index for cooling towers and spray ponds as other means for cooling circulating water.

Note: Fig. 5-1 shows an industrial power plant that is entirely staffed with radial-type diesels.

FIG. 5-1. Power plant employing diesel engines of the radial type. (*Courtesy Nordberg Manufacturing Company.*)

- Refer to these sections for guidance on specifying storage tanks for diesel engines.

(p) ***Data to Be Submitted by Bidders on Engine***

- See itemized form, Table 5-2, Physical and Performance Data.

5-2. TOTAL ENERGY WITH DIESEL ENGINES

(a) ***Diesel Engines for Total-energy Installations***

Note: The application of diesel engines advantageously to total-energy projects became practical with the general acceptance of summer cooling as a modern necessity, plus the recognition of absorption cooling as an economical way of salvaging what previously had been considered waste heat.

The ratio of waste heat from a diesel exhaust to the shaft horsepower developed by the engine is about 1½ to 1.

By harvesting this heat to generate steam at low pressure, and by then delivering this steam to an absorption machine, cooling is obtained at a relatively low cost. The net effect, under favorable conditions of occupancy and building services needed, is an installation of lighting, power, heating, cooling, and pumping that often is more economical to own and operate than is a comparable complex of purchased power, steam boilers for heat and process, and motor-driven cooling compressors.

Guide specifications for the diesel engine total-energy installation may be checklisted as follows.

(b) *Diesel Engine for Total Energy*

- For the engine classification and details, see Sec. 5-1(a)–(b). Evaluate each of these subheads for applicability to the total-energy project. Particular attention should be given to "space available" in 5-1(a), since the plant is likely to be located in a building of diverse character, as a shopping center, a high-rise office building, or a school.

(c) *Electrical Components*

- These are not within the scope of this Handbook, and their specifications should be obtained elsewhere.

(d) *Absorption Refrigeration System*

- Refer to Division 16, all sections and subheadings.

(e) *Machinery Room Cranes*

- Refer to Division 4, Sec. 4-20(a)–(e).

(f) *Emergency Standby Power Generators*

Note: Since the total-energy concept is intended to produce an installation that is self-sufficient, good design anticipates possible interruptions to the operation of the basic diesel engine unit or units. To pick up the load immediately can be very important in some environments, as in a department store where all lights out might initiate a panic; consequently, customary practice is to provide either an emergency connection to the local utility power lines or a standby generator. If the standby generator is driven by another diesel engine, refer to all the previously identified sections that apply to diesels. If the generator is driven by a gas engine or gas turbine, see Secs. 5-3 to 5-5.

(g) *Electric Generator Characteristics*

- Not within the scope of this Handbook. However, certain generator data should be given to the diesel engine bidder, and for guidance on these data which must be obtained from the electrical specifier, refer to Sec. 4-15(a).

(h) *Reduction Gears*, if used, are to be described in detail by bidders. Detail drawings are required showing dimensions, weights of heaviest pieces, ratio of reduction, kind of gears, etc.

- Either specify, or require method of lubrication to be described by bidders.
- Specify kind and model of couplings, as flexible, hydraulic, etc.

(i) *Diesel Engine Control Systems*

- Specify if controls are to be actuated by electric service, electronic, pneumatic, or a combination of these.
  - For electric and electronic, state service available, as _______ volt, _______ phase, _______ cycle, _______ wire.
  - For pneumatic, state compressed-air service available. If none available, specify an auxiliary air compressor to be used for the pneumatic controls exclusively.
- Specify each automatic control required, as (1) to start the engine, indicating if start is to be based on time, load increase beyond capacity of other units, or an emergency loss of load by another unit; (2) to synchronize the machine with others if the generator is alternating current; (3) to put the unit on the line automatically when it reaches speed and synchronization; (4) note if cooling system pumps are to go into operation automatically or shall start manually before the engine is started.
- Specify shutdown controls, as to stop the engine on (1) a failure of the lubricating oil system; (2) a failure of the cooling system; (3) a failure or the onset of a dangerous condition in the generator, such as overheating. If there is doubt of the dependability of the engine governor or emergency trip to stop the engine when overspeeding, an additional control may be specified for this purpose.
- Specify any form of alarm required, as a whistle, bell, flashing light to warn of a dangerous condition of high temperatures, excess speeds, low air flash pressure, fire or overheating in the generator windings.
- Require bidders to submit complete data on the control and alarm systems they plan to provide, including their designed electric or compressed-air working needs.

(j) *Operating Instructions*

- Require this contractor to provide in duplicate the necessary instructions, drawings, and diagrams to enable the owner to operate the diesel engine safely and efficiently.

(k) *Division of Energy*

- Refer to Sec. 5-6(a) for the division of energy into shaft horsepower and reclaimable heat in Btu per hour.

5-3. GAS ENGINES

(a) *Engine Classification and Details*

- State service in which engine is to be engaged, as driving a gas line compressor, air compressor, generator, etc.
- Stipulate if engine is direct-connected or drives through a reduction gear.
- Specify type of engine, as (number) cylinders; (2)(4) cycle; whether power cylinders are horizontal in line, vertical, or V shape.
- Specify kind of ignition, as magneto, individual cylinder coils.
- Specify brake horsepower required, using gas fuel as specified, and with a piston speed maximum of ________ fpm.
- Specify gas fuel to be furnished, as natural, manufactured, by-product, sewer, etc. Provide proximate analysis of gas. Specify pressure at which gas is supplied to the gas-air proportioning valve, as ________ oz per sq in. Specify pressure, ________ psi, as injected into 2-cycle scavenging engine.
- Specify compression range of power cylinder as (90 psia) (150 psia)(250)(400), etc.
- Stipulate space available, noting limitations, if any.
- Specify any special materials, shapes of piston, or particular arrangement of piston rings, if considered desirable.
- Specify materials for bearings, connecting rods and main.
- Specify method of bearing lubrication. Specify the form of lubricating oil cooler, as shell and tube, etc.
- If lubricating oil is to be centrifuged, refer to Sec. 5-1(n) for specification data.
- State maximum and minimum temperatures of water to oil cooler, both winter and summer, ________ F. Provide water analysis with particular reference to corrosive elements present, if any.

- Specify engine cooling system, as (closed) (open) with air-cooled radiator, etc.
- Specify maximum temperature rise acceptable in engine water jacket, _______ F. (*Note:* 12 to 15 F is a conventional maximum.)
- Specify maximum water pressure available and minimum, if furnished from an independent system; _______ psig.
- If water cooling is in a heat exchanger, specify kind, as shell and tube, materials, etc. Stipulate who will furnish the exchanger.
- If cooling radiator incorporates fans, stipulate location, the cfm of the fans, the temperature range of the ambient atmosphere, the noise level acceptable, in decibels.
- Require bidders to furnish as appropriate: Full details, specifications, and drawings as necessary to illustrate the design and construction of the engine and of all its auxiliaries.
- Require bidders to fill in a data and performance form approximately as shown in Table 5-3, this form being expandable or reducible as seems expedient to the specifying engineer.
- State elevation, _______ ft above sea level.

(b) *Engine Starting System, Storage Battery*

- Refer to Sec. 5-1(b) for items to include in the specification.

(c) *Engine Starting System, Other*

- If engine is to be started by a starting motor, electric or pneumatic, or an auxiliary starting engine, so stipulate.
  - Specifying who is to furnish and install the starting equipment.

(d) *Air-intake System and Air Filters*

- Refer to Sec. 5-1(d) as appropriate.
- For intake mufflers, refer to Sec. 5-1(e).

(e) *Engine Exhaust System*

- Refer to Sec. 5-1(f) for details of items to specify as appropriate to the job in hand.

(f) *Engine Speed Governors*

- Refer to Sec. 5-1(g).

(g) *Engine Foundations*

- Refer to Sec. 5-1(h).

(h) *Gas Engine Instruments*

- Refer to Table 5-1, which lists instruments for a diesel installation. Delete the instruments not required for a

Table 5-3. Physical and Performance Data for Gas Engines

Job ____________ Engine make ____________
Fuel ____________ Model ____________
Number of units ____________ Driving ____________

Item	*Information by bidder*
Number of power cylinders	
Operating cycles (2-scavenging)(4-scavenging)	
Cylinder arrangement (Horizontal)(In Line)(Vertical)(V Type)	
Cylinder diameter, in.	
Piston stroke, in.	
Rpm at full load	
Piston travel, full load, fpm	
Weight of engine, operating, lb	
Gas pressure to cylinder, 2-cycle engine, psia	
Gas pressure to proportioning valve, 4-cycle, oz	
Space required ________ Ft high, ________ Ft wide, ________ Ft long	
Brake mean effective pressure, full load, psi	
Brake horsepower, full load, bhp	
Fuel consumption,* cu ft/hp-hr	
Compression pressure, psia	
Temperature of gas entering cylinder, F	
Temperature of exhaust, F	
Lubricating oil cooling water, gpm	
Lubricating oil cooling water temperature rise, F	
Pressure drop in cooler, psi	
Engine cooling system, gpm	
Temperature, water into engine jacket, F	
Temperature, water leaving engine jacket, F	
Pressure drop in heat exchanger (cooler), psi	
Number of fans, radiator type of air cooler	
Output of each fan, cfm	
Speed of fan, rpm	
Noise level, full-load operation of fans, decibels	
Temperature of air entering radiator, F	
Temperature of air leaving radiator, F	
Fan motor characteristics ____________ Hp, ____________ Rpm	
Voltage	
Amperage, running	
Amperage, locked rotor	
Permissible overload ____________ Per cent for ____________ hours	
Oil pilot fuel (if used), lb or gal per hp-hr	

* Fuel consumption is based on a fuel supply of 1000 Btu per cu ft, calculated on the lower heat value of the gas.

gas engine, and change those provided for an oil fuel to instruments suitable for gas fuel. Also add a velometer or anemometer of a portable type, for use in measuring air flow through the radiator.

(i) *Gas Engines and Diesel Fuel*

- Refer to Sec. 5-1(k), Dual-fuel Diesels. Write the gas engine specification from the viewpoint of gas being the basic fuel and diesel oil the emergency or temporary fuel.

(j) *Gas Engine Fuel Systems*

- Specify who is to furnish and install all the required gas piping, pressure controller, meter, etc.
  - For specifying the gas piping, see Sec. 4-10(a) and (b). For valves and fittings, refer similarly to the sections in this Handbook covering these items.
- For natural gas handling, refer to Sec. 4-10(a).
- For the storage, piping, safety, pumping, and delivery of liquefied petroleum gases, refer to Sec. 4-10(b).

5-4. GAS ENGINES FOR TOTAL ENERGY

Note: A gas engine fits into the total-energy plan as well as a diesel engine, but the exhaust of a gas engine makes available much more heat than is characteristic of the diesel. The production of heat is in the ratio of 2½ to 1 to the useful work of the engine, or shaft horsepower. This means that in some projects, where cooling demands are greater than power demands, a gas engine might conceivably be preferred to the diesel.

(a) *The Gas Engine Classification and Details*

- See Sec. 5-3(a) and prepare the specification in accordance with the items given there. For items relative to gas engine installations, follow as appropriate the subheadings of 5-3(b)–(j).

(b) For units of a total-energy installation in addition to the gas engine, refer to Sec. 5-2(c)–(h).

(c) *Gas Engine Control Systems*

- Refer to Sec. 5-2(i). Adopt items from this section as they are applicable to the gas engine.

Note: In a total-energy installation, for either a gas engine or a diesel engine, the interlocking of controls may become quite complex, particularly where alterations in various loads are to be handled automatically. To illustrate, if a basic light and power load falls off to such an extent that in a multiunit installation there is not enough heat to maintain the air conditioning requirements, the automatic starting of a mechanical refrigeration compressor to make up the air conditioning needs is a common expedient. A complete detailed diagram of all control stations and their interconnections is essential in obtaining a satisfactory installation.

(d) *Gas Engine Testing Code*

- Specify that the gas engine shall be tested in accordance with the ASME Test Code, PTC 17, "Internal Combustion Engines."
- This is applicable to all gas engines, whether part of a total-energy installation or not.

(e) *Division of Energy*

- Refer to Sec. 5-6(a) for specifying information on the division of the gas engine energy into shaft horsepower and reclaimable heat.

5-5. GAS TURBINES

Note: The specification writer, particularly if he is not the installation designer, should realize that gas turbines come in various combinations of combustors, compressors, inter- and aftercoolers, reheaters, and equipment for waste-heat recovery. The basic arrangement of the cycle is shown in Fig. 5-2, using

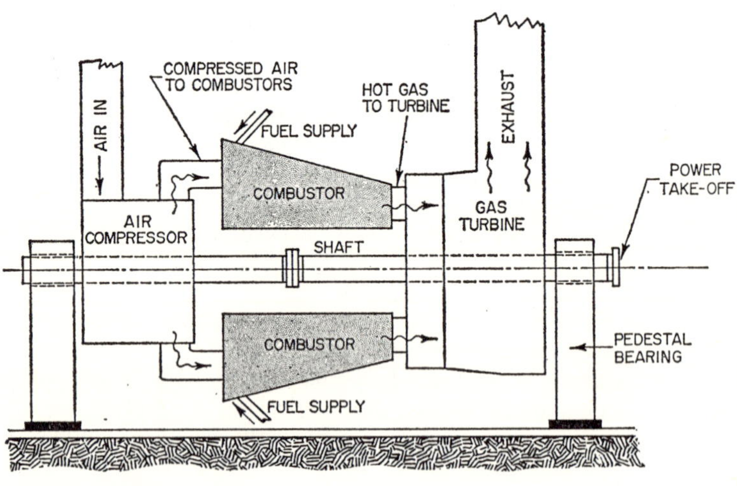

FIG. 5-2. Basic gas turbine. *Note:* A single combustor may be used, or half a dozen may be placed around the shaft (no scale).

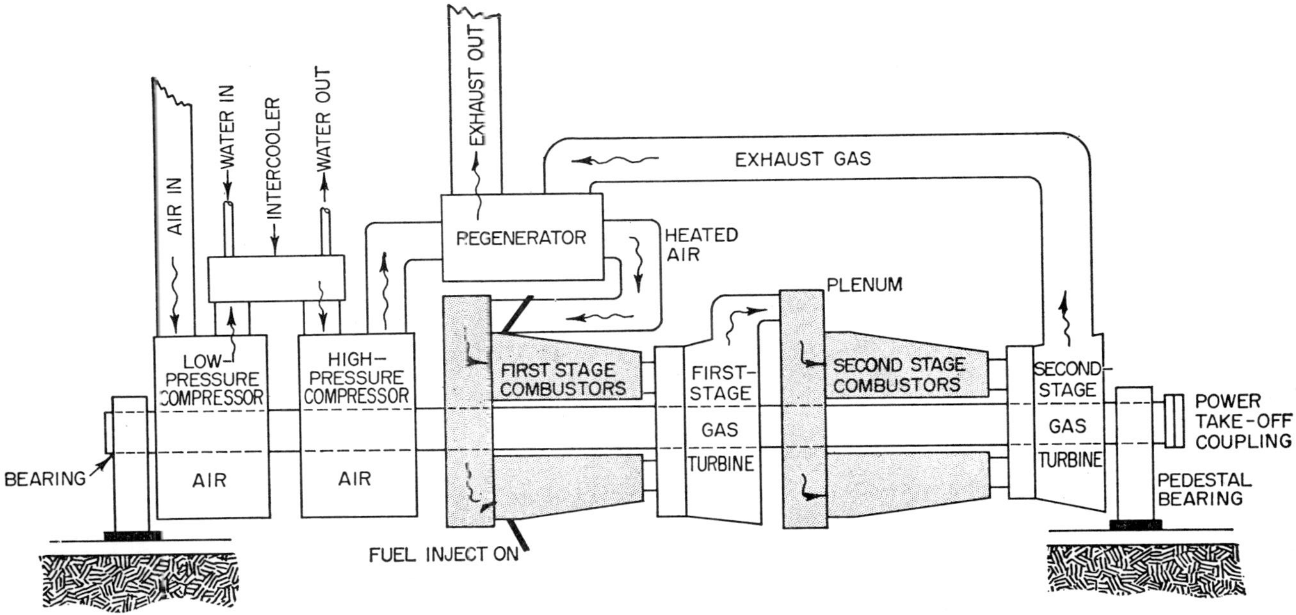

Fig. 5-3. Compound arrangement single shaft. *Note:* Intercooling reduces power to compressors; regenerator increases heat energy to first-stage combustors. Insulate ducts as specified. Locate thrust bearings as appropriate.

only the three essential components of compressor, combustor, and turbine.

An expansion of this simple cycle, incorporating an intercooler, regenerator, and stage compression for two turbines on a common shaft, appears in Fig. 5-3. Finally, the complications of a two-shaft installation are illustrated schematically by Fig. 5-4.

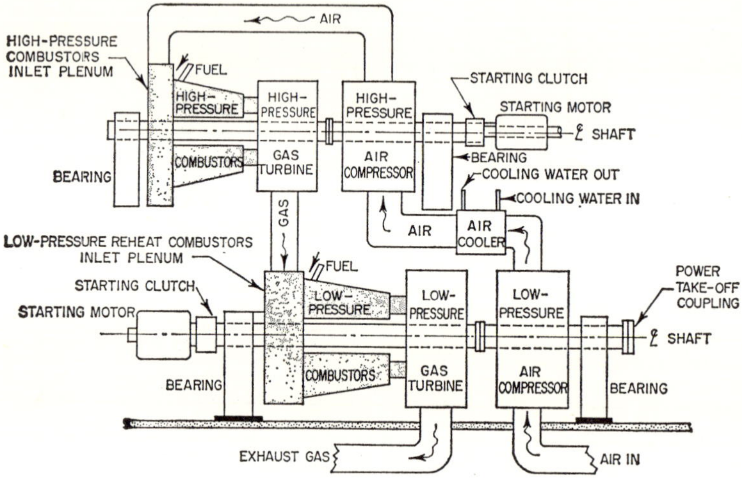

FIG. 5-4. Two-shaft reheat arrangement. *Note:* (1) Sketch is schematic to show air and gas sequence. (2) Location relationships of two shafts may be selected for convenience and space restrictions. (3) Starting motors may be replaced by pneumatic or internal combustion engines to suit conditions.

In any specification that is issued for a gas turbine, the complexity of the arrangement, with attention called to each of the components, must be set forth clearly; see Tables 5-3 and 5-4. Another consideration not to be overlooked is the counteraction provided against noise. Unless effectively muffled, the howl of the rotors revolving from 40,000 to 50,000 times per minute becomes intolerable.

(a) ***Gas-turbine Design and Arrangement***

- Specify the required shaft horsepower to be available at the power takeoff coupling.
- Specify an open-cycle, closed-cycle, or semiclosed-cycle

Table 5-4. Physical and Performance Data for Gas Turbines

Job ______ Turbine make ______

Fuel (*Kind of Gas*) ______ Model ______

Number of units ______ Driving ______

Item	*Information by bidder*
Brake horsepower at full load, bhp	
Fuel consumption,* full load, cu ft/hp-hr	
Rpm at full load	
Pressure of gas entering combustor, psig	
Temperature of gas entering combustor, F	
Pressure of air entering combustor, psig	
Temperature of air entering combustor, F	
Pressure of combustion gas entering turbine, psig	
Temperature of combustion gas entering turbine, F	
Pressure of turbine exhaust gas, psia	
Temperature of turbine exhaust gas, F	
Temperature, exhaust gas leaving regenerator, F	
Temperature, air entering compressor, F	
Temperature, water entering intercooler, F	
Temperature, water leaving intercooler, F	
Starting motor horsepower	
Amperage, locked rotor	
Noise level, full load, decibels	
Overall dimensions, ft and in. ____ Height, ____ Width, ____ Length	
Weight, operating, lb	
Number of compressors	
Number of combustors	
Number of turbines	
Number of shafts	
Lubricating oil, pressure, psig	
Lubricating oil, filter, kind	
Lubricating oil, maximum safe temperature, F	
Combustion temperature in combustor, F	
Weight of exhaust gases, lb/sec	
Starting method (electric)(pneumatic)(gas)	
Maximum hours continuous operation between regular maintenance	
Required time, start-up to full load, minutes	
Guaranteed and expected thermal efficiency, %	

* Correct fuel consumption to the basis of 1000 Btu per cu ft calorific value of the fuel gas, calculated on the lower heat value of the fuel gas. Correct atmospheric air conditions to 80 F and 14.17 psia.

machine. Also stipulate the general arrangement, as single or multiple shaft, and whether intercoolers and regenerator heat exchangers are to be included.

- Specify the number of combustors and their materials of construction.
- Specify the number of compressors and the number of stages in each. Also specify the materials of shaft and wheels.
- Specify the number of turbines and the materials of shaft and blades in each.
- Specify if machine is to be located indoors, or is to be furnished as a package unit suitable for outdoor placement. If a package unit, specify minimum clearances inside the casing and the means and methods of entry.
- Stipulate who is to furnish machine's foundation. If required of gas-turbine contractor, specify size and construction of the foundation, together with the mix of concrete. Specify the curing time for the concrete. Specify the number, kind, and size of the holding-down bolts. Provide the contractor with soil information, as the kind of clay, sand, loam, etc., and its load-bearing ability in tons per sq ft.
- Specify space limitations, if any.
- Name the driven machine, as a generator, blower, etc., and specify the kind of connecting coupling.
- Specify form of waste-heat recovery, as a waste-heat boiler, heat exchanger, or other means.
- Specify the kind of intake muffler desired, and decibel rating.
- Specify the kind of exhaust muffler desired, and decibel rating.
- Require bidders to submit complete drawings and photographs as necessary to illustrate the design and arrangement of the gas turbine.

(b) *Turbine Starting System, Electric*

- For a motor starter, specify electric service available; ________ volts, ________ phase, ________ cycle, ________ wire.
- Specify kind of declutching arrangement desired for cutting out the motor after the gas turbine is up to speed and self-supporting.
- Specify ignition method, as spark or hot bulb.

- Require bidders to furnish electrical equipment in accordance with NEMA standards.

(c) *Turbine Starting System, Pneumatic*

- Specify pressure of available compressed air, and give source, as air flasks, central plant compressor, etc.
- Refer to Sec. 5-1(c), Engine Starting Equipment, Compressed Air for details of the pneumatic system.

(d) *Turbine Starting System, Other*

- If an auxiliary engine or steam turbine is to be used for starting, so stipulate.
- In specifying an auxiliary diesel engine starter, refer to Sec. 5-1(a) and subsequent appropriate sub-sections, as "Engine classification and details."
- In specifying a gas engine starter, refer to Sec. 5-3(a), Gas Engine Classification and Details, and all appropriate subsections.
- If specifying a gasoline engine starter, stipulate capacity, _______ hp at _______ rpm when burning gasoline of _______octane rating.
  - Specify number of cylinders and arrangement, as in line, vertical, V type, horizontal.
  - Specify declutching, as automatic or manual.
- If stipulating a steam turbine as the starter, specify the steam pressure, _______ psig; if saturated, give quality, _______ per cent; if superheated, note superheat, _______ F.
  - Specify back pressure on turbine if not condensing, _______ psia. For condensing operation, give vacuum, _______ in. Hg.
  - Specify turbine hp, _______ at _______ rpm.
  - For details of steam turbines, refer to Sec. 4-15, Steam Turbines for Generating Power, and to the indicated auxiliary references for turbine drives.
  - As with other forms of starters, specify the kind of declutching arrangement desired.

(e) *Gas-turbine Bearings*

- Specify kind of shaft bearings, as babbitt lined, rollout; ball bearings, roller bearings.
- Specify kind of thrust bearing, single or multidisk, roller or ball thrust.
- Require bidders to describe method provided for the adjustment of axial thrust, and the range of clearance so adjustable.

(f) *Combustors*

- Specify form of combustor, as a burner tube angular to the shaft, or vertical and independently mounted.
- Specify kind of igniter desired and, optionally, the method of igniter retraction.
- Specify maximum solids in the fuel gas, ______ ppm by weight. Also specify percentage of water in the gas.
- Specify range of fuel flow from light-off to full flow.
- If burning oil in the combustor instead of gas, specify the kind and amount of contaminants, as vanadium, sodium, sulfur, ash, etc.

(g) *Air-intake System and Intake Muffler*

- These are approximately the same as for the diesel and gas engines. For details to be specified, see Sec. 5-1(d) on ducts and filters and 5-1(e) on mufflers.
- *Air filters* are included in these two sections also.

(h) *Exhaust System*

- For details to be covered in the specification, refer to Sec. 5-1(f), Exhaust System, for diesel engines.

(i) *Gas-turbine Foundations*

- Refer to Sec. 5-1(h), Engine Foundations.

(j) *Jet Engines*

Note: The airplane jet engine has been adapted to the gas-turbine field. In this application, the engine functions as a gas generator and is not mechanically connected to the power turbine. It has its own separate shaft which carries one or more turbine wheels and a multistage air compressor. The exhaust from the combustion chamber passes in part through the turbine, thereby driving the air compressor on their common shaft. The remaining exhaust, plus the discharge from the jet engine turbine, is introduced into the nozzle box of the power turbine by way of a connecting duct. This power turbine is sometimes called a "free" turbine, since its shaft is separate and independent from the jet engine's shaft, and their speeds of rotation may be quite different.

- Specify the number of jet engines desired to serve the power turbine. Since the jet exhaust is into a common collector duct that discharges its volume into the turbine nozzle box, any practical number of jets may be adopted from the viewpoint of cutting them in and out of service as needed to meet the load on the power turbine.
- Specify the size of each jet engine generator, as ______ hp with a fuel rate of ______ Btu per hp-hr.

- Specify the fuel to be used, giving a proximate analysis.
- Specify the minimum acceptable thermal efficiency, per cent, and require bidders to state the efficiency guaranteed.
- Specify maximum space available.
- If unit is to be installed as a package outdoors and within its own housing, so specify. Require bidders to furnish detailed and dimensioned drawings.
- For details of items to be specified as these relate to starting, bearings, foundations, refer to the appropriate subheadings under this section, since these are equally applicable to installations using gas for the power turbine from a jet-engine type of gas generator.

(k) *Gas-turbine Instruments*

- The instruments tabulated in Table 5-5 are subject to change, increase or decrease, depending on the kind of gas turbine to be monitored. The designing engineer's decisions will shape the specifications as issued.

Table 5-5. Gas-turbine Instruments

Area of measurement	Kind of measurement	Kind of instrument
Fuel gas to combustor	Volume, flowmeter	Recording
	Pressure, psig	Recording
	Temperature, F	Recording
Air entering compressor	Temperature, F	Recording
Air leaving compressor	Pressure, psig	Recording
Air entering combustor	Temperature, F	Recording
Gas entering power turbine	Pressure, psia	Recording
	Temperature, F	Recording
Exhaust gas, power turbine	Pressure, psia	Indicating
	Temperature, F	Recording
Exhaust gas leaving regenerator	Temperature, F	Indicating
Atmospheric pressure	Barometer	Indicating
Combustor, interior, each	Temperature, F	Indicating*
Lubricating oil	Pressure, psig	Indicating*
	Temperature, F	Indicating*
Exhaust gas, power turbine	Volume, cfm or lb/sec	Recording
Power turbine, speed	Rpm	Indicating
Jet engine shaft speed	Rpm	Indicating†

* With alarm to signal the onset of a dangerous condition.

† For each jet engine gas generator if design incorporates these units.

(l) *Controls*

- Refer to Sec. 5-2(i).
- In addition to controls indicated as applicable in that section, specify how the speed of the power turbine is to be maintained within tight limits. Also how the speed of the jet engine form of gas generator is to be varied to meet the load demands of the free turbine using the jet engine output of gas.
- Because of the effect of extreme combustor temperatures on the material life of the combustors, specify monitoring alarm pyrometers as considered suitable. Specify if alarm is to be accompanied by an automatic shutdown of the machine. Set alarm temperature as ________ F.
  - Specify nature of alarm, as whistle, bell, flashing light, red light, or a combination of these.

(m) *Gas-turbine Tests*

- Specify that the gas turbine shall be tested in accordance with the ASME Test Code, PTC 22, "Gas Turbine Power Plants."

(n) *Duct Insulation*

- Specify kind and thickness of insulation to be applied to exhaust ducts and air-intake ducts wherever the conservation of heat within the duct is desired.
  - Specify mode of application, kind of joints, joint lap, use of vapor barriers, etc.

(o) *Existing Building*

- If gas turbine is projected for installation in an existing building, describe facilities for weight handling and entry clearances.

(p) *Operating Instructions*

- This contractor shall provide, in duplicate, the necessary instructions, drawings, and diagrams to enable the owner to operate the gas turbine safely and efficiently.

(q) *Water-cooled Heat Exchangers*

- If the design of the gas-turbine plant includes series air compressors with one or more intercoolers between the compressors, these coolers must be covered in the specifications.
- Specify the kind and material of the intercooler, as shell and tube, Admiralty tubes, etc.
- Provide bidders with cooling-water data, as minimum temperature, winter, and maximum temperature, summer. State source and content, if any, of contaminants.

Also pressure, _______ psig. Require bidders to submit pressure drop, _______ psi, and temperature rise, _______ F, at the design flow of _______ gpm.

- Specify air temperature reduction in cooler, _______ F, under the full-load design conditions.

5-6. GAS TURBINES FOR TOTAL ENERGY

Note: Basically, specifying the type, form, and arrangement of a gas turbine for a total-energy project is covered in detail in Sec. 5-5(a)–(q). However, the following item must be introduced into the specification.

(a) ***Division of Energy***

- Specify the percentage of the gas turbine's total capacity of useful energy that is converted into shaft horsepower, and the percentage that becomes available for process steam, hot water, absorption refrigeration, direct air heating, etc.
- Specify that shaft horsepower is to be expressed in terms of brake horsepower, and the reclaimable heat in terms of Btu per hour. All figures to be based on full-load operation.

DIVISION 6

Power Plants, Hydroelectric

6-1. HYDROELECTRIC PLANTS, GENERAL

(a) *Classifications*

- Specify if operation is as a run-of-river installation or a storage pond, without or with off-peak pumping.
- Specify whether it carries the base load or comes on the line just for peaks.
- Note if the plant is isolated or is convenient to a community.
- Specify design as high head, or over 500 ft; low head of 100 ft or under; or medium head, in between.
- Specify if control will be automatic and remote, or if it is to be manual.

Note: Head classifications are arbitrary. Some engineers extend medium head to 800 ft; some depress low head to 50 ft.

(b) ***Waterwheel Details, High Head***

- Specify design head, ________ ft.
- Specify full-load horsepower.
- Specify kind of wheel, as vertical, reaction (Francis), horizontal, impulse (Pelton).
- Specify efficiency at full load. Request efficiency curves from bidders showing range of efficiencies at various loads.
- Provide bidders with drawings showing location of waterwheels and the draft tube characteristics and dimensions.
- Specify kind of speed governor desired, giving the per cent of speed rise acceptable if the turbine drops its load.
- Specify kind of bearings wanted, as ball, roller, or sleeve line bearings, and Kingsbury, Gibbs, or other form of thrust.

- Specify bearing lubricating system, as pumped oil, pumped water. Require bidders to state pressure required, ______ psig.
- Identify machine to be driven by the turbine, as an a-c electric generator, direct-connected, with an on-the-shaft exciter. Generator to deliver ______ kw at ______ volts, ______ per cent power factor, etc. Obtain from the designing electrical engineer the detailed specification for the electric generator driven by the turbine, and incorporate in the waterwheel specification. If the installation is to operate as a generating-pumping unit, with the turbine connected to a dynamo capable of running either as an electric generator or as a driving motor, the electrical specification shall so stipulate, providing available current and voltage characteristics on the motor cycle, and the desired characteristics on the power-generating cycle. The electrical engineer also shall specify the unit controls. Bidders shall provide complete information on the control equipment being furnished by them, as makes and ratings.
- Require bidders to submit complete drawings showing details of design and construction and giving all key dimensions. Also weights, ______ pounds, on thrust and step bearings.
- Specify materials to be used in the runner, including the crowns in the Francis design and the disks of the Pelton.
- Specify desired speed range, ______ rpm.

(c) ***Waterwheel Details, Low Head***

- Specify design head, ______ ft.
- Specify full-load horsepower.
- Specify kind of wheel, as (Kaplan) propeller type, with (fixed) (adjustable) blades.
- Specify additional requirements as given for high-head turbines in Sec. 6-1(b).

(d) ***Waterwheels, Medium Head***

Note: Usual practice specifies the Francis or Pelton design for medium heads and high heads. At low heads the Kaplan adjustable blade wheel maintains relatively flat efficiency curves but is not well suited to heads in excess of 100 ft. For this reason the other two designs dominate the medium- and high-head operation.

(e) ***Lubricating Systems***

- If turbine is one of several, it may be preferable to

pump the lubricating oil from a central pumping and filtering station. In these instances, specify as follows:

- Pressure ________ psig, of oil supply to bearings.
- Temperature of oil leaving cooler, ________ F.
- Flow of oil, ________ gpm.
- Describe kind of oil filter projected, as basket, cloth, mesh, etc. State cleanliness of oil leaving filter in percentage of contaminant particles in oil by ________ ppm. State size of particles removed.

- Specify the kind of oil pressure failure alarm to be furnished with each turbine.
  - Specify the kind of excess-temperature alarm to be furnished for the lubricating system of each turbine.
- In specifying oil pumping equipment, require pump motor to meet requirements of NEMA and the National Electrical Code.
- Stipulate kind of pump, as rotary, screw, reciprocating, etc. State if pump and motor are to be mounted on a common bedplate.
- State who is to furnish pump, filter, piping, and other equipment in the central-station lubricating oil system.
- For single-unit installations, where no separate lubricating system is planned, bidders shall furnish complete details of the lubricating oil system for their equipment, including operating temperature, ________ F, operating pressure, ________ psig, oil failure devices, alarms, oil purifying and cooling method, general arrangement, etc.

(f) *Turbine Control System*

- Specify the manner of controlling the flow of water from the penstocks into the turbine runner, as adjustable wickets, etc.
- Specify if water admission to the runner is to be automatically or manually controlled.
- Specify if adjustment of the water admission is synchronized with the load, so that it responds to load changes.
- Require bidders to furnish drawings and specifications of their control system, thus illustrating how movements of the wickets, or propeller blades or buckets, are implemented.
- Require information on the sensitivity of the control, that is, time expired in seconds before the flow adjustment is accomplished following a change in load.

(g) *Relief Valves and Hammer Protection*

- Operation of valve permits water to flow from the penstock into the draft tube or tailrace by bypassing the turbine when the load falls or is dropped.
    - Specify that valve is to be synchronized with the load control so that a falling of load opens the valve proportionately and an increase of load causes the valve to close, also proportionately.
- Specify the rate of opening and closing over full range, _______ seconds.
- Require bidders to show how the valve is connected to the gate operating ring of a reaction turbine, or the needle valve of the impulse turbine.
- State if combined discharge of turbine and relief valve is to be a constant over all ranges of load.
- Specify if governor movements are to be slowed and by how much, as a preventative of penstock hammer.
- On impulse turbines only, a form of jet deflector is frequently provided and installed between the water nozzle and the wheel buckets. Since this deflector is intended to waste water without any work, it maintains the rate of flow through the penstocks at the full-load rate, diverting flow away from the turbine buckets during low loads.
    - Specify the provision of a jet deflector if this form of control is to be adopted.

(h) *Surge Tanks*

- Specify the kind of tank, that is, the simple cylindrical tank or the Johnson differential.
- Fabricated of (riveted) (welded) steel plate.
    - In accordance with the ASME Code for Unfired Pressure Vessels.
- Specify volume of tank, cu ft.
- Provide drawing or sketch showing how tank is connected into penstocks.
- Specify painting of tank inside with _______ coats of _______________.
- Specify painting outside with _______ coats of _______________.
- Require bidders to furnish drawings that illustrate design and construction of tanks being offered. Also dimensions and other details, as weight of tank full, _______ pounds.

- Specify the calculated height of the surge, _______ ft average; _______ ft maximum.
- Specify maximum velocity, _______ fps, in the penstocks.

6-2. PENSTOCKS AND DRAFT TUBES

(a) *Penstocks, Steel*

- Specify construction of penstocks, as steel plates (riveted) (welded). Thickness of steel to be not less than _______ gage, or _______ in.
- Specify design, as uniform thickness, cross-section, etc.
- Specify if penstock is to be buried or entirely exposed.

  Note: Full exposure is generally preferred, to facilitate inspections and repairs.
- Specify the theoretical head, _______ ft, and the effective head, _______ ft. If some item of head loss is of substantial value, as in the valves or trash racks, call attention to this for the information of bidders.
- Specify kind of valves to be installed at the top of the penstocks, as gate, butterfly, needle, etc.
  - Specify materials of construction and maximum working pressure of valve. Specify degree of tightness required, and if any special sealing arrangement is to be provided. Specify kind and placement of vacuum breaker on penstock.
  - Specify how valves are to be operated, as by electric motor, manually, etc.
  - Specify kind and location of expansion joints to absorb effects of temperature changes.
  - Specify working and test pressures, _______ psig.

(b) *Penstocks, Concrete*

- Specify dimensions. Provide drawing showing details of construction, as thickness of concrete, steel reinforcing, finish of interior surfaces.
- Specify maximum velocity of water in penstocks, _______ fps.
- Specify maximum working pressure. Specify test pressure.
- Specify kind of valves to be installed at head of penstocks and elsewhere, stipulating materials, working pressures, test pressures.
- Specify operation of these valves, as by motor or manually.

- Specify the mix of the concrete to be used, and give the curing time, days. Obtain detailed concrete design and pour directions from the civil engineers engaged in designing the dam and spillway. These data may be included with either the mechanical or civil specifications as decided by the designing engineers.

(c) *Draft Tubes*

- Provide a drawing showing contour of draft tubes as projected. Show dimensions for steel and concrete, and the arrangement of all components.
- Specify height of turbine above tailrace, _______ ft.
  Note: This is important to obtain the most efficient shape and location of the draft tube.
- Specify who furnishes, or constructs, the draft tubes.
- Specify form of draft tube desired, as spreading, elbow, straight cone, hydraucone, etc.
- If any special precautions are taken to prevent cavitation, specify what these are.

6-3. IRRIGATION PUMPS

(a) *Hydroelectric Pumps, Water Supply*

Note: In some areas where irrigation is of prime importance, the hydraulic turbines drive pumps as a prime duty, with the generation of power dependent on how much water is left over from the irrigation needs. In plants of this kind, one or more penstocks carry water from the dam to the hydroelectric pumps, and one or more penstocks provide similar service for the power generators. An illustrative plant is the Chandler Power and Pumping Plant, in the Kennewick Division of the Yakima, Washington, Project.

Writing specifications for this kind of plant follows the guide subjects given in Sec. 6-1(a)–(h) for the turbine, excepting that the driven machine shall be identified as a water pump in place of the electric generator named in Sec. 6-1(b). Specifications for the pump shall be provided for the turbine bidders for their information, although the project plan may call for the pump to be built and installed by another contractor. In this circumstance, pump specifications need not be more detailed than the statement of capacity, _______ gpm; discharge head, _______ ft; suction head, _______ ft; weight of pump at full load, _______ pounds; kind of pump, as centrifugal, ________ stages; kind of coupling that connects turbine to pump; and by whom furnished and installed. Also, if pump is fitted with its own step or thrust bearing, or if the weight and thrust will be imposed on the turbine bearings. Also speed, rpm.

- Specify pump details of design and construction for guidance of pump builder in addition to items already named above, as follows:
  - Pump rotor shall be of the (open) (closed) type, and constructed of ________________ (material).
  - Pump to be arranged for (horizontal) (vertical) placement.
  - Casing to be of ________________ (material) and (horizontally) (vertically) split. In either case withdrawal or lifting of the rotor shall be accomplished without disturbing the lower half of the casing. Specify gasket, if any.
  - Pump suction shall be (single) (double).
  - Specify if pump is to be fitted with a balance chamber or any other device for balancing the thrust.
  - Specify material of shaft sleeves.
  - Specify kind and shape of stuffing boxes and packing material required.
  - Specify design pressure of the casing and the expected maximum working pressures, ________ psig.
  - Specify kind, size, and location of the pump discharge valve. Specify how bonnet is to be packed under pressure.
  - Specify extension stem if valve operating platform is high above the pump discharge pipe.
  - Specify materials of valve, body, seat, gate, or disk.
  - Provide information note as to who furnishes and installs the valve and the discharge piping that connects to both the pump and the valve.

6-4. ICE CONTROL

(a) *Ice Prevention Methods, Compressed Air*

- Provide drawing showing arrangement of compressed-air piping at the dam.
- Specify pressure of air available, ________ psig.
- Specify kind, size, and number of air nozzles.
- Specify materials of compressed-air piping and the material of the nozzles.

- Specify source of compressed air. If air storage tanks or some other reservoir is not available, so state. If tanks are available, state capacity and pressure, _______ psig.
- Specify actuating control, as air thermostat, water thermostat, or both. Specify how the control shall actuate the air valves and how and when to close them.

(b) *Ice Prevention Methods, Steam*

- As with air, provide a drawing showing the layout of the steam piping, valves, and nozzles.
- Specify steam pressure, _______ psig.
- Specify materials of piping, valves, and fittings; also the steam nozzles.
- Specify sizes and working pressures of all components of the system.
- Control the emission and cutting off of steam by thermostat as stipulated in Sec. 6-1(a) for Compressed Air.
- Specify source of steam, as direct-fired boiler, electric boiler, etc. Specify the weight of steam available for this service, _______ pounds per hour.
- If steam is not wasted into the water, but heats the area by means of submerged coils, specify the method of collecting the condensate and its return to the boiler or boiler-feed pump. Specify capacity of the heater, in Btu per hour, the capacity of the return pump, and the capacity of the condensate storage tank if this is part of the system. Specify drive of return pump, giving electric service characteristics, as voltage, phases, cycles, etc.
- Specify kind of return pump, as centrifugal, rotary, reciprocating, etc. Specify suction conditions, as _______ psig or _______ in. Hg column lift, as applicable.
- Specify kind and thickness of insulation on steam and condensate return piping. Specify kind and size of automatic air vent valves and vacuum breakers as they may be needed, particularly on condensate storage tank.
- Specify number of steam heaters required.

(c) *Ice Prevention Methods, Electrical*

- Specify the kind, number, and placement of the electrical resistance heaters to be used.
- Specify their capacity, each, _______ kw per hour at _______ volts, _______ cycles, _______ phase, _______ wire service.

Note: Details of these heaters and the specifying of the wiring and control components should be provided by the electrical engineers assigned to the hydroelectric plant project. This is also true for any electrical methods selected to prevent the formation of ice in the penstocks between the pond and the runner of the hydraulic turbine.

- Specify the activating temperatures for cutting in and cutting out the electrical heaters, ________ F.

6-5. FLOW MEASUREMENT

(a) ***Equipment for Flow Measurement in Conduits or Penstocks***

- Specify whether flow measurement is to be determined by the salt velocity method, the salt solution method, the Gibson method, or any other approved technique.
- Specify the number and locations of the salt injection valves.
- Specify the time of passage of the salt, ________ seconds, at full load. (*Note:* 9 seconds may be considered a minimum.)
- Specify the length of the test section of conduit, ________ ft.
- Specify use of a recording chart to show the continuous flow over the selected period of time.
- If desirable, specify the installation of a motor-driven turbulator upstream from the first measuring electrode, to ensure proper mixing of the water and brine solution.
- Specify kind of meter to be used, as propellor, and degree of accuracy required, as ________ per cent, plus or minus ________ per cent.
- Specify placement location. Require that meter rotating element can be removed from the casing or housing for repairs or cleaning, without disturbing the flow through the conduit.
- Require bidders to submit drawings and complete detailed data on the make, model, and characteristics of the measuring equipment being offered.

6-6. REGENERATIVE PUMPING

(a) ***Pumped Storage Plants***

Note: The pumped storage hydroelectric plant becomes practical when a surplus of power from other sources is available during those hours or seasons in which the normal storage pond becomes depleted from scanty rainfall and inadequate stream flow. Reported efficiencies of the system approach 80 per cent, meaning that for every 100 kw used for pumping back to the pond, 80 kw of new power will be realized. Since the pumping power is surplus and involves only incremental costs, the balance is quite favorable to the regenerative pumping operation. This may be particularly attractive if the extended pond availability meets or helps to meet recurring peaks.

- Specifying the mechanical equipment for such a regenerative scheme does not differ extensively from the requirements of a hydroelectric pumping station for water irrigation, as detailed in Sec. 6-3(a), Hydroelectric Pumps, Water Supply. The water must be lifted from the tailrace to the storage pond and the pumping must be accomplished by means of outside power. For power-generating stations, the designer aims at providing water turbines that can function also as pumps, with the electric generator on the regular power-generating cycle able to double as an electric motor on the pumping cycle.

For specifying the two-purpose electrical machine (termed a dynamo), refer to Sec. 6-1(b). For details of the modifications involved in the water turbine, consult the manufacturer. In addition to the items listed in Sec. 6-1(a)–(h), in the event a separate pump and driver are to handle the regenerative pumping, include the following:

- Specify the difference in elevation the pump must overcome, ________ ft static head.
- Specify the suction lift, ________ ft; the total discharge head, ________ ft.
- Specify the construction of the pump, design, kind, materials, etc.
- Specify kind of drive. For motor drives, refer to Sec. 4-12(a), Raw Water Pumps.
- For steam-turbine drive, refer to same section's steam turbine paragraphs.
- For a diesel engine drive, see Sec. 5-1(a) and subsequent subheadings as appropriate.
- For gas engine drives, see Sec. 5-3(a)–(j).
- Gas-turbine drives are detailed in Sec. 5-5(a)–(q).

- Whether a single turbine-generator–pump motor unit is required, or whether the turbine-generator and pump motor are separate units, require bidders to submit complete data, including dimensions, arrangements of units, operating performance data, drawings, weights, materials of construction, and other pertinent information.
- Specify if the changeover from generation to pumping is to be accomplished automatically or manually, and at what intervals of time or specified times. Require full details of the automatic control systems, if these are to be used.
- Specify the characteristics of the outside power to be provided for the pumping cycle, as _______ volts, _______ phase, _______ cycles.
- Specify minimum and maximum seasonal water temperatures, F.

6-7. TERMS AND TESTS

(a) *Hydroelectric Plant Nomenclature*

- Specify that all terms used in the specification by the owner, and all terms used by bidders, shall be in accordance with the latest Standards Publication of the National Electrical Manufacturers Association (NEMA) for "Hydraulic Turbines, Governors and Accessory Equipment."
- Specify also that equipment offered shall be in accordance with the requirements of these standards.

(b) *Testing*

- For field testing of an installation to verify the guaranteed horsepower and efficiency, specify that tests shall be conducted in accordance with the ASME Power Test Code No. 18 for "Hydraulic Prime Movers."
- Alternatively, the owner may accept a laboratory test on a model runner, stepped up in horsepower in accordance with established procedures and corrections.

6-8. WATER LEVEL RECORDERS

(a) *General Application*

- Water level recorders are to be used to maintain a continuous record of changes in water level as these occur in the streams feeding the hydroelectric plant storage pond.

- Specify the streams in which the instruments will be placed, and indicate their exact locations.
- Specify the components of each recording station, as clock mechanism, stage mechanism, float, counterweights, pencil holder and pencil, carriage, base, cover, etc.

(b) *Key Elevations*

- Specify the following data for each instrument and for each location:
  1. Elevation of instrument table, as ________ ft.
  2. Level of extreme high water, ________ ft.
  3. Level of extreme low water, ________ ft.
  4. Maximum distance of float travel, ________ ft (distance between high and low water).
  5. Normal distance of float travel, ________ ft.
  6. Elevation of bottom of gage house well, ________ ft.
  7. Shape and internal cross section of well, dimensions.
  8. Length of float cable for above conditions, ________ ft.

(c) *Arrangement for Operation*

- Describe operation of equipment, as chart bearing record cylinder rotates from the actions of the falling weight of the clock mechanism. Meanwhile, water level is recorded on the chart by a pencil connected to the float, so that the rise or fall of the float is reproduced on the chart. Clock is rewound by hauling up on weights.
- Specify that the design and workmanship must be such as to eliminate backlash, slippage, excessive friction, or any other condition that would affect adversely the efficiency and accuracy of the device.

(d) *Materials*

- Specify that float, cables, metal tapes, and all other metallic components shall be fabricated of noncorrosive materials, suitable for the area and atmosphere in which they will function.
- Specify the kind of chart graduations desired, as feet, tenths of a foot, hundredths of a foot; time scale to be specified also, as days, hours, etc.
- Chart shall be of adequate scope to record continuously for ________ days between chart changes on the cylinder.

(e) *Information by Bidders*

- In addition to submitting drawings, photographs, and sketches as necessary to fully illustrate the design and performance of the recorder, bidders shall supply the following information:

1. Rate at which the clock weights descend, _______ in. per day (or ft per day).
2. Recommended diameter of float pipe, so float may move up and down in the pipe with a minimum of friction or interference.
3. Operating time, days, between changes of the chart.
4. Travel of chart per day, _______ in.
5. Float travel, _______ in., corresponding to 1 in. on chart.
6. Minimum time and stage graduations on the chart.
7. Full dimensions and weights of the equipment.

(f) *Installations at Site*

- Specify who furnishes the ladders, platforms, housings, and other appurtenances required by a stream level gaging station.
- Specify if actual site work is to be accomplished by the instrument supplier, the owner, or some other contractor.
- If installer is different from the instrument supplier, specify relationship of instrument supplier and this installing contractor to avoid any time loss from a lack of coordination of their work.

6-9. PITTING FROM CAVITATION

(a) *Defined*

- According to the NEMA "Standards for Hydraulic Turbines, Governors and Accessory Equipment," pitting refers to the removal of metal from the hydraulic runner of a Francis-type wheel and from the discharge ring of a propellor turbine, as well as the runner of a propeller turbine, as a direct result of cavitation. Erosion and corrosion, the presence of injurious elements in the water, or surface markings from chipping and grinding in an area to be repaired, *are not included* in pitting by cavitation.

(b) *Excessive Pitting from Cavitation*

- For purposes of guarantees against loss of metal by cavitation, excessive pitting from this cause is set by formulas as follows:

1. Francis turbines and comparable designs, $W = D^2$
2. Propeller turbines, $W = 1.5D^2$

W = pounds of metal removed during 8,000 hours of operation
D = discharge diam of the runner, ft

- In specifying pitting limits, the resistance of the metal in the considered turbine will affect the extent of pitting likely to be experienced. The formulas are based on mild steel; some alloys may be more resistant. For this reason, the time of 8,000 hours may be increased or decreased as seems proper to the engineer and specification writer. Otherwise, losses in excess of the formula-established values may be regarded as unsatisfactory.

Table 6-1. Physical and Performance Data for Hydraulic Turbines

Job ________ Turbine make ________
Head, ft ________ Model ________
Number of units ________ Driving ________

Item	*Information by bidder*
Horsepower output	
Head of ____*____ ft	________ Hp
Normal speed	________ Rpm
Efficiency, at ___*___ per cent of maximum horsepower,	________ per cent
Efficiency, at ___*___ per cent of maximum horsepower,	________ per cent
Efficiency, at ___*___ per cent of maximum horsepower,	________ per cent
Hydraulic thrust plus turbine rotor, lb	________
Diameter, largest part going through generator stator,	____ Ft ____ In.
Governor, make and type	________
Elevation, bottom of draft tube, ft	________
Total net weight of unit, lb	________
Weight, heaviest piece, lb	________
Time for opening gates, seconds	________
Time for closing gates, seconds	________
Lubricating oil system (unit)(central)	________
If furnished, state method of flow measurement	________
Guaranteed rate of pitting from cavitation, lb/8,000 hr	________
Life of guarantee, years	________

* Figures to be inserted by purchaser before calling for bids.

- Specify how long the guarantee against excessive pitting from cavitation shall be in effect, ________ years. Provided the turbine is not operated during that period for more than ________ per cent of the time at less than minimum horsepower outputs. Or alternatively, at no more than _______ per cent of the time at more than the maximum specified horsepower output.

- Refer to Table 6-1, Physical and Performance Data.

6-10. SPARE PARTS FOR HYDROELECTRIC TURBINES

(a) Specify the NEMA Standards Minimum List of Spare Parts as follows:

1. For Francis or propeller turbines:
   A complete set of replaceable protective devices for gate operation.
   A complete set of packing.
   One bearing liner for the main shaft.
2. For impulse turbines:
   One needle tip.
   One nozzle seat.
3. One set of governor packing.
   One set of governor springs.

Note: These are minimum. Other items recommended for consideration are detailed in the NEMA Standards list. Refer to that standard in preparing a list of spare parts, and stipulate as seems expedient.

6-11. HYDROELECTRIC PLANT INSTRUMENTS

(a) ***Location Drawings***

- In some areas of measurement, multiple instruments of the same kind are placed to obtain uniformity of information. For example, Pitot tubes may be set at right angles in a penstock, while in the same penstock several piezometer openings may be spaced equally around the circumference. By subtracting the piezometer head from the Pitot head, the velocity head is obtained. This is readily translated into flow velocity.

(b) ***Summary of Instruments***

- Refer to Table 6-2. This list and the natures of the individual instruments are subject to change at the discretion of the designing engineer.
- For details in specifying particular instruments, both indicating and recording, refer to Sec. 4-21(a) and all subsequent sections as appropriate to a hydroelectric plant.

Table 6-2. Instruments for Hydroelectric Plants

Area of measurement	Kind of measurement	Kind of instrument
Water entering penstocks	Temperature	Indicating
	Elevation or level	Indicating
Water flowing in penstocks	Flowmeter	Indicating
		Recording
	Velocity	Indicating
		Recording
Water in tailrace	Elevation or level	Indicating
		Recording
Turbine governor	Oil pressure	Indicating*
	Oil temperature	Indicating*
Governor	Gate position indicator	Indicating
Tachometer	Turbine speed	Indicating
Head pressure	By Pitot tube	Indicating
Velocity head	By piezometer (with Pitot tube)	Indicating
Miscellaneous	Pressure	Indicating
	Temperature	Indicating

* Instrument is fitted with contactors to energize an alarm signal if pressure drops to a dangerous low or temperature rises to a dangerous high.

Note: Remote readings should be specified if desired.

DIVISION 7

Power Plants, Nuclear

7-1. GENERAL CLASSIFICATIONS, NUCLEAR REACTORS

Note: At present there are operating, or ordered, at least 12 forms of nuclear reactor intended for, or now generating electrical power, producing heat for process applications, propelling ships, or serving as instructional or research facilities.

Among these 12 forms, there are individual differences to be observed in the choice of fuels, the primary coolant, controls, core and structural materials, temperatures and pressures, as well as moderators and reflectors.

Plant efficiencies likewise vary, with no one design so far established as being consistently the best, or highest. Statistical reports show efficiencies ranging from a low of 19 per cent to a high of 33.5 per cent. The average over about 50 plants calculates to 24.8 per cent.

Fuel elements include uranium sulfate solutions; highly enriched, slightly enriched, and fully enriched uranium oxide; also U^{235} in concentrations of 20 per cent or 90 per cent. If not liquid, elements are designed in the form of plates, rods, and tubes. Plutonium may be found in breeder reactors. Moderators in service are identified as ordinary water (H_2O), heavy water (D_2O), graphite, polyphenyl, and terphenyl. Both ordinary water and heavy water not only are good moderators of the reaction, but double as primary coolants as well. In this latter role they are pumped, or operate by natural convection, through a heat exchanger, where the reaction heat is converted into steam or alternatively raises the temperature of a process fluid.

Other coolants observed are liquid sodium, which is pumped as long as its temperature stays above 208 F (it solidifies at lower temperatures), and some suitable gas, usually either helium or carbon dioxide. Circulation of the gas through the heat exchanger and back to the reactor is accomplished by means of a blower.

Since the design of nuclear reactors is as flexible as these facts indicate, the business of the specification writer is to confer with the probable suppliers for the job in hand, and shape up the specification accordingly. In short, the specification for a nuclear reactor becomes basically a closed one, with most, if not all, the requirements settled with the pre-

Table 7-1. Information by Bidders on Nuclear Reactors

Job ________	Builder of reactor ________
Kind of reactor ________	Capacity ________
Function ________	Number of units ________

Item	*Information by bidder*
Kind of fuel	
Number of fuel elements	
Degree of enrichment	
Fuel inventory, kg or tons	
Fuel critical mass, kg or tons	
Breeding factor (conversion factor)	
Refueling time or cycle	
Burnup, %	
Moderator	
Reflector	
Shielding, material and thickness	
Primary coolant	
Circulation, natural convection or pumped, gpm	
Secondary loop, fluid	
Temperatures, primary coolant leaving reactor, F	
Pressure, primary coolant leaving reactor, psia	
Pressure drop in heat exchanger for primary coolant, psi	
Temperatures, primary coolant returning to reactor, F	
Temperatures, steam or process fluid leaving heat exchanger, F	
Pressures, steam or process fluid leaving heat exchanger, psia	
Controls, kind and number of rods	
Plant efficiency, %	
Structural materials used in core	
Element heat flux, average, Btu/sq ft per hr	
Element surface temperature, F	
Neutron flux (n/cm²/sec)	
Fast	
Thermal	
"Scram" time, seconds	
Maximum normal operating temperature, full load, F	

sumed successful supplier before the specification is written. It thus becomes a technical record of the reactor details, rather than a vehicle for bidding.

Nevertheless, a nuclear reactor checklist will be helpful to the specification writer in making sure that the design selected will have all its parts, and that an important aspect, or performance item of significant value, is not overlooked. For that reason a specifying guide is given in subsequent sections of this division for each of the 12 established forms of reactor.

A general suggested form to be used by bidders in the submission of information concerning the equipment they propose appears as Table 7-1.

7-2. PRESSURIZED WATER REACTORS

(a) *Stated Characteristics*

- Specify operating pressure and temperature desired in primary coolant leaving the reactor.

  Note: Working examples of this type of reactor generally function from 1,000 to 2,000 psig. These pressures are necessary to obtain steam generation at the pressures and temperatures required.
- Specify service in which reactor will function, that is, in the generation of electric power, the supplying of heat to process fluids, in research, or in instruction. State if reactor is to be fixed or portable.
- Specify the minimum breeding factor; also maximum.
- Specify minimum acceptable burnup percentage and minimum refueling cycle under full power.
- Specify moderator, reflector, and primary coolant.
- Specify temperature and pressure of steam to be generated in the heat exchanger, when supplied with feedwater at a temperature of ________ F and pressure of ________ psia.
- Specify the weight of steam to be generated, ________ pounds per hour.
- Specify kind, model, and materials for the primary coolant pump.
  - Stipulate maximum horsepower.
  - Specify kind of pump driver. If by electric motor, state service available, as ________ volts, ________ phases, ________ cycles, etc.

- For gas-cooled reactors, specify kind of gas, as carbon dioxide, helium, or other.
  - Specify kind of gas blower and its driver. If electric motor, state service available. If by steam turbine, state steam conditions as pressure, _______ psig; temperature, _______ F; quality, _______ per cent or superheated.
  - Specify blower capacity, _______ cfm; discharge pressure, _______ in. W.G. or psia.
  - Specify seals for pump and blower and degree of sealing integrity required.
- Specify minimum plant efficiency, _______ per cent.

(b) *Safety Features*

- Specify means desired to stop the reactor automatically in the event speed and rate of fission rise dangerously. Example: Control rods, if mounted vertically, may be arranged to drop into the core by gravity under predetermined conditions. Horizontal rods are power-inserted.
- Specify the kind and thickness of shielding.
- Specify the means and methods of refueling.
- Specify the "scram" time as _______ seconds in the emergency shutdown of the reactor.
- If subatmospheric pressures are maintained within an air-space pressure vessel surrounding the shielding, specify pressure desired, _______ psia or _______ in. W.G.
- Specify that reactor materials and construction must satisfy ASME Pressure Vessel Code, Section III, for Nuclear Vessels.

(c) *Information by Bidders*

- Require complete drawings and specifications showing the internal arrangement of the reactor, all key dimensions, weights, and structural materials used in the core and in the reactor other than in the core.
- Require bidders to fill in requested information as itemized on the form, Table 7-1.
- Require bidders to furnish complete data on the number, function, and kind of reactor operating and monitoring instruments. These will include neutron detectors, located usually in the core of the reactor, and with individual characteristics suitable for measurements at reactor start-up, shutdown, low-power, and high-power conditions.

(d) *Coolant Instruments and for Steam*

- Specify flowmeters to record on charts the flow of coolant from reactor to heat exchanger and return, ________ gpm, or cfm, as appropriate.
- Specify temperature and pressure recording instruments for service on the coolant circuit, stipulating the required accuracy and range of temperatures and pressures to be sensed and recorded.
- Specify steam flowmeters with recording charts. Also call for recording gages on the temperature and pressure of the feedwater to the heat exchanger and of the steam leaving the heat exchanger.

(e) *Automatic Alarms*

- Specify kind and placement of alarms to warn of the onset of hazardous conditions. Specify whether signals are to be visible, audible, or both.

7-3. BOILING WATER REACTORS

Note: This is a direct-cycle reactor with the water coolant flashing into steam inside the reactor, and thence proceeding directly to the point of use, as a steam turbine. From the turbine the steam passes into a condenser. The water of condensation is pumped back to the reactor, and the cycle is repeated.

Advantages of this boiling reactor lie in its relative simplicity, and in being suitable for the generation of steam at both medium and high pressures. Units delivering steam at 1,000 psig have operated successfully.

Disadvantages include the radioactive contamination of the steam, although reports indicate that no turbine damage has been attributable to this contamination. However, radioactive deposits on the turbine blading introduce a personnel hazard during times of cleaning, and effective precautions must be taken.

There is also a possibility of the reactor being extremely sensitive to impurities in the water, and requiring that the water be treated for demineralization. This adds to the cost of the installation.

Water inside the reactor may be circulated by pump, or alternatively, permitted to move by natural convection. The pump provides more uniform cooling, but natural convection may be satisfactory in some cases.

(a) *Application and General Data*

- Specify the function of the reactor, as power generation, research, breeding, etc.
- Specify the power output, or heat output, or both.
- Specify the fuel, as U^{235}; Pu^{239}; whether enriched, percentage of enrichment, etc.
- Specify limiting dimensions of the site, if any.
- Specify kind of reflector, water, ________ in.
- Specify the fluid in the secondary circuit, as water, sodium, other.
- Specify the inventory mass and critical mass of the fuel, ________ kg or tons, as appropriate.
- Specify if internal circulation is natural or pumped.

(b) *Operating Requirements*

- Specify the temperature and pressure of the steam leaving the reactor, ________ F and ________ psia.
- Specify the temperature and pressure of the condensate returning to reactor, ________ F and ________ psia.
- Specify volume of condensate return, ________ gpm.
- Specify the weight of steam generated, ________ pounds per hour at ________ psia, from feedwater at ________ F and with coolant temperatures to the heat exchanger of ________ F.
- Specify average circulation inside the reactor, ________ gpm.
- Specify kind, number, and efficiency of steam separators in the reactor.
- Specify kind, and capacity in pounds per hour, of the steam superheater, if any.
- Specify purity of coolant water to be maintained by demineralization, or in any other manner.
- Specify minimum acceptable efficiency, ________ per cent.
- Specify refueling cycle, weeks, months, years, etc.

(c) *Pumps, Circulating and Condensate Return*

- For each pump, specify the kind, materials of construction, and design details; also the suction and discharge heads.
- Specify the drivers. If electric motors, require compliance with the standards of NEMA to the extent practicable.

- Specify if pump and driver are to occupy a common bedplate and stipulate kind of coupling.
- State electric service available, as _______ volts, _______ phases, _______ cycles, etc.
- If possible, submit to bidders a sketch showing location of the pumps with relation to the reactor and to the condenser hot well.

(d) *Structural Features*

- Specify materials to be used in the structure of the reactor, as steel, aluminum, zirconium, etc.
- Specify kind of foundation, and by whom furnished.
- If furnished by reactor contractor, specify nature and bearing capacity of the soil, _______ tons per sq ft.
- Specify the kind, schedule number of water piping, and welded joints.
- Specify certification of welders. State any particular qualifications that seem necessary.
- Require compliance with Section III, ASME "Code for Unfired Pressure Vessels."

(e) *Safety Features*

- Refer to Sec. 7-2(b), Safety Features, for pressurized water reactors, and follow this as suitable to the boiling water reactor.
- Also specify warning signals to function automatically on the approach or development of hazardous conditions.
  - Specify dangerous temperatures, pressures, velocities, etc. State locations of alarms desired.
  - Specify if signals are to be audible, visible, or both.

(f) *Instruments*

- In addition to the instruments normally required for the control and monitoring of a boiling water reactor, specify such flowmeters as seem desirable for evaluating the performance of the heat exchanger in generating steam. Also for measuring the condensate flow back to the reactor. Specify that all flowmeters are to be recording.
- Specify recording temperature and pressure gages for service on the steam and condensate.
- Specify the graduations and life period of the record charts.
- Specify which of these instruments, if any, is to be fitted with a danger alarm.
- Specify range of accuracy as ± per cent. Accuracy must be unaffected by radioactive contamination.

(g) *Information Required from Bidders*

- Require bidders to submit drawings, cuts, photographs, and specifications as necessary to thoroughly illustrate and make clear the reactor being offered.
- Require also that head-capacity-efficiency curves shall be furnished for each pump.
- Require bidders to submit the filled-in data form, Table 7-1.
- Require bidders to furnish complete information on the number, kind, and function of the reactor monitoring and control instruments.

7-4. SWIMMING POOL REACTORS

Note: This design operates at atmospheric pressure or a little higher. Although its basic purpose is experimental in most installations, with internal water cooling by natural convection, an increase of power output is sometimes accomplished by circulating the coolant water through an outside heat exchanger. Power produced seldom exceeds 1,000 kw in this form of unit.

Coolant temperatures leaving the reactor may range up to 140 F, seldom higher. This means that a substantial volume of coolant must be circulated through the exchanger in order to recover heat in quantity.

(a) Specify the desired power rating, in kwhr.

(b) Specify the operating temperatures desired of the coolant leaving the reactor and leaving the heat exchanger (if an exchanger is used).

(c) Specify coolant pressures, leaving the reactor and leaving the heat exchanger.

(d) Specify how reactor is to be used, that is, experimental and research, for process operations, power, etc.

(e) Specify location and space available for reactor, giving available length, width, and height accommodations.

(f) Specify total filling of water, gal.

(g) Specify burnup, per cent.

(h) Specify fuel; kind and degree of enrichment.

(i) Specify if coolant is to circulate solely within the reactor, and if this circulation is by natural convection or by pump. If by pump, specify kind of pump, capacity, ________ gpm, and discharge head.

(j) For moving coolant through an outside heat exchanger, specify kind, capacity, and discharge head of the pump.

(k) For any pump, specify characteristics of electric service available, as voltage, cycles, phases, etc. Specify motor features required, including adherence to NEMA standards if feasible.

(l) Specify minimum acceptable plant efficiency.

(m) Specify refueling period at full power.

(n) Specify number and placement of ports.

(o) ***Safety Features***

- Specify means and methods for containing the radioactivity of the water circulating pumps, heat exchanger, and piping.
- Refer to Sec. 7-2(b), Safety Features, for the pressurized water reactor. Use as applicable to the swimming pool reactor.

(p) ***Information by Bidders***

- Require complete drawings and specifications, showing the internal arrangements, all key dimensions, and structural materials used in the core and in other parts as well. Also request bidders to fill in the form, Table 7-1, as applicable to the swimming pool design.

(q) ***Instruments***

- Specify all necessary operating and monitoring instruments required by this design of reactor.
- Specify, in addition, recording flowmeter for water circulation through heat exchanger.
- Specify recording pressure gages and recording thermometers for placement on the outside coolant circuit. If all circulation is internal, by natural convection, provide immersion bulbs in the hottest water of the pool, these to be connected to recording charts.
- Specify that all instruments, meters, gages, columns, etc., are to be unaffected by any degree of contamination. Specify tolerances in all cases, as plus or minus or as percentages.

7-5. TANK-TYPE REACTORS

Note: Also termed in some installations a "low-intensity test reactor," this design has been built primarily for research and materials testing. As with the swimming pool reactor, to which it is closely related, it operates mostly at atmospheric pressure or at a low positive pressure.

Coolant temperatures likewise remain low, ranging from 100 to 140 F.

The design provides no breeding or conversion factor and operates with water serving as both moderator and coolant. In capacity, it tends to be larger than the swimming pool arrangement, with the coolant circulating rate in some instances reaching 20,000 gpm.

Reflectors used include beryllium, graphite, water, or a combination of these.

(a) *Application and Details*

- Specify function of reactor, if for research, test, systems testing, etc.
- Specify number and kind of fuel elements, also cladding.
- Specify kind of fuel, U^{235}, enriched, or percentage of enrichment from 20 to 90 per cent.
- Specify inventory and critical mass, kg.
- Specify core structural material, as aluminum, aluminum alloy, stainless steel, etc.
- Specify refueling cycle, days, months, etc. Also specify burnup, ________ per cent.

(b) *Operating Requirements*

- Specify temperature and pressure of coolant leaving reactor, if design calls for such an outside circuit.
- Specify temperature and pressure of coolant returning to reactor, at full load.
- Specify surface temperature of elements, ________ F.
- Specify volume of water circulated, ________ gpm.
- Specify volume or weight of secondary fluid passing through heat exchanger, ________ gpm, pounds per hour, etc.
- Specify increase in temperature of secondary fluid in exchanger, ________ F. Specify entering temperature.

(c) *Reactor Structure*

- If designed to operate at a positive pressure, no matter how low, specify reactor must be designed in accordance with the ASME "Code for Unfired Pressure Vessels," Section III.
- Specify materials of construction of noncore parts, as steel, stainless steel, zirconium, etc.
- Specify kind and schedule number of piping for handling coolant between the reactor, the heat exchanger and return, if this external circuit is used.
- Specify welded pipe joints, and require welders to be certified.
- Specify kind of foundation and by whom furnished.

- Provide specification information on the nature of the soil at the reactor site, and give its bearing capacity, tons per sq ft.

(d) *Safety Features*

- Refer to Sec. 7-2(b), Safety Features, for the pressurized water reactor and adapt to the tank reactor as appropriate.
- Also specify what precautions are to be taken to prevent personnel injury from contamination of the external coolant circuit.

(e) *Instruments*

- Refer to Sec. 7-3(f), Instruments, for boiling water reactors, and follow as appropriate to this low-pressure reactor.

(f) *Pumps*

- If external pumping of the coolant is incorporated by design, specify the pumps as appropriate, using Sec. 7-3(c), Pumps, Circulating and Condensate Return, for boiling water reactors.

(g) *Information by Bidders*

- Refer to Sec. 7-3(g), Information Required from Bidders, of the boiling water reactor specification guide.

7-6. AQUEOUS HOMOGENEOUS REACTORS

Note: The difference between homogeneous and heterogeneous designs is in the arrangement of the fuel with respect to the moderator.

In the former, the fuel, whether solid or liquid, is evenly distributed throughout the moderator, of which this aqueous design is an example. So are the boiling water designs, unless the fuel is latticed into a heterogeneous arrangement as in some of the swimming pool reactors. The heterogeneous reactors also may employ fuel lumps or rods embedded in the moderator. The graphite reactors are other examples of this heterogeneous design.

In the aqueous reactor, the fuel is in solution, and both removal of waste and recharging usually can be accomplished simply by drawing off some solution periodically and replacing this with fresh. The result is continuous operation.

Power and heat recovery are achieved by circulating the fuel solution through an external heat exchanger, with a secondary fluid picking up the heat in the exchanger. A notable feature of aqueous design is a relatively easy control. By varying the temperature of the fuel solution returning to the reactor from the heat exchanger, fission is controlled so effectively that reactors of this kind have been operated without any need for the usual control rods.

Starting and stopping of the reactor is controlled by changing the concentration of the fuel solution. Either ordinary water or heavy water fuel solutions may be used in this reactor.

Disadvantages which must be considered in writing a specification include the degradation of piping and heat exchanger metals by the fuel solution. This solution may be strongly corrosive at high temperatures. Also, the solution is highly radioactive during operation of the reactor, and contaminates the exchanger, secondary fluid, piping, pumps, and other materials contacted. Especial pains must be taken to provide adequate shielding of personnel.

Another possibility is the decomposition of the water, both ordinary and heavy, as produced by radiation effects. The hydrogen, deuterium, and oxygen gases released by the decomposition may create an explosive atmosphere in the reactor.

(a) *Reactor Application and Size*

- Specify the function of the reactor, as electric power, process heating, research, development, medical therapy.
- Specify capacity, kw per hour, Btu per hour, or both.
- Specify solution, whether ordinary water or heavy water, and per cent enrichment of the solution. Also chemical nature of the solution, as uranium sulphate, uranium oxide, etc.

(b) *Operating Characteristics*

- Specify temperature, ______ F, of solution leaving reactor for the heat exchanger.
- Specify the pressure, ______ psia, of the solution leaving the reactor for the heat exchanger.

Note: Some aqueous homogeneous reactors have operated at 3,500 psia; others at atmospheric pressure. Temperatures reported vary from 100 F up to 800 F at the 3,500-psia operation.

- Specify the refueling cycle, as months, years, continuous.
- Specify rate of circulation of solution through the heat exchanger, ______ gpm.
- Specify temperatures of secondary fluid entering and leaving the heat exchanger, ______ F.

- Specify the probable fluctuations of load, per cent of full load, and frequency of fluctuations.
- Specify pressure, ________ psia, of secondary fluid in the heat exchanger.
- Specify efficiency desired, ________ per cent.
- Specify manner of control, as temperature, rods, etc.

(c) ***Pumps and Structures***

- Specify the kind of pump to be furnished, if necessary giving name and model.
- If pump is driven by an electric motor, state electrical service available, as ________ volts, ________ cycles, ________ phases, etc.
- Specify that motors must satisfy the requirements of the NEMA standards to the extent that the pump can be related to these standards.
- Provide a drawing showing placement of pump in the circulating system relative to the reactor and the heat exchanger, together with significant details.
- Specify that reactor design and construction must conform to Section III, ASME "Code for Unfired Pressure Vessels," latest edition, plus modifications as issued subsequently.
- Specify piping materials, test pressures, welding standards, etc.
- Describe kind of heat exchanger to be furnished, relative to design, materials, pressure loss at full load, etc. Specify working and test pressure of the exchanger. If exchanger is to be furnished by others than the reactor supplier, so state, with limitation of authority and responsibilities for each of the contractors involved.
- Specify materials used in construction of the reactor, as aluminum alloys, steel, etc.
- Specify manner and means of refueling and of withdrawing waste solution, number of ports, etc.

(d) ***Safety Features***

- Specify shielding of the reactor, kind and thickness.
- Specify the shielding for the solution circulating pumps, the heat exchanger, the piping connecting these components, and procedures to be taken for avoiding or limiting the contamination of the secondary fluid.
- Specify the automatic-stop controls which are to take over in the event that the reactor should run away, that temperatures should rise beyond control, or even that a gas explosion should occur.

- Specify the "scram" time as ______ seconds for the emergency shutdown.

(e) ***Information by Bidders***

- Require complete drawings and specifications showing the internal design and construction of the reactors and giving all key dimensions, number of ports, control equipment weights, etc.
- Require bidders to fill in and submit form shown as Table 7-1, or one prepared similarly.
- Require bidders to submit details of all monitoring, control, and safeguarding instruments to be furnished with the reactor.

(f) ***Instruments***

- In addition to the instruments normally expected and needed for an aqueous reactor, specify recording flowmeters to be used on the circulation of the fuel solution through the heat exchanger. These meters should register on a chart so fluctuations and checks against the load can be made over definite time elements.
- Specify thermometers and pressure gages, also recording, for use with the circulating fuel solution, plus another set showing conditions in the heat exchanger of the secondary fluid.
- Specify frequency of chart changes desired, ______ days.
- Specify that all instruments must function with their accuracy unaffected by contamination from the radiation present in or around the media being measured.
- Specify what instruments are to be fitted with alarms to show the approach of hazardous conditions. Specify if alarms shall warn by lights, bells, whistles, etc.

7-7. GRAPHITE REACTORS

Note: These are heterogeneous reactors, with the fuel loaded as lumps or pellets, slugs or rods, into cladding of one kind or another. Many of them use natural uranium, others enriched U^{235}.

Primary coolant may be air, water, carbon dioxide. This is a reactor of large physical size; consequently satisfactory housing may be difficult and expensive.

(a) ***Applications and General Data***

- Specify function of this reactor, as power generation, research, etc.

- Specify maximum dimensions of space available.
- Specify maximum weights that can be accommodated. If this contractor pours foundations, specify the kind of soil and give, if possible, load-bearing capability in tons per sq ft.
- Specify the output capacity of the reactor in kwhr or Btu per hour, as indicated by the function.
- Specify primary coolant, as water, air, gas, etc.
- Specify the converting or breeding factor if this is a desired factor.

(b) *Operating Requirements*

- Specify the temperature and pressure of the coolant leaving the reactor.
- If air-cooled, specify weight of air to be circulated, ________ pounds per hour.
- If water-cooled, specify volume of water to be circulated, ________ gpm.
- If gas-cooled, specify weight of gas to be circulated, ________ pounds per hour.
- For all coolants, specify the temperature and pressure at return to the reactor from the heat exchanger.
- Specify the refueling cycle, as weeks, months, years, etc.
- Specify the temperatures of the secondary fluid entering and leaving the heat exchanger.
- Specify pressure, ________ psia, of secondary fluid entering and leaving the heat exchanger.
- Specify the minimum efficiency acceptable, ________ per cent.
- Specify manner of control, giving materials, cladding, and other information pertinent to the control rods.
- Specify the probable fluctuations in load, per cent of full load. Give expected cycle of fluctuations.
- Specify the element heat flux, Btu per sq ft, per hour, average.

(c) *Pumps and Blowers for Coolant*

- For water circulation, specify kind of pump and form of drive.
  - For electric drive, specify electric service available, voltage, phases, cycles, etc.
    - Require adherence to NEMA standards as appropriate.
  - Require bidders to submit complete data on pump, including horsepower, maximum capacity, ________

gpm, ______ maximum discharge head, ______ ft, or pressure, ______ psia.

- Require bidders to submit drawings showing incorporation of pump in the coolant circulation.

- For air or a gas to be used as a coolant, specify kind of blower and form of drive.
  - If drive is by electric motor, specify the electrical service available.
  - Specify that motor is to meet requirements of NEMA motor standards as far as may be practical.
  - Specify that blower is to meet tests and standards of the Air Moving and Conditioning Association, Inc. (AMCA).
- For both pumps and blowers, specify bearings and lubricating systems.
- Specify that both pumps and blowers must be unaffected in their performance by the expected degree of contamination.

(d) *Structures*

- Specify that design and construction of the reactor must be in accordance with Section III of the ASME "Code For Unfired Pressure Vessels."
- Specify kind and source of the graphite, and stipulate the maximum operating temperature it must withstand.
- Specify all piping or duct materials forming components of the coolant circulating system.
  - Specify welding standards, and certification of welders.
- Specify the structural materials, as aluminum, zirconium, steel, etc.
- Specify manner in which waste is to be withdrawn from the reactor. Stipulate time allowable for refueling.
- Specify kind of foundation and who provides it.
  - If concrete, specify mix, curing period.
- Specify kind, material, and capacity of heat exchanger.

(e) *Safety Features*

- Specify shielding, giving materials and minimum thicknesses for heat, biological, gamma rays, etc.
- Specify how circulating piping or ducts, carrying coolant to the heat exchanger and back to the reactor, are to be shielded.
- Specify means and methods for limiting the contamination of the secondary fluid in the heat exchanger.

- Specify the provision of automatic controls to stop the reactor in the event of a reactor runaway, the development of excessive pressures or temperatures. Locate the sensing elements for fast response in the hazardous areas of the reactor.
- Specify the "scram" time as _______ seconds for the emergency shutdown to act.
- Specify safeguards to be provided for operators during refueling periods.

(f) *Instruments*

- In addition to the standard instruments provided for a graphite moderated reactor, specify recording flowmeters to detail continuously the flow of coolant through the coolant circuit.
  - Specify type of flowmeter for liquid coolant and for the gas coolant, that is, whether orifice, venturi, weir, Pitot tube, piezometer, etc.
  - Specify the graduations on the recording chart.
- Specify recording thermometers and pressure gages, also for service on the circulating coolant, and to record the conditions of the coolant leaving the reactor and reentering therein.
- Specify chart life, that is, the frequency with which charts must be changed.
- Specify instruments that are to be fitted with alarms designed to warn of the approach of hazardous conditions. Specify form of alarms; a light, whistle, bell, etc.
- Specify that all instruments and their appurtenances must function with undiminished accuracy despite the development of a specified level of radioactive contamination. Figure 7-1 shows a control room.

(g) *Information by Bidders*

- Require submission of complete drawings, cuts, photographs, etc., as may be considered necessary to show the internal design and construction of the reactor, and likewise giving all key dimensions, weights, etc.
- Bidders should fill in and submit an information chart similar to that shown in Table 7-1. Additions and deletions to the form may be made to suit the particular conditions in hand.
- Bidders shall provide detailed information on the system of monitoring being furnished, the number and forms of operating controls, and the details of the instruments.

FIG. 7-1. Control room of Philadelphia Electric Company's Peach Bottom Atomic Power Station. Large circular panel at the far end of the room (above the seated operator) is the control rod monitor which displays all control rod positions and programs. This is a helium-cooled, graphite-moderated 40,000 kw unit. (*Courtesy Philadelphia Electric Company.*)

7-8. HEAVY WATER REACTORS

Note: Natural water contains about 0.015 per cent of deuterium oxide, or about 1 part in 6,500. Heavy water is obtained by a concentration of this deuterium, the procedure being usually by electrolysis, chemical exchange, or distillation. From any viewpoint, production of heavy water (D_2O) is expensive. However, D_2O has notable advantages as a moderator and reflector; consequently a smaller, heterogeneous natural uranium reactor is possible in comparison with a graphite reactor, and will require less fuel for equivalent performance.

If the heavy water is used in a homogeneous arrangement, together with natural uranium, the combination can become critical; this is not possible with graphite and natural uranium, and can be used to simplify design. Control is achieved by varying the concentration of the heavy water, injecting either ordinary water or more heavy water as necessary to meet load requirements.

Heavy water units circulate the D_2O at pressures from atmospheric to 1,000 psia and at temperatures from about 100 to 900 F. In pressurized reactors, if heavy water is the moderator instead of ordinary water, pressures up to 2,000 psia or even higher are to be considered.

(a) *Applications and General Data*

- Specify function, as power generation, research, experimental, etc.
- Specify capacity, kwhr, Btu per hour, or both.
- Specify location, space available, limiting dimensions, if any.
- Specify maximum weights that can be accommodated. If reactor contractor is to supply and place the foundation, provide information on the nature of the soil and its load-bearing capacity, tons per sq ft.
- Specify the mix, placing, and curing of the concrete foundation to the responsible contractor.
- Specify converting or breeding factor if this is a desired function.

(b) *Operating Requirements*

- Specify pressures and temperatures of heavy water on leaving the reactor and on returning thereto.
- Specify volume to be circulated, ________ gpm.
- Specify concentration at full load, ________ per cent D_2O.
- Specify refueling period, days, weeks, months, etc.
- Specify the pressures and temperatures of the secondary fluid for both entering and leaving the heat exchanger.
- Specify the probable fluctuations in load, both in degree (per cent of full) and in frequency.
- Specify the element heat flux, Btu per sq ft per hour, average.
- Specify the minimum acceptable operating efficiency.

(c) *Pumps for Coolant*

- Specify as outlined in Sec. 7-7(c) for pumps.

(d) *Structures*

- Require design and construction to accord with the ASME "Code for Unfired Pressure Vessels," Section III.
- Specify materials of structure, steel, aluminum, concrete, etc.
- Specify the materials, weight, and schedule numbers (if appropriate) for the circulating piping, the connections for ordinary water dilution and heavy water makeup to the reactor.

- Specify welding standards, certification of welders, the kind of flux, etc.
- Specify the kind, materials, and capacity of the heat exchanger.

(e) *Safety Features*

- Refer to Sec. 7-7(e) or graphite reactors, and follow as appropriate for heavy water.
- Also specify protective measures to be provided for varying the concentration of the heavy water, so the desired results are obtained without danger to the operator.

(f) *Instruments*

- Follow guidance of Sec. 7-7(f) as appropriate to a heavy water reactor.

(g) *Information by Bidders*

- Refer to Sec. 7-7(g) for graphite reactors. Require the submission of information as appropriate for the heavy water reactor.

7-9. SODIUM COOLED REACTORS

Note: The use of liquid sodium instead of water allows reactor operation at low pressures, or atmospheric, while at the same time reportedly giving a better overall efficiency than water. High temperatures at low pressures are readily reached, since the specific heat of sodium is 0.253. In several sodium cooled reactors on which data are available, the temperature of the sodium at atmospheric pressure is given as between 900 and 1,000 F. This permits the establishment of secondary-fluid temperatures, depending on the fluid, up to 800 F and pressures equally high, again depending on the characteristics of the secondary fluid.

Sodium introduces several hazards, since it reacts violently with some substances and may begin to burn at once if exposed to the atmosphere. Because of these characteristics, the specification for a sodium reactor must incorporate safety precautions beyond the usual radioactivity shielding of the water reactors.

Purity of the sodium for heat transfer is also essential. Foreign substances in small measure will reduce the effectiveness of the heat pickup.

Sodium cooled reactors are moderated by graphite and consequently might also be classified as graphite reactors. For this reason, references will be made to graphite reactor specifying guides as appropriate, in addition to the particular sodium phase requirements.

(a) *Application and General Data*

- Specify function of reactor, as power, experimental, research, etc.
- Specify maximum space available, giving dimensional limitations, if any.
- Specify the reactor output in kwhr, Btu per hour, or both.
- If area places limitations of weight that can be accommodated, give maximum loading permissible, tons per sq ft.
- If reactor contractor is to furnish and install foundations, provide full information on kind of soil and its bearing capacity, tons per sq ft.
  - Also specify concrete mix, and give curing time.

(b) *Operating Requirements*

- Specify temperature and pressure of sodium coolant leaving the reactor, and its temperature and pressure at reentry to the reactor.
- Specify refueling cycle, weeks, months, years.
- Specify temperatures, ________ F, of secondary fluid entering and leaving heat exchanger.
- Specify pressures of secondary fluid entering and leaving heat exchanger.
- Specify minimum acceptable efficiency, ________ per cent.
- Specify the element heat flux, Btu per sq ft per hour, average.
- Specify the probable fluctuations of load, in per cents of full load. Also indicate expected frequency of these fluctuations.
- Specify how reactor is to be controlled, giving materials in control rods and other pertinent data.

(c) *Sodium Pumps*

Note: The major problem in a liquid sodium pump is to provide a complete seal, so that the sodium does not leak out, nor does contamination by dirt or other foreign materials leak in. For this reason, various kinds of pumps are being used with varied success. A thorough discussion with the reactor supplier is certainly essential before deciding on a type and design. Choices are indicated by the following types:

- Specify that pump is to be electromagnetic, canned, centrifugal, positive displacement.
- Specify how sealing is accomplished, as by means of an inert gas in which the pump assembly is submerged.

- Require pump supplier to state frequency with which sealing other than gas, as carbon rings, plates, etc., must be renewed.
- For canned pumps, require supplier to submit data on means provided for cooling the motor and bearings. Since liquid sodium is at high temperatures, this cooling problem is serious.
- For electromagnetic pumps, specify if design is to be of the conduction or induction type.
  - Specify electric service available, as _______ volts, _______ phases, _______ cycles, etc.

  Note: Both alternating and direct current can be used with this form of pump. Specify which is wanted.
- For all pumps of whatever type, specify the capacity, _______ gpm. Also the pumping head as _______ ft or _______ psia.
- Specify the temperature of the sodium entering the pump, _______ F.
- For motor-driven pumps, specify that motor shall meet the requirements of the NEMA standards to the extent such conformation is practical.
- Specify that pump and driver must function without the development of lowered efficiency as the result of radioactive contamination from the pumped sodium.
- Require bidders to submit complete data—drawings, cuts, photographs, etc.—as necessary to illustrate the design features, materials, dimensions, etc., of the pump being furnished.
  - Also require bidders to submit pump head, flow, and efficiency curves. (Efficiencies may be unacceptably low for a satisfactory economic picture.)
- Specify the kind of inert gas used to seal the pump assembly, and the pressure to be maintained on the gas.

(d) ***Reactor Structures***

- Refer to Sec. 7-7(d), Structures, for the graphite moderated reactor, and use those items applicable to the sodium coolant reactor.
- Specify whatever extra precautions are to be taken to make the heat exchanger leakproof.

(e) ***Safety Features***

- Refer to Sec. 7-7(e), Safety Features, for graphite moderated reactors and specify accordingly.
- Also specify means to be provided for preventing or ex-

tinguishing fire resulting from the escape of sodium to the area. *Note:* Shutting off sodium combustion by smothering with nitrogen or proprietary powders has been successful. However, since the sodium is radioactive and would contaminate these materials, hazards to personnel must be avoided. This might involve additional shielding. Method chosen should be as approved by the reactor supplier.

(f) ***Instruments***

- In addition to the standard instruments provided for a graphite moderated reactor, specify recording flowmeters for service on the sodium coolant circuit. Because of problems of leakage, temperature maintenance, vapor pressure, and the tendency of sodium to precipitate impurities in relatively cool spots, metering accuracy is difficult to maintain. Results with orifices and Bourdon tubes have not been altogether satisfactory. Diaphragm loaded instruments generally have been reasonably satisfactory, but not always. Discussions with both the reactor builder and the instrument builders are essential before final decisions on instrument types and details are made. General features of the instrument specification include:
  - Specify accuracy required, as ± percentage.
  - Specify range of performance, as _______ gpm for flowmeters, from _______ to _______ F for temperatures, and _______ to _______ psia for pressures.
  - Specify the time chart and the graduations thereon.
  - Indicate which instruments are to be recording and which indicating.
  - Specify frequency of change for the recording charts.
  - Specify what instruments are to be fitted with alarms designed to show the gradual or sudden development of dangerous conditions.
  - Specify the forms of the alarms, as light, bells, whistles, or a combination. Specify where the alarms are to be located, as in the office, reactor room, etc.

(g) ***Information by Bidders***

- Refer to Sec. 7-7(g), Information by Bidders, for graphite moderated reactors, and specify accordingly as may be appropriate for the sodium coolant reactors.
- In addition, require bidders to state the total weight

of sodium for the complete charge in an operating system.

7-10. ORGANIC MODERATED REACTORS

Note: The organic materials consist of polyphenyl or terphenyl. These materials presently are considered resistant to becoming radioactive, which was not always the case.

Their thermal characteristics are good, permitting operation at high temperatures. Circulation is by pump, with primary discharge through a superheater (where superheated steam is wanted) and thence through the saturated steam generator.

The organic materials function both as moderators and as coolants.

Fuel may be fully enriched uranium oxide, or an enriched percentage of uranium.

A combination of heavy water as the moderator and an organic fluid as the coolant has been proposed and is expected to extend the life of the fuel. This would place the organic liquids in an intermediate position.

(a) *Applications and General Data*

- State function of reactor as power, experimental, research.
- Specify space availabilities, giving maximum or limiting dimensions, if any.
- Specify reactor output in kwhr, Btu per hour, or both.
- Specify moderator, whether polyphenyl, terphenyl, D_2O, or other.
- Specify kind of coolant, polyphenyl or terphenyl.

(b) *Operating Requirements*

- Specify temperature and pressure of organic coolant leaving the reactor, and its temperature and pressure at reentry, ________ F and ________ psia.
- Specify volume of coolant being circulated, ________ gpm.
- Specify superheat, ________ F, to be added in superheater when passing ________ pounds of steam per hour.
- Specify temperature, pressure, and quality of saturated steam entering superheater.
- Specify temperature and pressure of coolant entering and leaving the steam generator (heat exchanger).

- Specify temperature and volume of feedwater entering the steam generator, ______ F and ______ gpm.
- Specify pressure of feedwater entering steam generator, ______ psia.
- Specify refueling cycle, weeks, months, years.
- Specify minimum acceptable efficiency, ______ per cent.
- Specify the element heat flux, Btu per sq ft per hour, average.
- Specify the probable fluctuations of load, in per cent of full load. Also indicate frequency of fluctuations.
- Specify how reactor is to be controlled, from start, through load, to shutdown. Stipulate materials in control rods, as boron-cadmium alloy, etc.

(c) *Organic Pumps*

- Specify kind of pump (usually centrifugal), giving number of stages, single or double suction, and form of driver.
- Specify the discharge head, ______ ft, or pressure, ______ psia.
- Specify suction conditions, as positive head, in. of vacuum, etc.
- Specify pump seals, kind, materials.
- Specify pump bearings and system of lubrication.
- If pump is motor-driven, provide data on electric service available, voltage, cycle, phases, etc.
  - Specify type of motor. Require compliance with the NEMA standards to the extent practical.
- Require bidders to furnish complete data on the pumps being offered, including drawings, cuts, and key dimensions. Also required is a head-capacity-horsepower curve, with efficiencies indicated at guarantee point and others as stipulated.

(d) *Reactor Structures*

- Specify that design and construction must be in accordance with Section III of the ASME "Code for Unfired Pressure Vessels."
- Specify piping materials to be used in piping coolant between the reactor and other components of the installation.
- Specify all welded joints, work to conform to welding standards, with certified welders.

- Specify structural materials as aluminum, steel, zirconium.
- If reactor builder also supplies the foundation, specify the concrete mix, curing time, and other pertinent requirements.
- Furnish contractor with data on the kind of soil present and its bearing capacity, ______ tons per sq ft.
- Specify design and materials of steam superheater. Describe for bidders the circulation of the coolant through the reactor and through the superheater.
- Specify design and materials of the heat exchanger; also its capacity in ______ pounds of steam per hour.

(e) *Safety Features*

- Refer to Sec. 7-7(e), Safety Features, for graphite moderated reactors. Use this section as a guide as appropriate in writing the safety specification for the organic moderated reactor.
- Specify alarms, their placement and activating hazards.

(f) *Instruments*

- In addition to the standard instruments for an organic moderated reactor, require recording meters for the flow of coolant. Also temperature and pressure recorders on the same circuit.
- Specify the graduations on the charts for each record.
- Specify the chart life, or frequency of chart change.
- Specify steam flowmeters for recording the output of steam, also pressures and total steam temperatures.
- Require that all instruments function without adverse effects from the expected degree of contamination.

(g) *Information by Bidders*

- In addition to information required for individual components, such as pumps, motors, etc., require bidders to submit detail drawings, cuts, and structural specifications, all showing clearly the design and organization of the reactor.
- Require complete information and details of the instruments being furnished. Also a schedule showing the number of each, and the proposed location of placement.
- For instruments that are to operate pneumatically, require information on air pressure needed, ______ psia.
- Require each alarm to be identified by kind, instrument, and component it serves.

- Prepare and issue to bidders a form similar to that illustrated in Table 7-1. Require bidders to return the form, with information entered, together with their bids.

7-11. BREEDER REACTORS

Note: The only natural fissionable material presently known is a constituent of uranium, identified by the symbol U^{235}. The quantity of U^{235} in natural uranium is extremely small, only a little more than 0.7 per cent. The remainder of the natural bulk consists of uranium 238, with U^{238} as its symbol.

U^{238} is not fissionable material, but is termed a "fertile" material, meaning that through a series of reactions and disintegrations it can be converted into the fissionable material plutonium, symbol Pu^{239}. The series begins with the absorption by U^{238} of a low-energy neutron, and thereafter produces U^{239}, which in turn by radioactive disintegration ends up as plutonium, the Pu^{239}.

One other fertile material is known, by name thorium 232, and this, like U^{238}, is found in nature. After going through the series of absorbing a neutron and decaying radioactively, the final product is the fissionable isotope of uranium symbolized by U^{233}.

Reactors built for the primary purpose of producing Pu^{239}, thus creating a new nuclear fuel, are termed breeder reactors. Thorium as yet is not being bred to generate U^{233} in significant quantities; performance of plutonium appears satisfactory enough to dominate the breeder reactions.

The effectiveness of the breeding is expressed in "doubling time," or the time needed, usually in days, for the fissionable material in the reactor to double itself. In construction, fertile material may be incorporated in the core of some breeders; in others it may be given the form of a blanket surrounding the core and thus perform reflector functions to a limited extent while being converted.

Although the basic purpose of a breeder reactor is generally considered to be the creating of new atomic fuel, the generation of power simultaneously from the heat of the reaction is not to be overlooked. Since these reactors are not moderated, high temperatures of the coolant are quite common. Reported conditions indicate an average range among different installations of 600 to 900 F. By exchanging coolant heat with a secondary fluid of sodium or sodium-potassium, pressures up to 1,200 psig with temperatures only slightly below those of the coolant have been achieved.

(a) *Applications and General Data*

- Specify the purposes or function of the reactor, as breeding, electric power, research, etc.
- Specify the breeding factor as 1.0 ± 0.04; 1.2 ± 0.01, etc.
- Specify the power or heat levels, or both.
- Specify the coolant, as sodium, sodium-potassium, etc.
- Specify the fuel, as 90 per cent U^{235}; plutonium; enriched uranium alloy, etc.
- Specify limiting dimensions of the reactor site, if any.
- Specify if neutron energy is to be fast or intermediate.
- Specify kind of reflector, as natural U; steel; graphite; depleted uranium.
- Specify the secondary loop fluid, as water, sodium, etc.
- Specify critical mass and inventory mass, kg, tons, etc., as appropriate.

(b) *Operating Requirements*

- Specify temperature and pressure of coolant leaving the reactor, and its temperature and pressure at reentry, ________ F and ________ psia.
- Specify volume of coolant being circulated, ________ gpm.
- Specify temperatures and pressures of secondary fluid entering and leaving the heat exchanger, ________ F and ________ psia.
- Specify the refueling cycle, weeks, months, years.
- Specify minimum efficiency of plant, ________ per cent.
- Specify the element heat flux, ________ Btu per sq ft per hour, average.
- Specify the probable fluctuations of load, in per cent of full load. Also indicate frequency of fluctuations.
- Specify how the reactor is to be controlled, from start, through load, to shutdown. Stipulate materials in control rods, number of rods, etc.
- Specify minimum temperature rise of secondary fluid in the heat exchanger, ________ F.

(c) *Coolant Pumps*

- If sodium or sodium-potassium coolant is to be used, refer to Sec. 7-9(c), Sodium Pumps, for sodium coolant reactors. Prepare this specification similarly.

- For other coolants, consult reactor builder, and evaluate the type of pump proposed.

(d) *Reactor Structures*

- Refer to Sec. 7-8(d), Structures, for heavy water reactors and observe items given as appropriate for this breeder reactor.
- Specify details of foundations and who furnishes them.

(e) *Safety Features*

- Specify shielding, giving materials and minimum thickness for heat, biological protection, gamma rays, etc.
- Specify how coolant piping, pumps, and heat exchanger are to be shielded.
- Specify the means and methods to be followed in limiting radioactive contamination of the secondary fluid and equipment on the secondary circuit.
- Specify the automatic controls needed to stop the reactor in the event of a reactor runaway, the development of excessive temperatures, pressures or both. Indicate where the automatic emergency sensing elements are to be inserted.
- Specify the "scram" time as not longer than ________ seconds.
- Specify protection for operators during refueling periods.
- Specify means for extinguishing sodium or potassium fires if these should result from a coolant leak.

(f) *Instruments*

- Refer to specifying guide in Sec. 7-9(f), Instruments, for sodium coolant reactors.
- Adapt as suitable to the instruments required for the breeder reactors.

(g) *Information by Bidders*

- Require bidders to fill and submit an information form similar to that shown in Table 7-1.
- Require complete drawings, cuts, photographs, and other data to illustrate the design and arrangement of the reactor being proposed.
- Require statement of total weight of sodium or other coolant under conditions of full load.

7-12. PORTABLE REACTORS

Note: The term "portable" means that the reactor is designed in units, each unit being of a size which can be transported and handled by ordinary equipment. Portable reactors have been designed for service in areas where the needed amount of conventional fuel is either unavailable or extremely difficult and expensive to obtain. Examples are found in the Arctic and Antarctic regions, their working functions being to furnish light, power, and heat.

For example, a project portable reactor for the U.S. Army will be put together of pieces no one of which exceeds 9 × 9 × 27 ft or weighs more than 10 tons.

This is a pressurized water reactor.

Power output is designed at 1,700 kw and 10,000 kw of heat. Water pressure is in the area of 1,200 psi. Steam generation is at 200 psia and 380 F saturated. Refueling is on a schedule of 12 months, and the fuel consumption about 8 pounds.

(a) *Application and Details*

- Specify the kind of reactor, as pressurized water, boiling water, etc.
- Specify projected function, as power generation, heat, light, etc.
- Specify maximum dimensions of any one piece.
- Specify the maximum weight of any one piece.
- Specify the dimensional limitations at the site, if any.
- Specify foundation at the site, giving nature and bearing capacity of the soil, ________ tons per sq ft.
- Specify refueling cycle in days, weeks, months, etc.
- Specify burnup, ________ per cent.
- Specify the power and heat levels, ________ kw.
- Specify the kind of fuel.
- Specify the kinds and materials of the reflector and moderator.
- Specify the secondary loop, as water, sodium, sodium-potassium, etc.
- Specify the critical mass and inventory, kg or pounds.
- Specify coolant.

(b) *Operating Requirements*

- Refer to Sec. 7-2(a), Stated Characteristics, for pressurized water reactors. Adapt these as necessary to the operating requirements of the portable reactor.

(c) *Pumping Requirements*

- Refer to the pumping guide for the boiling water reactor, Sec. 7-3(c).

(d) *Structural Features*

- Specify materials, dimensions, thickness, and other aspects of the reactor that must be adapted to portable service.

- If welding is to be done at site, specify qualified and certified welders.
- Specify packaging, crating, and slings necessary for handling the various components. State particular requirements for the safe transportation of the fuel elements.
- Provide bidders with information on the kind and location of the conveyances to be used in moving the reactor.

(e) *Safety Features*

- Refer to Sec. 7-2(b), Safety Features, for the pressurized water reactor, adapting this section to the portable reactor as appropriate.

(f) *Instruments*

- Refer to Sec. 7-3(f), Instruments, for boiling water reactors. Follow this guide as appropriate.
- Specify kind and placements of the dangerous-conditions alarms. Indicate whether warning signals are to be visible, audible, or both.
- Specify the temperatures, pressures, and other abnormal conditions that shall actuate the warning signals.

(g) *Information from Bidders*

- Refer to Sec. 7-3(g), Information Required from Bidders, for the boiling water reactor.
- Also require bidders to submit a schedule of each piece, giving its dimensions and weight. In addition, bidders are to provide details on the packing and crating, showing sealing procedures used, reinforcing, and other safeguarding data for the parts in transit.

Note: In preparing a specification for a portable reactor, useful information and guidance might be obtained from the U.S. Army Corps of Engineers, Ft. Belvoir, Virginia.

- Another source of mobile-unit information is the U.S. Navy Bureau of Yards and Docks, Arlington, Virginia.

7-13. REACTOR DATA, U.S. ATOMIC ENERGY COMMISSION

(a) The Atomic Energy Commission has not issued standard purchase specifications comparable to specifications standardized by other departments of the government. Reactors for government projects have been designed and built as tailor-made operations; in other words, one of a kind. All commission requirements have been shaped primarily for safeguarding the health and lives of the public.

Part 50 of the Atomic Energy Commission Rules and Regulations covers these aspects of the reactor projects:

LICENSING OF PRODUCTION AND UTILIZATION FACILITIES

Sections 50.1–50.3	General Provisions
Sections 50.10–50.12	Requirement of License, Exceptions
Sections 50.20–50.24	Classification and Description of Licenses
Sections 50.30–50.39	Applications for Licenses, Etc.; Ineligibility of Certain Applicants
Sections 50.40–50.45	Standards for Licenses and Construction Permits
Sections 50.50–50.59	Issuance, Limitations and Conditions of Licenses & Construction Permits
Section 50.60	Allocation of Special Nuclear Material
Sections 50.70–50.71	Inspections, Records, Reports
Sections 50.100–50.103	Revocation, Suspension, Modifications, etc.
Section 50.110	Enforcement

(b) ***Appendix A: Guide to Technical Specifications for Nuclear Reactors***

This is primarily a checklist of items to be specified. Although the preceding 12 sections in this division of the Handbook give details of data to be specified and to be furnished by bidders on nuclear power plants, specification writers are urged to obtain a copy of the Atomic Energy Commission list and to apply the information therein to the requirements of the reactor being projected.

(c) ***Design Criteria for Nuclear Power Plant Construction Permits***

The Atomic Energy Commission has issued tentative design criteria for guidance in preparing nuclear power projects for submission to the Commission in requesting construction permits.

While these criteria are subject to change as experience and knowledge of the art increase, they introduce new standards of thinking in providing safety for both nuclear plant operators and the general public. If nuclear designers and specification writers do not already have these criteria, they are urged to obtain copies of this and all later revisions.

Of the total of 27 given criteria,

Nos. 1–3 apply to reactor facilities and performance standards.

Nos. 4–16 apply to the reactors and their fuel.

Nos. 17–23 are concerned with engineered safeguards.

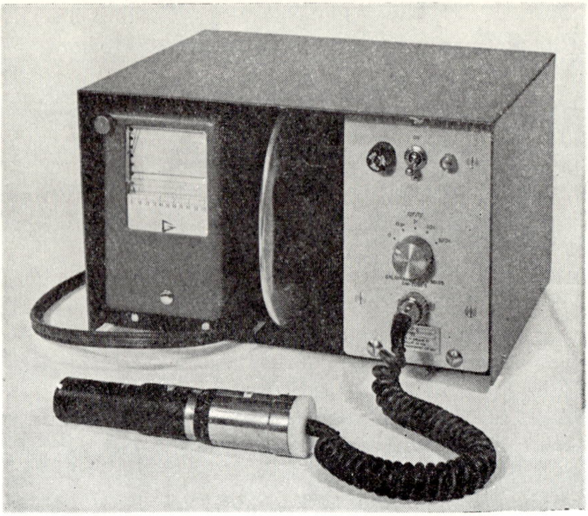

FIG. 7-2. Recording radiation rate meter. Continuous record over 31-day time interval. Adaptable for gamma, and beta-gamma detection. (*Courtesy Gelman Instrument Company.*)

Nos. 23–27 are directed to radioactivity control. See Fig. 7-2.

(d) ***Source of Commission Rules and Regulations***

- Address the United States Atomic Energy Commission, Washington, D.C., 20545.

7-14. DISPOSAL OF ATOMIC WASTE

(a) ***General Notes***

- The term "waste" is applicable to any material, solid, gaseous, or liquid, that is disposed of by discharge into surrounding spaces or areas. The most intense radioactive waste is produced by the chemical processing of spent reactor fuel, this processing being outside the scope of this Handbook. Radioactive waste is also a direct production in the form of coolants, both liquid and gaseous; also ventilation air and various liquids and cleaning materials used in connection with the operation of the reactor.

  In reactors using liquid fuel, gaseous products of fission include krypton, xenon, and iodine. These are all radioactive, and a safe place to receive them must be found.

Specification writers should obtain all knowledge possible about disposing of the particular waste that is expected, and should shape the specification accordingly. The present methods will be described briefly to indicate the basic problem of each. Providing specification details is not practical because the wide variations in the characteristics of the waste materials, plus possible recovery or purification actions prior to the final discharge, make each case a one-of-a-kind problem, generally to be solved by specialists in the field.

(b) *Disposal by Dispersal, Liquid*

- Radioactive liquid wastes may be retained in settling basins or tanks as long as considered necessary to permit a decrease of radioactivity to an acceptable level. The waste may then be diluted with large volumes of water by releasing a carefully controlled discharge into a river. Regular checks of this river-and-waste mixture are obligatory to make sure the radioactive level is held below the point of hazard to fish or other organisms living in the river.

  Discharge to dry wells has been considered, but the possible ultimate contamination of underground water supplies tends to discourage this plan.

  Concentration and evaporation have been used to reduce the bulk of the waste, the residue then being disposed of by burial either in the earth or at sea. Reports indicate this method gives successful results, although finding land burial areas might become a problem in time.

(c) *Disposal by Dispersal, Gas and Air*

- Common practice is to allow at least 100 days for cooling of the radioactive spent fuel, and then to release the gases to atmosphere from a high stack.
- This procedure calls for continuous checking of weather conditions to make sure that temperature inversions, rain, or fog do not appear and return the discharged gases to the earth. So long as a return does not occur, there is no hazard.

  An alternative method is to treat the gases chemically, converting them into solids and liquids that may be disposed of by burial, feeding into a river, storage, or other nonhazardous procedures.

  In some instances, electrostatic precipitation of particulates in contaminated air is practiced, in order to prevent these particulates from ultimately escaping back to the earth.

(d) *Disposal of Solids*

- Contaminated equipment and filter material in some instances may be decontaminated and reused. Stainless steels particularly may be reclaimed. Mild steel is less responsive to the usual decontamination procedures of acid treatments.
- When reclaiming is not possible or economically feasible, the material customarily is disposed of by burial.
- Regular burial procedures require periodic testing of the soil and water in the area to ascertain if contamination is spreading, and if the development of a hazard seems likely.
- Disposal of combustible material by incineration is possible but introduces a danger from contaminated smoke.

7-15. STANDARD COMPONENTS OF NUCLEAR POWER PLANTS

Note: Steam turbines, condensers, pumps, valves, fittings, piping, etc., as a rule will be the same in a nuclear plant as in a conventional power plant. Some nuclear shielding around the turbine may be required if contaminated steam is being introduced into the turbine. This is extraneous to the turbine, however, and in no way invalidates the guide specifications for turbine-generators, Sec. 4-15, Steam Turbines.

Biological shielding should be specified as directed by cognizant health and medical authorities.

All other components of the power plant not involved or in any way connected to the fission process, may be specified in accordance with the guides and checklists itemized in the appropriate sections of this Handbook.

DIVISION 8

Power Plants, Miscellaneous Equipment

8-1. AIR COMPRESSORS

(a) *Air Compressors, Reciprocating*

- Specify design arrangement as vertical, horizontal, L, or V mounted cylinders.
- Specify capacity in ______ cfm of free air compressed to ______ psig.
- For single-stage compressors, specify the aftercooler (if necessary), describing design and arrangement and giving desired materials, method of rolling tubes into tube sheet, and desired temperature drop, ______ F, when supplied with ______ gpm of cooling water at a maximum temperature of ______ F and a pressure of ______ psig.
- For compound or multiple-stage compressors, specify both aftercoolers and intercoolers, with details of design, materials, and capacity, as listed above for the aftercooler.
- Specify the driver, as electric motor, or gasoline or other internal combustion engine.
  - Specify kind of electric motor (as synchronous), and give details of voltage, cycles, phases, and other data required by the bidder.
  - For an internal combustion engine drive, refer to Sec. 5-1, Diesel Engines; Sec. 5-3, Gas Engines, and Sec. 5-5, Gas Turbines. Follow the guidance given in these sections as it is applicable to the driver in this case.
- Specify method of connecting driver to compressor, as direct-connected (with motor rotor on compressor crankshaft), or by belt.
  - If belt, specify whether flat or V desired. Give method of fastening. If idler is required, state size, kind, and location.
- Specify if compressor and driver are to be mounted on a common bedplate. Require base or bedplate as appropriate to be bolted or grouted to the floor.
- Specify kind and size of bearings.

- If compressor or motor is to have a maximum speed, so specify, giving rpm for each.
- Describe starting equipment and controls to be furnished by vendor.
- Specify size, type, and decibel rating of intake muffler if required.
- Specify the internal diameter, _______ in., and the stroke, _______ in., for each cylinder.
- Require that cylinder wall thickness must be sufficient to permit reboring in the future, the reboring to increase the internal diameter up to _______ in., without hazard.
- Specify materials of cylinder, and minimum acceptable strength, _______ psi.
- If crankcase, cylinder heads, and cylinders are to be water-jacketed, so specify, giving temperature range of available water supply, _______ F in summer and _______ F in winter. Give analysis of water with particular reference to foreign or material in suspension that might be left in the passages.
- Specify cross bracing for frames of cross-compound units so that low-pressure and high-pressure cylinders are kept in proper alignment.
- Specify the material and finish of the crankshafts, as forged steel, statically and dynamically balanced.
- Specify material of connecting rods, and methods of take-up for wear.
- Specify kind of lubricating system desired, as automatic, manual, grease cups, etc.
- Require indicator connections to be provided on each cylinder so that performance can be ascertained and measured.
- Require bidders to fill in the blanks of the form, Table 8-1.
- Specify kind of valves wanted, as plate or disk, poppet, or mushroom.
- Require bidders to submit performance curves showing the volumetric efficiency, shaft horsepower, actual capacity, and compression efficiency. State on what basis the compression efficiency is to be computed, isothermal or adiabatic.
- Specify if compressor is to be governed for a constant pressure discharge with fluctuating capacity, or a constant capacity at fluctuating pressures.

Table 8-1. Air Compressor Data by Vendor

Vendor ____________ Job ____________
Number of units ____________ Model ____________

Item	*Information by bidder*
Number of cylinders	
Internal diameter of cylinders (give for each)	
Stroke, in.	
Piston displacement, full load, cfm	
Volumetric efficiency, full load, %	
Rpm at full load	
Horsepower, full load	
Amperage, full load	
Overall length, ft and in.	
Overall width, ft and in.	
Overall height, ft and in.	
Air inlet, diam, in.	
Compressed-air outlet, in.	
Total weight, lb	
Weight of flywheel, if used, lb	
Decibel rating, without muffler	
Decibel rating, with muffler	
Half load, rpm	
Half load, volumetric efficiency, %	
Half load, driving horsepower	
Half load, total piston displacement, cfm	
Water jacketing: cylinders, crankcase, or both	
Water quantity required, gpm	
Temperature rise, F	
Maximum acceptable inlet temperature, F	
Required inlet water pressure, psig	
Pressure drop in compressor, psig	
Cylinder lubricating oil, base*	
Cylinder lubricating oil, flash point, open cup, F minutes	
Cylinder lubricating oil, viscosity, furol, seconds	
Cylinder lubricating oil, pour point, F	
Cylinder lubricating oil, carbon residue (Conradson), % maximum	
Cylinder lubricating oil, consumption per cylinder, drops/minute	
Kind of silencer: Snubber, filter silencer, etc.	

* Lubricating oil characteristics may be specified if desired, but bidders should fill in these blanks so that any divergence from the specifications becomes obvious.

Note: See Fig. 8-1 for a photographic illustration of a vertically designed compressor.

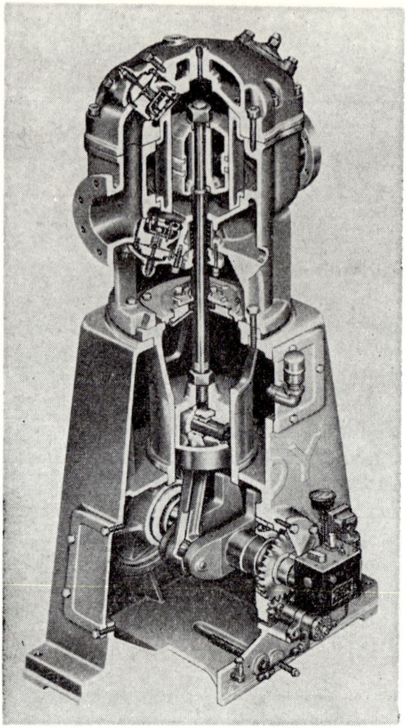

FIG. 8-1. Single-cylinder vertical air compressor, showing internal arrangement and construction. (*Courtesy Joy Manufacturing Company.*)

- Indicate if control shall be by varying the compressor speed, or by stop-and-start cycles.
- Specify altitude of site, and air density correction therefore if necessary.
- Specify gage board and location, and list the instruments to be mounted thereon, giving their operating ranges.

(b) *Air Compressors, Rotary*

Note: Rotary compressors are commonly used for establishing a vacuum, and for delivering air or gas at pressures from a few pounds up to 100 psi without the use of liquid seals. By using liquid seals, the unit may establish vacuums up to 28 in. Hg column, or approximately double that of an unsealed compressor. Discharge pressure will be as specified.

- Specify the operating characteristics of the compressor, as free air or vacuum on the suction side, and delivery pressure, ________ psig.
- Specify the required capacity under the conditions specified, ________ cfm.
- State service on which compressor is to be used, as air, gas, vapors, etc.
- State if compressor is to be sealed or unsealed.
- Require bidders to furnish complete information on design and materials incorporated in the compressor.
- Require bidders to fill out a form similar to Table 8-1, modified to suit the rotary design desired.
- Refer to Sec. 8-1(a), Air Compressors, Reciprocating, for data on the compressor drive to be specified, and for other items appropriate for a rotary unit.
- Require bidders to submit performance curves showing relative capacities, speeds, and efficiencies; also shaft horsepowers as various discharge pressures.
- Specify the gage board location, as on wall or mounted on compressor. Also specify the instruments to be mounted on the board, giving range of operation for each, and specifying the construction in design and materials.

(c) *Air Compressors, Vaned, Rotary*

- Specify air conditions, capacity, etc., as for other rotary compressors.
- Specify number of stages.
- Specify the general construction of the compressor and materials used for casing, vanes, or blades, and for shaft.
- Specify how vane rubbing against the casing is to be controlled, as by oil film, etc.
- For other details of the vaned rotary, refer to Sec. 8-1(b), as appropriate to this form of compressor.
- Require complete drawings and information from bidders, showing construction, clearances, weights, lubrication, etc.
- Require bidders to submit a form similar to that shown in Table 8-1 for rotary compressors.

(d) *Air Compressors, Centrifugal*

Note: Centrifugal compressors differ from a fan in their purpose, the former being designed to achieve air pressure at the point of delivery, the latter being concerned with the delivery of volumes of air at relatively low pressure. In calculating compressor performance, the laws of thermodynamics rather than hydraulics apply.

- Specify the capacity of the compressor, as _______ cfm of free air at _______ psig, with intake air at an elevation above sea level of _______ ft and a temperature of _______ F.
- Specify the number of stages if close control of design is desired; otherwise manufacturer may be directed to determine the number for the pressures required.
- Specify the construction materials, as shaft of _______ _______ steel; casing of _______________ iron or steel; kind and arrangement of gas seals.
- Specify the materials in the impellers as _______________ steel; with diaphragms of _______________ steel.
- Specify how impellers are keyed to shaft, and how diaphragms are anchored in the casing.
- Specify method of maintaining the axial position of the rotor, and maximum longitudinal movement allowed.
- Specify minimum and maximum clearances between impellers and casing, and impellers and diaphragms.
- Specify bearings, materials, lubrication, and kind of thrust compensation desired, as by a balance piston.
- Specify if impellers are to be of the backward curved type to avoid overloading.
- Specify driver, as internal combustion engine, electric motor, hydraulic turbine. For specifying directions concerning internal combustion engines, refer to Secs. 5-1, Diesel Engines; 5-3, Gas Engines; and 5-5, Gas Turbines. Note that in most instances, the compressor operating with a gas turbine is of the axial type rather than centrifugal.
    - For electric motor drives, state service available in voltage, cycles, phases, etc. Specify kind of motor wanted, synchronous, induction, or other special. Specify kind and models of starting equipment for the electric motor. Require all electrical components to satisfy the NEMA Standards as applicable, and the National Electrical Code, plus local codes.
- If it is desired to connect a centrifugal compressor to a waterwheel (hydraulic turbine), refer to Sec. 6-1(a)–(h), Hydraulic Turbines.
- Specify how compressor is to be connected to the driver, as direct, through a flexible coupling; or by means of a belt.
- Specify if compressor and driver are to be mounted on a

common bedplate. In any case, state how compressor and driver are to be anchored, by bolting, grouting, etc.

- Specify who provides the foundation on which compressor and driver are to be mounted.
- Specify maximum compressor rpm.
- State if compressor intake is to be equipped with a silencer. Specify maximum acceptable decibel rating.
- Require bidders to submit detailed drawings showing internal arrangement of the compressor and all key dimensions.
- Require bidders to submit relative performance curves, showing delivery pressure, ________ psig; shaft horsepower, and efficiency, ________ per cent, based on free air at the inlet of ________ F temperature, at an elevation of ________ ft above sea level.
- Require bidders to submit data as called for by the chart, Table 8-2.

Table 8-2. Centrifugal Compressor Data by Vendor

Vendor ________ Job ________

Number of units ________ Model ________

Item	*Information by bidder*
Number of stages	
Diameter of impellers	
Speed, full load, rpm	
Delivery volume, full load, cfm	
Delivery pressure, full load, psig	
Horsepower, full load	
Tip clearance over impellers, in.	
Clearance between impeller and diaphragm, in.	
Overall length, ft and in.	
Overall width, ft and in.	
Overall height, ft and in.	
Total weight, lb	
Decibel rating, without silencer	
Decibel rating, with silencer	
Diameter, air intake, in.	
Diameter, air discharge, in.	
Ratio of compression	
Compression efficiency, adiabatic, %	
Compression efficiency, isothermal, %	
Direction of rotation	

- If an aftercooler is to be provided, refer to Sec. 8-1(a) for the construction, materials, and requirements of the cooler.
- Specify interstage cooling nozzles for casing, if intercooling is desired.
- Specify location of instrument board and list instruments to be mounted on the board, giving operating ranges and the design and construction of the instruments.

(e) *Axial Flow Compressors*

Note: Axial compressors have been developed primarily in connection with gas-turbine units, but their characteristics of high efficiency, large-volume capacity, and smaller space requirements than rotary or centrifugal compressors of equal capacity, have expanded their application to general air compressing service.

- Specifications as shown in Sec. 8-1(d) on centrifugal compressors are applicable to axial compressors as well. However, add the following items:
- Specify how the inlet guide vanes are to be adjusted.
- Specify means of adjusting stator blades while the compressor is operating.
- Require bidders to submit drawings and material data illustrating the design and showing all key dimensions, clearance, and weights.
- Require bidders to submit performance curves showing output in cfm, and relative pressure and horsepower at the indicated speeds and for the air inlet conditions specified.
- Specify cooling means and methods, as water, air, etc.

(f) *Steam Jet Compressors*

Note: Available in standard single-stage and two-stage arrangements. Because of the presence of steam vapor in the discharge, use of these units for exhausting condensers and creating vacuums in other vessels is more frequently specified than for the production of high-pressure air. The following guide items recognize this application for producing a vacuum.

- Specify the volume of air to be handled, as ________ cfm at a maximum vacuum of ________ in. Hg. Discharge pressure to be ________ psig.
- Specify the steam conditions motivating the ejector (the prime ejector in a two-stage compressor), giving steam pressure, ________ psig, temperature, ________ F, and

quality, as dry saturated, _______ per cent moisture, or _______ degrees of superheat, F.

- Specify the design and construction of the intercooler (two-stage compressor), as _______ passes, with shell of _____________ material and tubes of _____________ material, designed for working pressures of _______ psig working water pressure. Tube sheet to be of _______ _______ material, with tubes expanded into the sheet.
- Specify the steam-air conditions for the second-stage ejector as _______ psia, temperature of _______ F with a total volume of _______ cfm. Specify discharge conditions as _______ psig and temperature of _______ F. State if discharge is against atmosphere, into a receiver, or meets other disposition.
- Specify the design and construction of the aftercooler (applicable to both single and two-stage units), following the details indicated for the intercooler.
- Specify condensing water available for the coolers, giving minimum and maximum temperatures, summer and winter; also the range of water pressure available, _______ psig.
- State source of water and the presence of any sediment or corrosion-creating element.
- Require bidders to submit drawings and specifications to illustrate the equipment being offered, and showing all key dimensions, weights, general arrangement, etc.
- Specify that inches of vacuum achieved is to be referred to a 30-in. barometer.
- Require bidders to submit performance curves showing the pounds of steam consumed per hour against the hourly delivery of evacuated air in _______ pounds, under the operating conditions specified. If desirable, call for curves for single-stage, two-stage, and three-stage operation, thus making an economic comparison possible in obtaining the best buy.
- Specify the thermometers, kind and range, to be mounted on the coolers' inlets and outlets, also on the ultimate discharge.
- Specify the pressure gages and mercury gages to be used for indicating air inlet absolute pressures and ultimate discharge conditions.
- For all instruments, specify the designs and calibrations.

(g) *Controls for Compressors*

Note: Since the number of control steps and the nature of

the compressor operation, as continuous or intermittent, affects the control, the following suggestions may be used in developing the kind of control desired.

1. FOR RECIPROCATING COMPRESSORS
   - Motor-driven: Steps vary from two (2) up. Specify automatic stop and start with automatic unloading on start-up. Provide changes in capacity by unloading or other acceptable means.
   - Steam engine or internal combustion engine: Provide a steam flow control or throttling valve for the steam engine. For internal combustion drives, either throttle fuel to the engine, or provide automatic cut-in and cut-out. Manual switches can be used also.
2. ROTARY COMPRESSORS
   - Generally the same as for the reciprocating units.
3. CENTRIFUGAL AND AXIAL FLOW
   - Motor-driven: Any number of steps up to infinity. Drive through a hydraulic coupling.
   - Steam or internal combustion engine: Throttle steam or fuel to the engine as appropriate.
   - On the discharge of centrifugal and axial flow compressors, provide a blowoff valve to protect the compressor against the possibility of unstable operating conditions.
4. STEAM JETS
   - Infinite number of steps as desired. Change operation by throttling steam to the unit.

(h) *Safety Devices for Compressors*

- Specify a pressure relief valve on the discharge side of each compressor. Valve to open automatically and dump excess pressure.
- Specify an oil pressure gage with contactors that will shut down the compressor if lubricating oil pressure drops below a safe value. Specify whether gage is to operate on line voltage or at reduced voltage under electrical control. For pneumatic control, specify the compressed-air pressure available for this purpose.
- Specify a similar pressure gage for mounting on the cooling water supply. Specify electrical or pneumatic operation, as for the oil gage.
- Specify a similar pressure gage to stop the compressor should the discharge pressure go above the preset safe limit.

- Specify thermostatic shutdowns for (1) excessive discharge temperatures; (2) excessive bearing or oil temperatures. Alternatively, a thermal shutdown may be specified instead of a pressure gage for the cooling water circuit, if so desired.
- In each case, specify the pressure or temperature at which the safety device is set to act.

(i) ***Receivers for Compressed-air Systems***

- Refer to Sec. 5-1(c), Engine Starting Equipment.

(j) ***Piping for Compressed-Air Systems***

- For high-pressure piping, specify conformance to "Code for Pressure Piping," American Standards Association, B31.1.
- Welders must be qualified under the same code; also the ASME Boiler and Pressure Vessel Code, Section IX, for ferrous and nonferrous piping.
- Specify that pipe hangers and supports shall meet the requirements of the Manufacturers Standardization Society, Standard Practice SP-58, latest edition.
- In an air-intake system, refer also to Sec. 5-1(d), Air-intake System, and use as appropriate. For mufflers and silencers, refer to Sec. 5-1(e), Mufflers.

8-2. FIRE DETECTION SYSTEMS

(a) *Note:* When they occur, which is not frequent, power plant fires usually appear in the electrical side of the plant, from the overloading of a live cable or bus, or from a short circuit. Generator fires have been almost eliminated by the permanent placement of excessive-temperature sensors buried in the windings, which signal the approach of dangerous conditions.

In the mechanical plant, fires generally follow some condition of careless housekeeping and originate in the building rather than in the equipment. Fires have resulted however, from the ignition of soot in air preheaters, from fuel oil or lubricating oil spilling on hot surfaces, and from spontaneous combustion in coal storage bins and in pulverized coal mills and chutes.

In industrial steam or power plants, the time of maximum danger is during the shutdown hours overnight and on weekends and holidays, when few if any persons are present to observe the birth of a fire, and the alarm is slow getting out.

Since fire extinguishing programs vary over an almost limitless choice of water, foam, carbon dioxide, dry chemicals, and soda-acid, available in fixed systems, portable systems using wheeled carts, hand extinguishers, automatic systems, and manually operated systems, every installation is a one-of-its-kind design, and a general specification is not practical.

For this reason, guide specifications are provided in this Handbook for detection systems only. Warnings are released by these systems as the result of smoke appearing to a smoke sensor, or from the development of excessive temperatures on a heat sensor, or as a result of both reactions.

(b) *Fire Detectors, Duct-mounted, Ionization Type*

- Specify position in duct circuit where the sensor is to be placed. Provide drawing, if possible.
- Specify the source of electric energy, giving voltage, cycles, phases, etc.
- State range of air flows through the duct, from a minimum of _______ cfm and temperature of _______ F through a maximum of _______ cfm and _______ F temperature.
- State pressure conditions in the duct, from a minimum of _______ psia to a maximum of _______ psia.
- State range of humidity conditions in duct, from a minimum of _______ per cent relative humidity to a maximum of _______ per cent relative humidity.
- State if automatic fire dampers are positioned in the duct, and if so, indicate locations.
- State if duct system is zoned, giving number of zones from which alarm signals will be required, and state location of central cabinet by which the signals will be received.
- Specify number and kind of alarm signals, as by light, bell, horn, or a combination of these.
- Specify if central alarm control station is to be connected to an outside fire department.
- Specify that having given an alarm signal, the system must be manually reset. The signal is to continue until manual action is taken.
- Specify that equipment and system, both in parts and in whole, must be in accordance with the requirements of the Underwriters Laboratories, or Factory Mutual Laboratories.

- List equipment that is to be actuated by the detector signal, as fans or blowers to be stopped or started, pumps to be stopped or started, dampers to be opened or closed, etc.
- Provide information on the horsepower of the various motors that will be actuated as a result of the warning signal.
- Specify that all electrical work must be carried on by qualified electricians. Also materials and design must be in accord with the National Electrical Code and applicable local codes.
- Require bidders to furnish full information on the system being offered, including drawings, material specifications, dimensions, weights, etc.

(c) *Fire Detector, Ceiling-mounted, Ionization*

- Specify number and position on ceiling of the detector heads in each room or space to be protected.
- Specify the electric service available, giving voltage, phases, cycles, etc.
- Specify number and position of zone cabinets when several rooms or spaces are to be protected. Indicate location of central alarm cabinet, to which each zone cabinet is connected.
- Specify that that all wiring, conduit, boxes, etc., must be in accordance with the National Electrical Code and local codes.
- Specify that all electrical work must be accomplished by qualified electricians.
- Specify the number and kind of alarm signals.
- Specify that alarm signals must continue until stopped by manual action.
- Specify that the system and its parts must satisfy the Underwriters Laboratories or Factory Mutual Laboratories.
- List all equipment that is to be stopped or started as a result of the alarm. State horsepower of motors affected.
- Require bidders to furnish full information on the design and arrangement of the detector heads, giving dimensions, weights, materials, etc.

(d) *Fire Detector, Rate-of-Rise Type*

- Specify the number and positions of the detectors in each space or room to be protected.

- Specify number and kind of alarm signals, and give their locations.
- Specify the maximum linear feet of tubing to be provided with each detector.
- Specify that all parts and the whole of the system must carry the approval of the Underwriters Laboratories or of the Factory Mutual Laboratories.
- List all equipment that is to be stopped or started as a function of the alarm signal. State horsepower of all motors affected.
- Specify if central control station is to transmit a signal to an outside fire department.
- Provide bidders with drawings or information concerning size and arrangement of spaces to be protected, including ceiling or roof heights.
- Specify that all electrical materials must satisfy the National Electrical Code.
- Require all electrical work to be accomplished by qualified electricians.
- Require bidders to submit full information on their systems, including key dimensions, weight, material specifications, etc. Also voltage for alarms.
- State electrical service available, as voltage, phases, cycles, etc.

(e) *Fire Detector, Sensitive-cable Type*

- Specify the rooms or spaces to be protected, providing a drawing, if possible, to show dimensions of these spaces. Indicate height of ceiling if only a plan is available.
- Specify temperature level at which detector is to operate, as 155 F, regular rating; 185 F, intermediate rating; and 275 F, high rating.
- Indicate location of sensing cable, as on ceiling, wall, over molding, etc.
- Specify that system as a whole and in all its parts must carry approval of Underwriters Laboratories or Factory Mutual Laboratories.
- Specify number and kind of alarm signals, and their locations.
- Specify what equipment is to be stopped or started as a result of the signal. Give horsepower of each motor affected.
- Specify if central control station is to be connected to an outside fire department.

- Specify that electrical materials must meet the National Electrical Code and applicable local codes.
- Specify that all electrical work is to be accomplished by qualified electricians.
- Require bidders to submit full information on their systems, including dimensions, weights, design, materials, etc. Also voltage for actuating the alarms.
- State current available, as voltage, phases, cycles, etc.

(f) *Smoke Detector, Duct-mounted, Photoelectric-cell Type*

- Refer to Sec. 8-2(a), Fire Detectors, Duct-mounted. Each condition detailed in that section applies equally to the smoke detector. Add this breakdown condition:
- Specify that the occurrence of a mechanical failure in the smoke detector system shall be signaled by the activation of a warning light, bell, buzzer, or other approved signal.
- Require bidders to state electric service desired, giving their design voltage, phases, cycles, etc.
- Require bidders to provide full information, including drawings and photographs, on design, arrangement, materials, and weights of the equipment being furnished.

DIVISION 9

Steam Distribution Systems

9-1. TRENCHING AND MANHOLES FOR UNDERGROUND PIPING

(a) *General Data to Be Supplied to Bidders*

- Provide general plan showing projected location of steam line, visible interferences, the beginning and end of the line, changes of direction, the location of manholes, expansion joints, branches leading from the main steam line, and other important items the contractor must consider. Also provide an elevation drawing, with sectional views where appropriate. Drawings to show dimensions.
- State the elevation of the water table.
- Describe nature of soil through which the line is to pass, as sandy, rocky, clay, marl, etc. Give bearing capacity in tons per sq ft.
- If easements are required, stipulate who shall obtain them.
- If necessary to open a city street or to obstruct traffic, stipulate who will obtain the necessary permits and licenses, and who will bear whatever charges are involved.

(b) *Trench Construction*

- Specify precautions contractor must exercise to avoid disturbances to properties adjacent to the trench. Particularly require control of vibration from trenching machinery that may otherwise cause damage to plaster walls, ceilings, and foundations.
- Require contractor to describe how he intends to prevent earth waves from blasting, if blasting should become necessary.
- Specify that contractor must carry property damage insurance, covering both direct and indirect damage, in the amount of $_______, total claims, and $_______ for each claim.
- Specifv the kind and extent of the sheathing, giving maximum depth of trench, _______ ft, for which sheathing is not required.

- Specify the method of continuous unwatering to be employed, and designate where pump discharges are to be taken.
- Specify techniques for projecting trench across underground piping, sewers, electric and telephone cables, and all other underground obstructions, if these are known to exist.
- Specify that contractor must assume full responsibility for interruptions to other services that result from constructing the trench.

(c) *Manholes*

- Provide contractor with detail drawings of manholes, showing dimensions, arrangement, and construction.
- State if manholes are to be built at the site of each, or if prefabricated units (Fig. 9-1) shall be installed.
- If steel, prefabricated, specify gage of steel, how treated

FIG. 9-1. "Instant" manhole of corrugated steel. (*Courtesy Ric-Wil Incorporated.*)

for corrosion resistance, as galvanized, coated inside and outside, and whether cathodic protection is to be included.

- Itemize equipment to be contained in the manhole, as line valves, pipeline anchors, expansion joints, drainage sump with float-operated sump pump, heat exchanger for cooling sump pump discharge if required by point of discharge, ventilating ducts or equipment, instruments, and any other necessary items. Figure 9-2 is the interior of a manhole with a minimum of enclosed equipment.

FIG. 9-2. Interior view of the steel manhole. (*Courtesy Ric-Wil Incorporated.*)

- Specify how manhole covers are to be lifted and how to be kept locked and tamper-proof at other times.
- Specify kind of drainage or sump pump, as rotary, centrifugal, etc. Specify capacity in ________ gpm against a discharge head of ________ ft when pumping water of ________ F temperature.
- Specify materials and construction of the pump, as cast-iron casing, bronze impeller, etc.
- Specify pump drive, requiring motor to be moisture-proof and to operate on ________ volt, ________ phase, ________ cycle current.
  - Show bolting of pump and motor to foundation and means of connection, as flexible coupling.

- Specify kind of contact switch to be used in the manhole atmosphere of _______ F temperature and _______ per cent relative humidity.
- If sump drainage is to be by ejector, specify if steam or air. Give pressure, _______ psig, and if steam, state quality, as dry saturated, _______ degrees of superheat, etc.
- Specify kind and capacity of trap and point of discharge.

(d) *Conduits, Wood*

- This is a prefabricated conduit, with usage generally limited to dry areas, and for low-pressure steam. Figure 9-3 illustrates a common design.

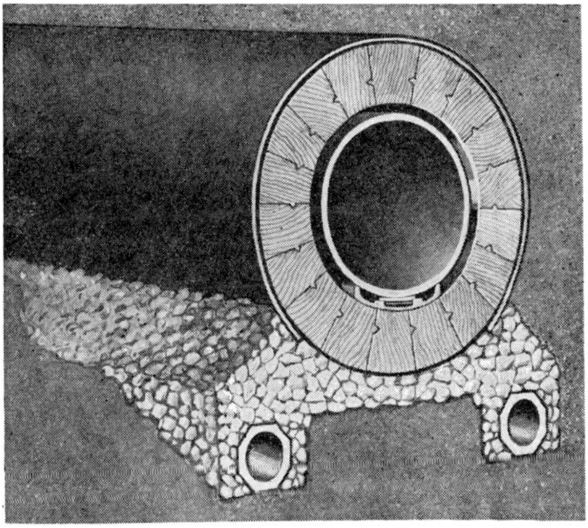

FIG. 9-3. Typical form of wood conduit, showing drain tiles in broken stone base.

- Specify kind of wood to be used, as pine, cedar, or other.
- Specify how joints are to be made and sealed.
- Specify sizes of stone bed under the conduit, and give sizes of drain tile.
- State pressure, _______ psig, and temperature, _______ F of steam.
- Specify kind and thickness of insulation.
- Specify kind and placement of pipe saddles.
- Specify means and disposal for groundwater accumulating in the drain tiles.
- Specify material and form of sealing jacket.

Note: Wood conduit may be difficult to obtain. Check supplier before issuing specification.

(e) ***Conduits, Tile***

- Prefabricated tile conduits, as shown in Fig. 9-4, are proprietary designs, furnished complete with baseplates, drains, support rollers for the pipe, and all other needed items.
- Specify number, sizes, and kinds of service for which enclosed pipes will be furnished. As ________ main steam pipe, ________ in. diam; ________ condensate pipe, ________ in. diam.

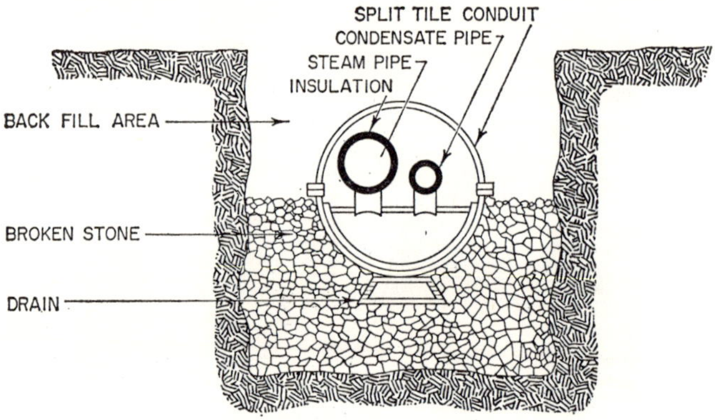

FIG. 9-4. Tile conduit. *Note:* In some forms of conduit, the interior is packed with insulation instead of wrapping the pipes as shown.

- Specify temperatures in each pipe and the kind of insulation to be provided with the conduit.

(f) ***Conduits, Masonry, Box Type***

- Figure 9-5 illustrates this kind of conduit.
- Specify if walls are to be poured concrete, or built up of concrete blocks.
- Specify the mix of concrete to be used, and stipulate the curing time to be allowed.
- Where reinforcing is used, specify thickness of rods, center-to-center distance, and placement.
- Specify insulation on pipe, spacing and kinds of pipe saddles, sizes of drains for groundwater.

Note: This form of conduit can be expanded into tunnel dimensions, since the construction is masonry. The sides and roof in some designs are formed into an arch.

- Specify how conduit is to be waterproofed, as by a membrane placed before backfilling, by using waterproofed concrete, etc.

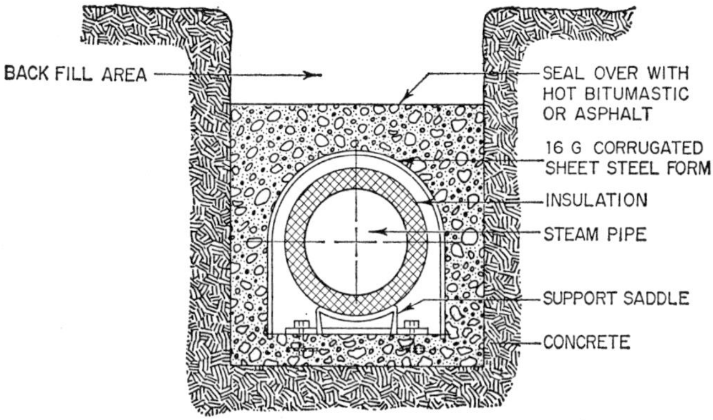

FIG. 9-5. Concrete conduit.

(g) *Conduits, Monolithic*

- These are formed by pouring over the pipeline, after it has been placed in position in the trench, a material that hardens around the pipe and is of a nature that provides the line with both physical protection and thermal insulation.
- In all cases, the material is proprietary, and the specification should be written to obtain the particular formula desired.
- If conditions indicate that a particular brand is preferable, state the brand name. If this is to be an open specification, add the phrase "or approved substitute."
- Specify that material is to be applied in strict accordance with manufacturer's directions.

(h) *Backfill*

- Specify the kind of material to be used for backfill, as clean gravel, sand, soil for topping, etc.
- Prohibit the backfilling with cinders, or any other material likely to be porous, to be acid-generating, or to have other undesirable qualities.
- Specify how the sheathing and concrete forms are to be handled, that is, pulled, stripped, left in place, or however.

- Specify the depth of the cover, _______ ft.
- Specify the finish paving. If concrete, specify mix; if bitumastic or asphalt paving, specify kind and methods of application. In every case, specify thickness.
- Specify method of tamping backfill, and density to be effected.

(i) *Special Precautions, Underground Projects*

- During the course of the work at the site, require contractor to provide barricades along entire length of work, with suitable lights or flares in place at night.
  - Specify the height, materials, and construction of the barricades.
- If work creates unusual hazards, require a watchman patrol to walk the job at night from sundown to sunrise.
- Where trench is more than six (6) ft deep, and at the same time oil, sewer, gas, or chemical lines are in the area, specify a daily check of the atmosphere in the trench before workmen are permitted to enter and work there continually.
  - Check to be made by a suitable instrument as manufactured by the Mine Safety Appliances Co., or other competent supplier.
- For tunnels, specify method and extent of ventilation, as by (portable)(permanent) blower with a capacity of _______ cfm when discharging air of _______ F temperature into the tunnel, and at a discharge pressure of _______ in. W.G.
  - Blower to be driven by electric motor (direct-connected) (belt drive) with electric service of _______ volts, _______ phases, _______ cycles.
  - Specify daily atmospheric safety check before workers are permitted to enter tunnel, by use of suitable instrument as previously indicated.
  - Specify tunnel lighting, giving the desired illumination as _______ watts per sq ft.
  - Specify that all wiring, conduit, switches, and other electrical materials must satisfy the National Electrical Code, the NEMA standards, and local codes applicable to such equipment.
  - Specify that all electrical work must be accomplished by qualified electricians.
- If excavated and construction materials may be piled up or stored along the line of the work, indicate where.

Require that suitable covers and/or warning signs shall be maintained on such piles or storage.

9-2. UNDERGROUND PIPING, STEAM

(a) *Main Steam Piping*

- Specify steam pressure, _______ psig; quality, _______ per cent saturated or superheat of _______ F.
- Specify the expected maximum flow rate, pounds per hour.
- Specify the kind of piping to be used, as carbon steel, alloy steel, wrought iron, etc.
- Specify schedule number, and indicate whether to have plain ends or threaded ends.
- Specify that pipe materials and construction must conform to the suitable ASTM standards as:
  - Seamless steel for pressures in excess of 400 psi, ASTM A-106 or A-158; fusion welded, refer to ASTM A-155.
  - Between 250 psi and 400 psi, lap welded or seamless, ASTM A-106; electric fusion welded, ASTM A-155; electric resistance welded, A-135. Seamless also may be ASTM A-53.
  - Under 250 psi, any of the ASTM standards for the higher pressure may be specified. Also ASTM A-134 or A-139 when fusion welded. For wrought iron, ASTM A-72. Brass pipe, ASTM B-43.
  - For pressures under 125 psi, ASTM A-120 applies to welded pipe.

(b) *Pipe Joints*

- Specify if welded. Require welders to possess certification by the National Certified Pipe Welding Bureau or other approved authority.
- State characteristics of electric service available to welders, or if welding machines must be provided in the field.
- If flanged joints, specify if companion flanges, Van Stone or other type of bolted faces.
- Specify type of gaskets.

(c) *Expansion Joints*

- For slip-type joints, specify if single or double slip. Specify travel of each, _______ in.
- If joint has base anchor, specify kind of anchorage, as to steel cross member as shown on the drawings, or to concrete block. Specify who supplies the anchorage, as piping contractor, or others.

- If joint is to have top takeoff for a branch pipe, specify size of takeoff. Specify joint material.
- For bellows-type expansion joints, specify the materials of the bellows and the number of the corrugations.
- For all joints, specify the method of connecting to the piping, as by flanges, welding.
- For all joints, specify the range of expansion to be absorbed, from a minimum of ______ in. to a maximum of ______ in.
- If expansion joint requires periodic lubrication, as for the slip, sleeve, or other, specify frequency acceptable.
- Require bidders to furnish information on packing, if any, as thickness, kind, number of rings, etc.
- Require bidders to submit drawings giving information on design and arrangement of the joint, dimensions, materials, operating data, etc.

(d) *Pipeline Valves, Gate*

- Specify materials of construction, as ______________ for body, ______________ for gate, ______________ for stem, ______________ for bonnet, ______________ for handwheel, ______________ for seat.
- Specify to be rising or inside stem; if outside, specify if yoke also is to be provided.
- Specify if valve is to be fitted with bypass, and if so, specify size and kind of valve in the bypass.
- Specify the classification of the valve as to working steam pressure, ______ psig, and working steam temperature, ______ F.
- Specify valve ends, as screwed, flanged, welded.
- Specify test pressures.

(e) *Pipeline Valves, Globe*

- Specify if valves have inside stem, outside stem, or outside stem and yoke.
- Specify if single- or double-seated.
- Specify materials of construction, as ______________ for body, ______________ for seat or seats, ______________ for disk, ______________ for bonnet, ______________ for handwheel.
- If bypass is to be provided, specify size, and kind of valve to be mounted in bypass.
- Specify classification of valve, as for ______ psi working steam pressure and ______ F working temperature.
- Specify valve test pressures.
- Specify valve ends, as screwed, flanged, or welded.

(f) ***Pipeline Traps***

- Specify number, kind, and various sizes of traps to be furnished.
- Specify capacities in pounds of condensate per hour.
- Specify range of steam pressures for each trap, from a minimum of ______ psig to a maximum of ______ psig.
- Specify back pressure on each trap, ______ psig.
- For traps with bypasses, state who furnishes the bypass together with the bypass valve.
- Specify materials of trap construction, as carbon steel, forged steel, alloy steel, etc.
- State position of each trap, as in manhole no. ______.
- Specify trap test pressure, ______ psig.
- Specify length of bare pipe to be provided on inlet side of trap to avoid presence of superheated steam or loss of trap seal, ______ in.

(g) ***Pipeline Insulation***

Note: This applies to pipelines that are independent of the conduit containing them.

- Specify kind and thickness of insulation, as 85 per cent magnesia, hi-temp magnesia, asbestos, silica, etc.
- Specify acceptable conductivity, and give temperature, ______ F, at which this conductivity is to be guaranteed.
- Require bidders to furnish a graph showing increase in conductivity in relation to increases of temperature differential across the insulation.
- Since the atmosphere of the conduit in which the pipeline is placed tends to be both hot and humid, specify limits on moisture-absorbing capacity of the insulation, ______ per cent by weight.
- Specify insulation jacket, as ______ oz duck, and the method of application, as (cemented) (sewed on). If asphaltic jacket, give weight.
- Specify how joints are to be made in the insulation, whether insulation edges are to be abutted, scarfed, or lapped.
- Specify kind and method of placing insulation on valves and fittings, as preformed sections, cement troweled on, etc.

Note: If actual amount of insulation is unknown at time of writing the specification, specify how prices are to be submitted. That is, $______ per ft for each size of pipe for the temperature range and temperature differentials to be met. Stipulate if valves and fittings are to be quoted similarly for each size and temperature, or if prices shall be based on equivalents. For example, 1 elbow shall be equivalent to 3 lin ft of pipe in that size; 1 valve shall be equivalent to ________ ft of pipe, etc.

(h) *Condensate Piping*

Note: In many metropolitan district steam systems, the cost of returning the condensate to the boiler plant is not economical, and the condensate from the users' premises is piped to the city sewer. Condensate removed by traps along the steam main is similarly handled, in most instances a heat exchanger being a legal requirement to avoid the possible creation of noxious vapors by the dumping of hot water into the sewer, although these sewers are of the storm or drain type, and not sanitary.

- Specify piping in accordance with Sec. 9-2(a), noting difference in pressures and temperatures.
- Specify kind and placement of rollers and saddles for allowing expansion movement of pipe.
- Provide drawings as necessary to show mounting of condensate piping in relation to main steam piping and other equipment in both the conduit and manholes.
- Specify the kind, model, and capacity of the heat exchanger if this form of equipment is required.
- Specify trap piping and valves to suit the pressures and temperatures of the trapping system.
- To specify pipe joints, refer to Sec. 9-2(b).
- For expansion joints, refer to Sec. 9-2(c).
- For valves, refer to Sec. 9-2(d) and (e).
- Traps to be as specified in Sec. 9-2(f).
- Insulation as in Sec. 9-2(g).
- If condensate piping is to be alone in occupying the trench or conduit, refer to Sec. 9-1, all subdivisions.

9-3. PIPING SUPPORTS, OUTDOORS

(a) *Supporting Structures, Wooden*

- State who shall furnish and erect the supporting structure, as (piping contractor) (another contractor) (owner), etc.
- Provide this bidder, whether the piping contractor or a separate contractor, with drawings showing the design, dimensions, and locations of the units from the source to the terminal of the piping run.

- State nature of soil in which supports are to be set, giving bearing capacity in tons per sq ft, also elevation of water table.
- Specify kind of wood and its treatment for preservation against weather and fungus.
- Specify pitch of pipe and direction thereof, giving relative elevations measured from an indicated datum or benchmark.
- Specify if self-supporting or guyed.

(b) *Supporting Structures, Steel*

- State by whom furnished and erected.
- Provide drawings showing the design, dimensions, and placement of the structures.
- Specify finish of structure buried below the surface of the ground, and aboveground, as red lead, bitumastic, galvanized, etc. Specify the number of coats of each, method of application, time allowed between coats, preparation of surface, etc.
- Provide bidders with soil information, kind, bearing capacity, tons per sq ft, etc.
- State elevation of water table.
- Specify the pitch of the piping, giving the relative elevations of the supporting structure.
- If structural legs are embedded in concrete, specify the size of the concrete blocks, the mix, distance below surface of ground, reinforcing curing time, etc.
- Specify how saddles and pipe rollers are to be placed and fastened to the supports.

(c) *Supporting Structures, Concrete*

Note: Concrete supports for aboveground piping generally take the form of concrete blocks elevating the pipelines from one to several feet above the ground.

- Furnish drawing showing shape of blocks desired and giving all dimensions.
- Specify the concrete mix to be used, noting if proportioning is to be based on weight or volume. *Note:* A mix by weight is not the same as a mix by volume.
- Specify kind and spacing of reinforcing; also method of anchoring pipe saddles.
- Provide full information on nature of soil, its load-bearing capacity, tons per sq ft, and the elevation of the water table.

- Specify pitch of piping, giving elevation at top of each concrete pedestal.
- State if piping on the concrete supports is to be steam, condensate, water, or all these. Show relative locations on top of the pedestals.
- Provide drawing showing location of blocks, including those supporting expansion loops, horizontal or vertical.

9-4. OVERHEAD PIPING, OUTDOORS

(a) *Steam Piping*

- Refer to Sec. 9-2(a)–(f) for guidance on piping, valves, fittings, joints, expansion joints, traps.

(b) *Insulation*

- Refer to Sec. 9-2(g) as appropriate.
- Since these overhead systems are exposed to the weather, specify kind and application of weather protection.
- If aluminum-finish jacket, specify whether finish to be smooth or corrugated.
- Specify vapor barrier if required, and require warranty that adhesive used to bond the vapor barrier to the metal will not attack the aluminum.
- Specify thickness of aluminum as ________ gage.
- Specify method of making aluminum joints, as screwed, riveted, bands, etc.
- Specify lap, ________ in., for all joints.
- Specify where insulation and aluminum jacket are to be preformed. Particularly applicable to valves and fittings.
- Require bidders to submit table of pipe sizes from ________ in. diam pipe to ________ in. pipe, and give a price per foot for each size.
- For preformed jackets and insulation for valves and fittings, require bidders to submit a table of unit prices for each size of valve or fitting specified.
- If covered with roofing, specify kind and thickness.

(c) *Pipe Rollers*

- Specify number of rollers required for each size of pipe.
- Request unit prices for each size.

(d) *Condensate Piping*

- For piping, traps, valves, etc., refer to Sec. 9-2(a)–(f). Follow as appropriate.
- For insulation and jacketing, refer to Sec. 9-4(b).
- If a heat exchanger is required, refer to Sec. 9-5(b).

9-5. STEAM DISTRIBUTION PIPING, INDOORS

(a) *Note:* An indoor system differs from an outdoor system mainly in having the piping suspended rather than mounted on top of a support, and in the absence of any need for weather protection. All the other considerations of piping schedule number, pitch, trapping, drainage, expansion and contraction movements, valves, fittings, joints, etc., are common to both inside and outside installations.

This is not to say outdoor suspension and indoor under supports are never used. They are, and will be observed in some circumstances. Outdoor suspension is seen particularly in the process and chemical industries, where one character of fluid may travel on top of a supporting structure, while another fluid hangs beneath the structure which is common to both.

The physical environment and interferences will dictate the piping locations.

- To specify the piping, valves, fittings, traps, etc., refer to Sec. 9-2(a)–(f).
- To specify insulation, refer to Sec. 9-2(g).
- Also specify the color identification to be used on the piping canvas jacket.
- Specify the piping hangers by type and sizes, as spring suspension, solid band, adjustable, nonadjustable, etc.
- Specify if hanger is to contain a roller on which the pipe moves.
- Specify the support to which the hanger is to be attached.
- Specify painting; color and number of coats.
- Specify minimum safety factor of (5) or other value.
- Specify material of hangers, as steel, galvanized, copper bearing, etc. Refer to appropriate ASTM standard.

(b) ***Condensate Piping, Indoors***

- As with the steam piping, indoor condensate piping differs from its outdoor counterpart mainly in the form of its support and in the absence of weather protection.
- For the specifying guide, refer to Sec. 9-2(h).
- Specify the painting of the pipe or its insulating jacket in accordance with the plant color code. Specify the number of coats of paint and the method of application, surface preparation, etc.

- If condensate is to run to the sewer, specify the maximum discharge temperature, _______ F.
- Specify the kind, style, and capacity of the heat exchanger or cooling radiator to be provided for cooling the condensate before discharge.
- If a condensate receiver is necessary, specify the size, capacity, materials of construction, and the location.
- Specify condensate pump taking suction from the receiver, giving capacity as _______ gpm when handling water at _______ F, and discharging against a head of _______ ft, and operating on a (suction lift) (suction head) of _______ ft.
- Pump to be driven (direct-connected) (through gears) by an electric motor of _______ hp, with electric service of _______ volts, _______ cycles, _______ phases.
- Stipulate if pump and motor are to be mounted on a common baseplate.
- Specify how pump is to be actuated, as by a float in the receiver, or other control.
- Specify that electric motor, starters, relays, etc., must meet requirements of the National Electrical Code, the NEMA standards, and any applicable local codes.
- Require that electric features of installation be accomplished strictly by qualified and licensed electricians.
- Specify pipe hangers in accordance with Sec. 9-5(a).

DIVISION 10

High-temperature–high-pressure Hot Water Systems

General Note. High-temperature–high-pressure hot water distribution has advantages over steam in being unaffected by changes of elevation of the ground over which the piping passes, and by the elimination of traps and condensate removal problems. Circulation is entirely by pump.

If the piping is run through conduits, or tunnels, or both, the problems of excavation, backfill, expansion and contraction safety precautions, and other aspects of the installation, are practically identical with those for a steam program. Since the water is circulated at steam pressures and temperatures, steam for various applications is provided at the point of use by introducing the water into a heat exchanger and taking off the steam generated therein at whatever pressure is required.

These components of the hot water system are examined and given guide specifications in the succeeding sections of this division. For trenching and conduits, see Sec. 9-1. For aboveground supports, see Sec. 9-3(a)–(c).

10-1. PIPING

(a) ***Materials and Classifications***
- Refer to Sec. 9-2(a)–(e).

(b) ***Pipeline Insulation, Underground and in Conduits***
- Refer to Sec. 9-2(g).

(c) ***Pipeline Insulation, Overhead, Outdoors***
- Refer to Sec. 9-4(b).

(d) ***Pipe Rollers***
- See Sec. 9-4(c).

10-2. HEAT EXCHANGERS FOR STEAM GENERATION FROM HOT WATER AND HOT WATER PUMPS AND CONTROLS

(a) ***Operating Data and Requirements for Bidders***
- For each heat exchanger, specify the temperature of the hot water entering, ________ F.

- Specify the volume of hot water to pass through the exchanger, _______ gpm.
- Specify the temperature, _______ F, of the water entering the exchanger and to be evaporated.
- Specify the volume of this water, _______ gpm.
- Specify the weight of steam to be generated as _______ pounds per hour at _______ psig and total temperature of _______ F.
- Specify the number of passes in the exchanger.
- Specify the construction of the exchanger, materials of shell to be _______________, of tubes to be _______________, and of tube sheet to be _______________.
- Specify test pressures, _______ psig.
- Specify that all aspects of the heat exchangers must satisfy the ASME "Code for Unfired Pressure Vessels" and bear the ASME stamp.
- Require bidders to furnish detailed drawings showing all dimensions, arrangements, tube supports, baffles, removable head or heads, pressure relief valves, weights, and other pertinent information.
- Specify the material and thickness of insulation and the finish jacket.

(b) *Heat Exchanger Condensate System*

- Refer to Sec. 9-2(h) for piping guidance.
- Refer to Sec. 9-5(a) and (b) for condensate handling indoors. This includes pumping, hangers, cooling radiators, etc.
- Specify painting of the heat exchanger condensate piping in accordance with the plant color code.

(c) *System Compression Tank*

- Specify number of expansion tanks.
- Provide drawing showing equalizing connections between multiple tanks. Show dimensions of both steam (over the water surface in the tank) and water equalizers. Also show location of tank or tanks with reference to pumps, and indicate elevation above pumps, _______ ft.
- Specify gage glass length as _______ in.
- Specify safety relief valves to conform to ASME requirements; also the drain and charging valves.
- Specify if pressure on liquid is to be provided by air or nitrogen.
- Require tank construction to satisfy all stipulations of ASME "Code for Unfired Pressure Vessels."

- Specify the kind, characteristics, and operating range of high- and low-water alarms.
  - Stating if alarms are to be visual, auditory, or both.
  - Require wiring materials and labor to meet requirements of National Electrical Code and any applicable local codes.
- Specify connecting piping, as ____________ (material), and Schedule No. ______, for a working pressure of ______ psig.
- Specify method of tank support and state who provides it.
- Specify temperature range of hot water entering the tank, from ______ F minimum to ______ F maximum.
- Specify capacity of tank, ______ gal.

(d) *Nitrogen Pressurizing*

Note: The tank for a gas pressurized system may be set either horizontally or vertically, but the latter is preferred as saving both space and the cost of elaborate supports for the horizontal tank.

- Specify the position and size of the tank.
- Specify the number and location of the nitrogen flasks that serve the tank.
- Specify the storage pressure as ______ psig, and the final delivery pressure to the tank as ______ psig.
- Specify the kind and number of pressure-regulating valves on the nitrogen flasks, as ______ first-step regulating valves reducing pressure from ______ psig to ______ psig, and ______ second-step regulating valves reducing pressure from ______ psig to ______ psig.
- Specify ______ safety relief valves set at ______ psig.
  - Safety valves to be in accordance with the ASME Code.
- Specify that storage flasks, or cylinders, must meet the requirements of the U.S. Interstate Commerce Commission and all applicable codes.
- Provide sketch showing placement of the nitrogen flasks, their interconnecting piping, means for exchanging charged units for empty flasks, control valves, etc.
- Specify that the expansion or pressure tank shall be constructed in accordance with the ASME "Code for Unfired Pressure Vessels."
- Specify excessive-pressure alarm, set to act at ______

psig, or ________ pounds pressure above the setting of the safety relief valves.

- State if alarm is to be visual, auditory, or both.

- For nitrogen compressor, if required, specify capacity, ________ cfm, at a pressure of ________ psig.

(e) *Water Circulating Pumps*

- Provide drawing showing location of the pumps with reference to the boiler and to the expansion tank.
- Specify capacity of pumps as ________ gpm when handling hot water at ________ F, and discharging against a total head of ________ ft, running at ________ rpm.
- Specify number and type of pumps, as (centrifugal) (rotary), ________ stages.
- Describe pump construction, as (close-grained cast iron) (steel), etc.
- Specify maximum working pressure, ________ psig, and test pressure of ________ psig.
- Specify horizontal or vertical split of casing, with discharge connection placed relative to the split. (For horizontal splits, pipe connections should be on bottom half for convenience.)
- Specify the design and materials of the impellers, as (double) (single) suction, bronze, etc.
- Specify shaft materials and require both static and dynamic balance at the factory.
- Specify the bearings wanted, as babbitt, roller, etc. Also thrust bearing.
- Specify that stuffing boxes must be water-cooled.
- Require bidders to submit detailed drawings showing the arrangement of the pump internally, methods of preventing leakage between stages, weights, etc.
  - Bidders also to state cooling water needed, ________ gpm, with water at ________ F minimum and ________ F maximum. Water to be supplied to the stuffing boxes at ________ psig.
- Specify if pump and driver are to be mounted on a common bedplate, and method of attachment to the foundation.
- Specify kind of coupling, as flexible, hydraulic, etc., or if pump is driven through a reduction gear.
- For data on motor drives, refer to Sec. 4-12(a).
- For data on steam-turbine drives, see Sec. 4-12(a).

- Require bidders to submit performance data in accordance with the form, Table 4-9, modified as necessary to suit this character of service. Both pump and turbine data are called for by this form.

(f) *Pump Controls*

- State general plan of system control, whether by temperature, pressure, or a combination of each.
- For temperature controls, specify placement of submerged bulbs, as in system return immediately before reentering boiler, or in expansion tank liquid.
- Specify the kind and design of the desired temperature sensor, including materials of construction, the nature of the expanding and contracting element, and the operating range from ______ F to ______ F, and calibrated in steps of ______ F.
- For pressure controls, specify position of sensitive element, as in liquid, nitrogen gas in the expansion tank, etc.
- Specify the number and locations of each control instrument.
- Specify how temperature or pressure sensing is to react on the pumps, as by stopping, starting, opening the bypass valve to permit the returns to be picked up by the pumps' suctions without going first to the boiler, etc. Specify if action is pneumatic or electrical.
- For pneumatic controls, specify the air pressure available, ______ psig. Identify source of the air.
- If special air compressor is required for the control system, so specify, giving capacity of compressor in ______ cfm of free air, at an entering temperature of ______ F and at a location of ______ ft above sea level.
- Specify the pneumatic tubing or piping required, as copper, brass, steel, wrought iron, etc.
- For electric controls, specify location of relays, and give characteristics of electric service available, as ______ volts, ______ phase, ______ cycles.
- Require all wiring and electrical components to satisfy the National Electrical Code, the NEMA standards, and applicable local codes.
- For air venting of the distribution system, specify the placement of air bottles or other suitable form of container, on high points of the system, and where there

occurs a change of elevation. Provide each air bottle with a manually operated vent valve.

- Valve to be of the (globe type) (gate type) and suitable for the working pressure, _______ psig, of the system at a temperature of _______ F. Specify sizes.
- Specify materials of air bottle, as seamless steel pipe, Schedule No. _______, or wrought iron, etc.
- Specify vent valve materials of body, seat, and disk or gate; rising or nonrising stem; screwed or flanged ends as appropriate.
- Specify total number and sizes of the air bottles required.

(g) *Accumulators*

- Specify number and placement on the system.
- Specify the capacity of each, as _______ gal.
- Specify construction, as welded, riveted, thickness of plates, heads.
- Specify construction to be in accordance with the ASME "Code for Unfired Pressure Vessels."
- Specify maximum working pressure as _______ psig.
- Painting to be specified in kind and number of coats.
- Specify insulation, as _______ in. thick, _______ _______ material, suitable for a maximum temperature differential of _______ degrees. Specify maximum temperature to be met by tank.
- Specify kind and design of foundations and supports.
- For instruments, gages, etc., see preceding section, Pump Controls.
- Specify the temperature and pressure instruments required, noting if they are indicating, recording, remote reading, etc.
- Specify how input and discharge of the accumulator is to be controlled, as by a thermostatic mixing valve or other means.

DIVISION 11

Premises Heating, Steam

11-1. DISTRICT STEAM SERVICE

(a) *General Notes*

• District steam companies have established requirements of piping arrangements with which the customer must comply. Therefore, before issuing a specification aimed at heating and piping contractors for work on the premises and at the expense of the customer, the specification writer should obtain a copy of the steam company's local rules and regulations, and incorporate these as appropriate into the specification.

To illustrate, Fig. 11-1 shows the service entry requirements issued as standard by the steam utility in a metropolitan center.

Particular attention should be given to the exact point at which the utility work ends and the customer's work begins.

Meters and their immediate connections usually are owned and installed by the utility, and the utility is responsible for their accuracy and proper functioning thereafter.

In some areas, and for small customers, so-called "demand limiters" may be provided by the utility, these consisting of a fixed orifice to be mounted between flanges in the steam supply. The orifice works on the principle that flow through it is limited by the pressure differential across the orifice, and in these cases, the utility adjusts the pressure-regulating valve ahead of the orifice to maintain a constant pressure entering the orifice, despite pressure fluctuations of the main steam entering the building.

The regulating valve is locked so that its setting can be changed only by representatives of the utility company. The customer then pays a demand charge based on the size of the orifice.

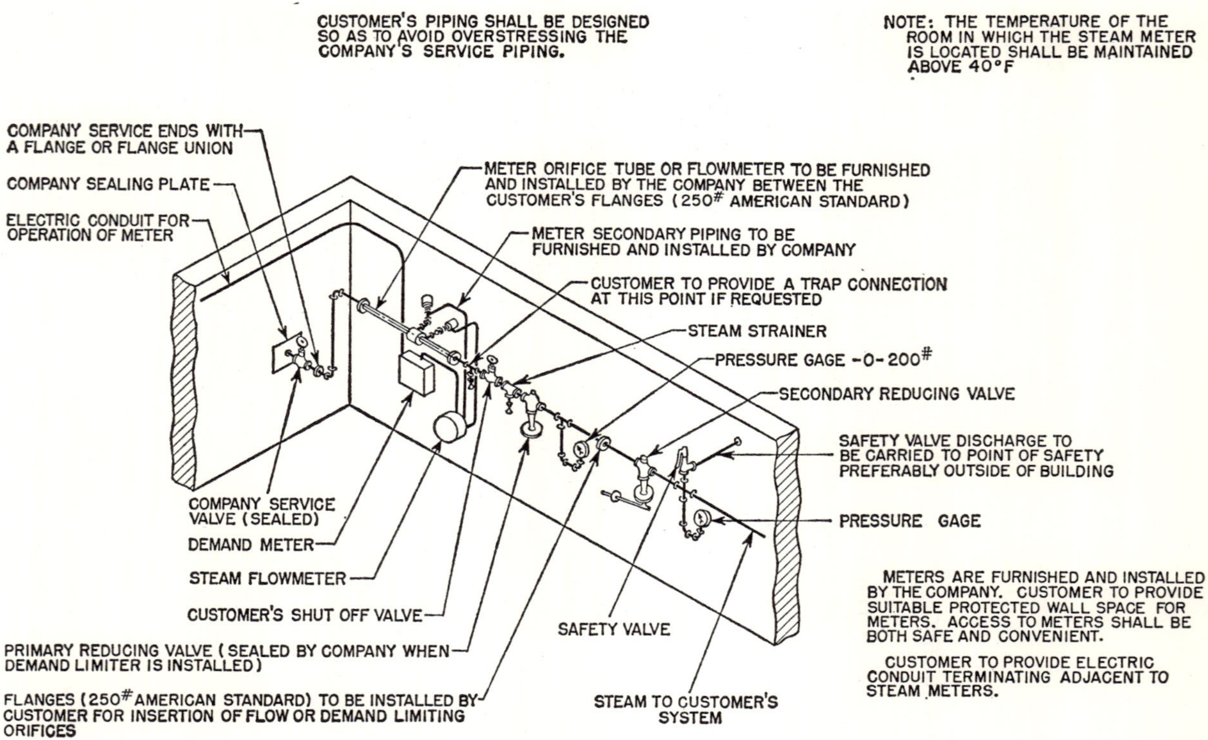

FIG. 11-1. District steam service connection.

For large purchasers of steam service, the utility company attaches a demand recording meter to the flowmeter, thus maintaining a record of the hourly, or half-hourly, withdrawal of steam from the supply mains. Demand charges are calculated on the records provided by these demand meters.

The specification writer should reach an agreement with the steam utility on the location of all meters, including the position of the small fixed orifice, and state these locations in the specification.

For those installations that do not require the return of the condensate to the utility steam plant, special attention must be given to the condensate disposal. Metropolitan regulations generally place a limit on the temperature of any discharge to the sewers; consequently, cooling radiators may be required to satisfy the local laws. In these circumstances, specify the kind and size of the cooling radiator, expressing size in square feet of radiation.

(b) *Piping*

- Specify the kind and working pressure of the steam piping, ________ psig. Give schedule number.
- Specify joints, as flanged, screwed, welded.
  - If flanged, specify kind of flanges, (cast iron) (steel) (raised face) (Van Stone), etc.
  - If screwed, specify threads, and say whether "pipe dope" is to be used on one thread only, or both. Specify kind of dope to be used.
  - If welded, specify kind of joint, as abutting, scarfed, etc.
    - Specify that welders must be certified by National Certified Pipe Welding Bureau or other approved authority.

(c) *Fittings*

- Specify materials, as cast iron, steel, malleable iron, bronze, etc.
- Specify working pressure, ________ psig, and temperature, ________ F.
- Specify ends, as flanged, screwed, welded.

(d) *Pressure-reducing Valves*

- Specify duty of valve, to reduce steam pressure from ________ psig to ________ psig, steam temperature to be _______ F (saturated) (or with _______ deg F of superheat).

- Specify type of valve, (spring loaded) (weight loaded) (diaphragm).
- Specify construction and materials, as iron body or bronze body, single or double seat, screwed or flanged ends.
- Specify number and sizes.
- Specify kind and size of steam strainer to be placed ahead of reducing valve. State if strainer basket or mesh is to be fabricated of copper, stainless steel, brass, etc.
  - Give strainer size.

(e) *Stop Valves*

- Specify type, as gate or globe. Give required materials of body, seat, and disk.
- Specify if valve is to have inside or outside stem, and if ends are to be screwed, flanged, or welded.
- Specify sizes, _______ in., and working pressure, _______ psig.

(f) *Safety Relief Valve*

- Indicate where relief valve is to be placed in the low-pressure piping, downstream from the reducing valve or valves.
- Specify size, and require valve to meet requirements of ASME for safety valves.
- Specify pressure at which valve is to be set to open.
  - Stipulate size of safety valve discharge piping, and state where excess pressure is released to atmosphere.
- Specify capacity of valve in _______ pounds per hour when relieving steam at a pressure of _______ psig.

(g) *Condensate Removal Trap*

- If steam trap is to be used to keep out surges of condensate from the main steam and from the branch into the building, specify kind of trap, as bucket, inverted bucket, etc.
- Specify capacity of trap in pounds of condensate an hour with an entering steam pressure of _______ psig and a back pressure on the trap of _______ psig.
- Specify drip connection to trap from steam piping to use Schedule No. _______ pipe of (steel) (wrought iron), etc.
- Specify stop valve to be provided on trap inlet, of suit-

able material and pressure, and a similar valve on the trap discharge.

- If several traps discharge into a common header, specify a check valve on the discharge of each.
- Provide sketch showing length of trap discharge piping, changes of direction, fittings, etc.

Note: Condensate must not go into a floor drain unless it is adequately cooled beforehand.

(h) ***Insulation***

- Specify insulation to be of _______ and _______ in. thick.
- Specify if insulation joints are to be abutted or lapped, and if lapped, specify the amount of lap as _______ in.
- Specify the material and weight of the jacket, as _______ oz duck, to be (sewn) (pasted) on the insulation.
- Specify the filler and paint to be used in coloring the line in accordance with the color code of the plant or building.
- Specify how valves and fittings are to be insulated, as with plastic insulation troweled on at the site, or by preformed insulation designed to fit the unit. This applies to flanges also.

(i) ***Pipe Hangers***

- Specify where pipe hangers are to be attached to the overhead beams or ceiling for the support of the steam line.
- Specify the kind of hangers to be used, (strap) (spring) (roller), etc.
- Specify the minimum acceptable load rating of the hangers.

(j) ***Gages and Thermometers***

- Specify the number, dial diameter, and calibration of pressure gages. State their locations.
  - Specify how the sensitive element is to function, as by Bourdon tube or spring, etc.
- Specify that each pressure gage must be provided with a pigtail connection and an isolating stop valve.
- For thermometers, specify if these are to be dial or stem type. Specify calibration and state locations. If dial, specify diameter. If stem, state desired length of stem and the submerged distance of the sensitive bulb.
- If bidders are to furnish thermometer wells, so specify, giving required material of the well.

(k) *Control Methods*

- State if control of steam into building system is to be automatic or manual.
- If automatic, describe sequences, noting if the action of the steam control valve is the result of impulses from an indoor thermostat, an outdoor thermostat, or a combination of both.
- Specify locations of the thermostats. Specify if these are to operate at line voltage or at reduced voltage.
- Specify whether the steam control valve is to be pneumatically or electrically operated.
- Specify if valve is to be of the modulating type or is to have only two positions, closed and open.
- If pneumatically operated, specify available air pressure, ________ psig, and name source.
- Electrically operated valve specifications must show characteristics of available service, as ________ volts, ________ phases, ________ cycles.
- Stipulate location of the control valve.
- Require all electrical material and labor to satisfy the National Electrical Code, the NEMA standards, and local codes.
- If control valve is to be manually operated only, specify type of valve, giving size, materials of the body and internals, whether ends are screwed or flanged.
- Specify if valve is to be furnished with rising or nonrising stem.
- Specify that valve must be suitable for fast and convenient operation. Require the addition of an operating chain if valve is in a high location, so valve can be controlled from the floor.
- Specify a position indicator visible from the floor if the open or closed position of the valve is not readily seen otherwise.

11-2. STEAM SOURCES OTHER THAN DISTRICT MAINS

Note: For convenience, the sources of low-pressure steam suitable for premises heating are specified in this section. In the later sections, when steam heating systems of various designs are being examined and specified, the source of steam to be included in the heating system specification should be lifted as appropriate from the details given herein. Thus a boiler, an exchanger, or an exhaust steam system can be coupled with any type of premises steam heating arrangement, and a complete specification developed accordingly.

(a) *Cast-iron Heating Boilers*

- Specify boiler to be of the cast-iron sectional header type, designed for a maximum working steam pressure of 15 psig and a temperature of ______ F.
- Capacity to be not less than ______ pounds of steam an hour at a gage pressure of ______ psi, when supplied with water at ______ F temperature.
- Specify that boiler must be constructed in accordance with the ASME "Code for Low Pressure Heating Boilers."
- Specify that safety valve must comply with the ASME requirements.
- Specify feedwater regulator of the ______________ type, together with a low-water alarm to function coincidentally with the low-water automatic shutdown of the firing equipment. Specify that firing equipment must be restarted manually.
- Specify boiler pressure gage, giving type, diameter of dial, ______ in. Also specify the calibrations, as from 0 to (15) psi in steps of ______ pounds each.
- Specify size of water makeup connection, ______ in. Also specify type of pressure-reducing valve on the makeup, and the manually operated stop valve.
- Specify kind and type of boiler blowdown valve, or valves, if a tandem arrangement is desired. Give size.
- Specify the number and size of the boiler steam outlet nozzles.
- Specify how boiler is to be fired (gas) (oil) (coal).
- Specify the calorific value of the fuel, as ______ Btu per pound, per gal, per cu ft, as appropriate.
- Specify the gage ______ of the breeching, and the kind of paint and number of coats to be used on the outside, the inside, or both.
- Specify the kind of barometric damper if installation is for natural draft.
- Specify the kind, size, and placement of a mechanical draft inducer if this is to be provided. Require motor to meet the National Electrical Code, the NEMA standards, and any applicable local codes.
- Specify size of gage glass and water column.
- Specify that boiler is to be jacketed as an enameled

steel package, with ________ in. of ________________ insulation between the cast iron and the inner surfaces of the steel jacket.

- Specify that boiler is to be installed and placed in operation strictly in accordance with the instructions of the manufacturer.
- Require bidders to provide detailed drawings of the boiler, giving dimensions, weight, and locations of all openings.

(b) *Low-pressure Steel Heating Boilers*

- Specify that boiler is to be of the return tubular, firebox type (locomotive), designed for a maximum working steam pressure of (15) psig and temperature of ________ F.
- Specify that boiler is to be delivered and installed as a package complete, with refractory in furnace. Furnace brick to be suitable for a temperature of ________ F when fired with (coal) (oil) (gas).
- Boiler is to be all welded in construction, in accordance with the ASME "Low Pressure Heating Boiler Code."
- Specify insulation to be of ________________ (material), and ________ in. thickness, all enclosed in a steel jacket or casing of ________ gage steel.
- Specify that tubes shall be of ________ gage steel, ________ in. diam, rolled in and beaded.
- Specify that boiler shall be provided with a surface blowdown connection ________ in. in size. Also washout or blowdown connections on each water leg.
- Specify that water column shall provide a glass not less than ________ in. long, and incorporate a low-water cutoff, low-water alarm, and feed pump control for introducing feedwater into the boiler.
- For draft inducer, barometric damper, breeching, and fuel data, refer to Sec. 11-2(a), Cast-iron Heating boilers.
- Require bidders to furnish detail drawings of the boiler offered, showing all dimensions, general arrangement and design, weights, and other pertinent data.
- Specify site of boiler installation, calling attention to any restrictions or difficulties of handling due to tight clearances, obstructions, storage, or other problems.
- Specify that boiler must be installed and lighted off in strict accordance with the manufacturer's instructions.

- Specify that safety relief valve or valves must comply with ASME requirements for such service.

(c) *Exhaust Steam*

Note: A supply of exhaust steam to be used for premises heating may or may not require a pressure reducing valve. If the exhaust steam comes from a low-pressure steam turbine, the need for a reducing valve is unlikely, since the lower the back pressure on the turbine, the higher its efficiency. However, if the turbine is designed for high pressure, with an exhaust at several hundred pounds, one or more stages of pressure reduction are essential before the steam can be used safely in the ordinary premises heating system.

This latter condition is met most frequently when the exhaust steam is needed for a process application primarily, and the process pressure is above that for heating.

A specification for the use of exhaust steam in heating includes the items listed in (d) below.

(d) *Exhaust Steam Piping*

- Provide bidders with a sketch showing the source of the exhaust steam and the projected method of tying the heating steam into this source.
- Specify the material of the piping for the exhaust steam, as steel, wrought iron, etc., and give the schedule number suitable for the pressures and temperatures to be handled.
- Specify the piping joints. Refer to Sec. 11-1(b).
- Specify the fittings as noted in Sec. 11-1(c).
- For the pressure-reducing valves, if needed, refer to Sec. 11-1(d).
- For stop valves, manually operated, see Sec. 11-1(e).
- For automatic control valves, refer to Sec. 11-1(k).
- For safety relief valves to be mounted on the low-pressure piping downstream from the final pressure-reducing valve, see Sec. 11-1(f).
- For trapping of the steam line, see Sec. 11-1(g).
- Insulation, pipe hangers, and instruments are specified in Sec. 11-1(h), (i), and (j), respectively.

(e) *Exchangers for Steam Generation*

Note: Because of the necessity for maintaining a uniform steam supply pressure for the steam heating system, use of the instantaneous flash exchanger (or convertor) must be adopted with caution. The storage-tank type, with the high-temperature water circulating through the coils, allows a storage of heat in the water that is helpful in meeting sudden steam demands.

- Specify the output capacity of the exchanger in pounds of steam an hour, when supplied with feedwater at ________ F, and with high-temperature, high-pressure water passing through the coil at a temperature of ________ F and a pressure of ________ psig, at the rate of ________ gpm.
- Specify steam pressure under the given conditions of generation to be ________ psig.
- Provide seasonal range, of feedwater temperatures from ________ F minimum to ________ F maximum. State if temperature fluctuations from causes other than seasonal are to be expected, giving frequency and extremes of temperature predicted.
- Specify construction of exchanger, giving materials of shell, tube bundle as (copper-nickel) (stainless steel) (Admiralty), etc.
- Require convertor (or exchanger) to meet the ASME "Code for Unfired Pressure Vessels."
- Specify the arrangement of the tube head, and the means for pulling the tube bundle from the tank.
- Specify the insulation on the tank, as ________________ (material) in a thickness of ________ in., and with a finish jacket of ________ oz duck (sewed) (stapled) (pasted) on. Or provide metal jacketing of ________ ________ (material) in ________ gage, joints to be (riveted) (bolted).
- Specify painting, giving pretreatment of surface and number of coats, of ________________ color.
- Specify the size and type of safety relief valve, to meet ASME relief valve requirements.
- Specify point where safety valve discharges to atmosphere as an area of no danger.
- Specify thermometers for placement on high-temperature water into exchanger coil, and on water leaving coil.
- Specify pressure gage to be mounted on steam side of exchanger, giving size, type, dial diameter, and calibration from ________ psig to ________ psig.

- Specify excess-pressure alarm to be set at ________ psig, and indicate if signal is to be visual, auditory, or both.
- Specify if signal is to be operated pneumatically or by electric relay. If the former, give available air pressure, ________ psig. If the latter, require compliance with National Electrical Code, NEMA standards, and local codes as applicable.
- Specify electric service available, ________ volts, ________ phase, ________ cycles.
- Require bidders to submit drawings showing dimensions, weights, and general arrangements.

(f) *Controls for Steam-generating Exchangers*

- Low-water cutoff for the high-temperature supply may be specified by some designers, omitted by others. The decisive factor should be looked for in the effects on the coil, tank, and surrounding materials as the result of being exposed to the full temperature of the hot water medium. If thermal strains or warping seem likely to result, the cutoff is desirable.
- Specify the high-temperature-water admission control valve by type, model, and capacity in ________ gpm. State range of operation from a minimum admission of ________ gpm to a maximum of ________ gpm.
- Specify the control valve actuating forces as the following: (1) Steam pressure in the steam line leading from the exchanger. This to be basic, normal operating control. (2) Excess pressure in the steam line to shut down the admission control valve. (3) A low-water alarm, if provided, to shut the admission valve in response to low-water conditions.
- Specify if control of admission valve is to be accomplished by pneumatic system or electric. Specify details of either system as suggested in the preceeding section, 11-2(e).

(g) *Steam from Hot Gas Heat Recovery Systems*

Note: Sources of hot gas from which heat may be recovered include diesel engines, gas engines, and gas turbines. This is particularly a practice in total-energy installations.

In design, the recovery apparatus usually consists of a heat exchange coil, mounted in the hot gas discharge duct, with a bypass duct and automatic control valves.

Pressure and temperature control of the steam may be accomplished by using the control valves to vary the weight of hot gas flowing through the coil.

In those instances where the amount of steam so generated fluctuates to an extent that produces a steam shortage from time to time, the hot gas may be led through a waste-heat boiler which includes a furnace and direct firing equipment capable of making up the deficit.

Specifications for either alternative should provide bidders with detailed drawings of the duct sizes, the arrangement of the coils or boiler, piping connections, etc. Other information to be furnished bidders is suggested in the following sections.

- Specify the kind, material, and arrangement of the steam-generating coils. Specify the total heat exchange, surface, _______ sq ft.
- Specify the size of the tubes and fins, also the spacing of the fins, as _______ fins to the linear foot of tube.
- Specify the volume of gas passing over the coils, as _______ cfm at a temperature of _______ F.
- Specify the range of hot gas volume fluctuations, as from a minimum of _______ cfm to a maximum of _______ cfm, and a temperature range from _______ F to _______ F.
- Specify the average weight of hot gas per hour passing over the coils as _______ pounds.
- Provide bidders with a proximate analysis of the gas.
- Specify the maximum friction loss acceptable over the coils, as _______ in. W.G.
- Provide bidders with drawing showing kind and dimensions of the duct valves used to divide the total volume of hot gas between the steam-generating coil and the bypass.
- Specify how modulating gas duct valves are to be operated, as pneumatic or by electric motor.
- Require bidders to furnish figures on steam generation at the following rates of hot gas flow over the coils, pounds per hour and pressure, _______ psig:

  1. 100% of gas flowing over coils.
  2. 75% of gas flowing over coils, 25% bypass.
  3. 50% of gas flowing over coils, 50% bypass.

  Any other ratios desired.

- Specify range of water temperatures supplied to the coils, from ______ F minimum to ______ F maximum.
- Provide an analysis of the water, showing hardness in terms of calcium carbonate, presence of sulfur, mud, organic matter, or contamination of any kind.
- Specify means desired for cleaning coils' external surfaces of soot, as by compressed air or steam soot blowing.
  - Require bidders to furnish information on frequency of blowing, pressure of air or steam necessary, ______ psig, efficiency of cleaning, ______ per cent.
- Specify if gas flow across coil is to be controlled by means of a pressure gage and relay on the steam header for steam leaving the coil. Indicate also if this gage opens, closes, or modulates the feedwater admission valve concurrently, or if another source of monitoring is to be provided.
- Specify the kind, type, and size of the feedwater admission valve.
- Specify kind, size, and number of coil drain valves.
- Provide duct drawings to give all dimensions, thickness of gage metal, kind of joints, method of making joints leak-tight, duct supports, and hangers.
- Specify painting, preparation of surfaces, kind and number of coats.
- Specify duct insulation to be of ______________ (material), ______ in. thick, with (abutting) (lap) (scarf) joints. Finish jacket to be ______ oz duck, (sewed) (pasted) (stapled) on. Or of metal of ______ gage with ______________ joints.
- Specify steam piping insulation as above, with preformed insulation for valves and fittings. Or other.
- Specify steam stop valves, manually operated, as noted in Sec. 11-1(e).
- Specify fittings as suggested in Sec. 11-1(c).
- Specify steam piping, Schedule No. ______, etc., as given in Sec. 11-1(b).

(h) ***Steam from Hot Gas Heat Recovery in a Waste-heat Boiler***

Note: The design and production of a waste-heat boiler is a more or less established practice with several of the large steam generator builders.

Included in the key information submitted by prospective purchasers of these units is the decision to order an auxiliary-fired or nonfired boiler. Specifying data in the following sections will consider both of these possibilities.

- Specify the capacity of the waste-heat boiler, as _______ pounds of steam an hour, at _______ psig pressure, evaporated from water at _______ F temperature, when supplied with hot gas at _______ F temperature in the amount of _______ pounds per hour (or cfm).
- Specify type of boiler as (water tube)(bent tube) (straight tube), _______ drum. Or fire tube, _______ pass unit, with dry pipe.
- Specify the velocity of the gas through the boiler, _______ fpm.
- Specify maximum acceptable draft loss through boiler, _______ in. W.G., if boiler is to operate on natural draft.
- If an induced draft fan is to be used, specify as itemized in Sec. 4-1(w), including the chart to be filled in by bidders.
- Specify if water circulation through the boiler is to be by natural convection or by forced circulation. (In many instances, forced circulation will be preferred.)
- For forced circulation, specify kind and type of pump and indicate its point of suction and point of discharge relative to the tube arrangement.
- Specify the pump capacity as _______ gpm when handling water at a temperature of _______ F and operating at a total dynamic head of _______ ft.
- Specify pump motor to meet requirements of National Electrical Code, NEMA standards, and local codes.
- State electric service available as _______ volts, _______ phase, _______ cycles.
- Specify the control for the pump, whether this is to be manual or automatically cut in and out as a result of the steam demand on the boiler. Example: Heavy demand, pump operates continually; light demand, natural convection may suffice, and pump remains still.
- Specify if special baffling or tubing arrangements are needed to avoid excessive slagging or fouling of the boiler passes in water-tube boilers.
- Specify how tube surfaces are to be cleaned, by compressed air or steam. Bidders shall state the pressure needed for each medium, _______ psig. Also the vol-

ume of steam or air needed, _______ psig, or the weight, _______ pounds, as appropriate.

- State range of feedwater temperatures, from a minimum of _______ F to _______ F maximum.
- For fire-tube boilers, specify mounting as horizontal, vertical, or inclined.
- Specify if an external steam drum is to be provided, and if so, state size and capacity desired.
- Specify that fire-tube boiler is to be of welded construction, with welding done by certified welders. Drums of water-tube boilers likewise to be welded by certified welders.
- Specify casing for the water-tube boiler, stipulating metal thickness as _______ gage, backed up by _______ in. of _____________ insulation.
- External surface of fire-tube boiler to be insulated by _______ in. of _____________ and jacketed by a _______ gage metal casing of _____________, and with joints bolted, screwed, etc.
- Specify painting and preparation of metal surfaces, and give number of coats and color.
- Specify if boiler foundation and settings are to be provided by the contractor or the owner.
- Specify safety relief valves in accordance with the ASME requirements, with discharge piping being carried to a safe terminal.
- Specify boiler blowdown system as necessary for both the water-tube and fire-tube units. For details of the blowdown valves, tanks, etc., refer to Sec. 4-6.
- For feedwater systems applicable to waste-heat boilers, refer to Sec. 4-12, modifying the provisions of that section as necessary to suit this waste-heat specification.

(i) *Waste-heat Boiler Deficit Firing for Steam Generation*

Note: Provision of auxiliary firing to make up for a steam shortage resulting from insufficient hot gas volumes entering the boiler passes, involves all the equipment expected to be needed for full-load firing.

Stokers, oil burners, gas burners, fuel storage and transfer facilities must be provided for the waste-heat boiler to meet peak demands. The main difference perhaps, will be in the amount of fuel storage necessary; this should be less for the auxiliary installation.

Therefore, in writing specifications for an auxiliary-fired

waste-heat boiler, reference is made to the direct-fired generators in preceding sections as follows:

- Specify coal handling and storage from Sec. 4-7.
- Specify fuel-oil handling and storage from Sec. 4-9.
- Specify gas fuel systems from Sec. 4-10.
- Specify ash-handling systems from Sec. 4-8.

- Specify auxiliary firing control system with diagram to show the interrelationships of the hot waste gas, steam pressure, the stop, start, and rate of firing of the firing equipment.
- Specify if these interacting components are to function by means of pneumatic controls or electric controls. Provide details of points desired in either system.
- Specify limit controls, as excess-temperature bulb in the breeching to shut down the firing units; also low-water bulb in the boiler to shut down the unit similarly.
- Specify the kind of automatic boiler feeder to be furnished.
- Specify all necessary gages and thermometers, giving kinds, the type of sensitive element wanted, the diameters of dial instruments, the steps of calibration.
- In connection with the unit shutdown as the result of excessive temperatures, excessive pressures, or low water, specify the kind of alarm desired, stating if the signal is to be visual, audible, or both. Specify that firing equipment after being shut down by emergency action must be restarted by hand.
- Specify that all waste-heat boilers, fired or nonfired, must be constructed in accordance with the ASME Boiler Code covering the particular pressures planned.

(j) *Superheated Steam from a Waste-heat Boiler*

- Specify degrees _______ F of superheat wanted, when superheater is handling _______ pounds of steam an hour, this steam entering the superheater at a pressure of _______ psig and a temperature of _______ F.
- Refer to Sec. 4-1(e).

(k) *Information by Bidders*

- Require bidders to furnish detail drawings showing dimensions, arrangements, and all pertinent data of the waste-heat boiler.
- Require bidders to furnish operating data as in Table 11-1. Specification writers are to add whatever other

Table 11-1. Waste-heat Boiler Performance

Condition I Unfired

Performance item	Maximum gas flow	75% gas flow
Hot gas per hour	Lb	Lb
Steam output, specified pressure and temperature, natural convection	Lb	Lb
Efficiency	%	%
Steam output, forced circulation	Lb	Lb
Efficiency	%	%
Draft loss	In. W.G.	In. W.G.
Gas temperature, leaving boiler	F	F

Condition II Auxiliary Fired

	Maximum gas flow	75% gas flow	50% gas flow*
Hot gas per hour	Lb	Lb	Lb
Fuel burned per hour	Lb	Lb	Lb
Steam output, specified pressure and temperature, natural convection	Lb	Lb	Lb
Gas per hour from fuel	Lb	Lb	Lb
Temperature of gas from fuel	F	F	F
Total weight of gas per hour through boiler	Lb	Lb	Lb
Temperature of total gas through boiler	F	F	F
Efficiency	%	%	%
Draft loss, total gas condition	In. W.G.	In. W.G.	In. W.G.
Steam output, specified conditions, forced circulation	Lb	Lb	Lb
Efficiency	%	%	%

* The 50 per cent hot gas flow is included with the auxiliary firing data since the auxiliary firing in combination with a low hot-gas supply might represent a practical operating condition. A 50 per cent hot gas flow would be of doubtful economy if not supplemented by heat from another source, in view of the resulting slow movement through the waste-heat boiler and the low efficiency of heat exchange under those conditions.

information they desire to this chart, for response by the bidders.

11-3. PREMISES HEATING EQUIPMENT

(a) *Steam Generation and Source Equipment*

- Refer to Secs. 11-1 and 11-2.

(b) *Steam Piping*

- For both one-pipe and two-pipe steam heating systems, specify pipe material as (steel) (wrought iron) (brass) (alloy), etc. to suit.
- Specify working pressures as _______ psig, and give appropriate Schedule No. _______.
- Specify pipe joints, as (screwed) (flanged) (welded) (soldered for copper tubing) (brazed), etc.
  - For screwed joints, specify thread and pipe "dope" to be used on threads. *Note:* Some dopes may leak out in time and stain the surroundings.
  - For welded, soldered, or brazed joints, specify the analysis of the rod or solder.
  - For flanged joints, specify material of flanges and type, as cast-iron screwed companion flanges. Specify gasket material to be used between flange faces. Note if flanges are to have raised faces, grooved faces, etc.
- Specify pitch of steam piping if not shown on the drawings, giving both direction and amount.
- Specify size, kind, and spacing of pipe hangers.

(c) *Steam Valves, Manual*

- Specify materials of bodies, seats, disks, gates, etc.; also valve ends, as screwed, flanged, welding, etc.
- Specify kind of valves, as gate, globe, rising or nonrising stems. If bypass is required, specify size of bypass, _______ in., and size of bypass valve. Specify bypass valves similarly as to material, ends, etc.
- Specify working pressures, _______ psig, and maximum temperatures, _______ F, for each valve.

(d) *Steam Valves, Motor-operated*

- Valve specification same as for the hand-operated units, except for modification needed by the pneumatic or electric operator.
- Specify the pneumatic or electric valve operating system desired, giving available air pressure, _______ psig, and electrical characteristics of electric service as _______ volts, _______ phase, _______ cycles.

- Specify if valve operator shall provide modulating movement of the valve (sometimes desirable for zone control and air duct installations) or shall take either a wide-open or closed position. If the latter, specify time elapsing between open and close, or vice versa, _______ seconds.

(e) *Steam Line Traps*

- Specify the kind, materials, and capacities, in _______ pounds of condensate per hour, of each line trap.
- Specify the pressure differential over the trap orifice, as _______ psig on the steam leg, _______ psig on the discharge side. Specify inlet steam temperature, _______ F.
- Specify feet of lift on discharge side.
- Specify valves for isolating the trap, kind, material, sizes, etc.

(f) *Steam Radiators, Cast-iron, Column Type*

- If not already appearing on the drawings, prepare a radiator schedule showing for each size the number of columns, number of sections, height, _______ in., and square feet of heating surface.
- Specify color and finish.
- Specify steam pressure, _______ psig, and temperature, _______ F.

(g) *Cabinet Convectors*

- As for cast-iron radiators, prepare a schedule of sizes, ratings, etc., unless such schedule appears on the drawings.
- Specify the kind of cabinet, whether freestanding, front or top discharge, slanting top or flattop, wall hung, etc.
- Specify cabinet metal as _______ gage; also color and finish.
- Specify the steam pressure, _______ psig, and temperature, _______ F.

(h) *Baseboard, Cast-iron*

- Specify height of cast-iron sections, _______ in.
- Specify placement with asbestos paper between the baseboard and the wall as protection for the wall against the formation of dust streaks.
- If allowed by the construction, insulate the wall behind the baseboard with _______ in. of _______________ insulation, and to a height of _______ in. minimum. *Note:* This is not necessary if the entire wall is insulated.

- Specify the heat output as _______ Btu per hour per linear foot of baseboard.
- Prepare a schedule of baseboard, showing the different heights, different thicknesses, and different heat outputs of the required sections, if this is not already on the drawings.
- Specify the color and finish of the baseboard.
- Specify the basis on which the baseboard capacity is rated, as I = B = R standards, or by an entering air temperature at the bottom of _______ F and a room temperature of _______ F measured _______ ft above the floor.
- Specify steam pressure, _______ psig, and temperature, _______ F.

(i) *Baseboard, Finned Tubing*

- If not on the drawings, prepare a schedule of the finned tube to be furnished, showing capacities, height, etc.
- Specify finned tube elements are to be assembled of annealed copper tubing, _______ in. diam, and bearing aluminum fins _______ to the ft; each fin to be _______ in. wide by _______ in. high.
- Specify how fins are to be attached to tubing, and specify working pressure, _______ psig.
- Specify the number of tiers high, and specify the thickness of the metal cover, _______ gage. Specify the means provided to compensate for expansion and contraction, as bellows-type expansion joint placed every _______ lin. ft.
- Specify how tubing elements are to be supported, by backplate brackets set _______ ft apart, etc.
- Specify dummy sections if these are needed to balance out the installation.
- Specify the pressure, _______ psig, and temperature, _______ F, of the steam circulating through the baseboard.
- Specify the basis on which the finned tubing is rated.
- Specify that output capacities are to be given in Btu per hour per linear foot.

(j) *Steam Unit Heaters*

- Provide bidders with a schedule, either in the specification or on the drawing, of the unit heaters required, stating the number of each, type of each, direction of blow and output capacity in _______ Btu per hour when supplied with steam at _______ psig and _______ F total temperature.

- Specify that ratings of unit heaters must comply with the standard test codes of the Air Moving and Conditioning Association.
- If area of application is favorable, specify that heater is to be fitted with adjustable outlet louvers.
- Specify that electric fan motors must be suitable for continuous operation and must comply with the NEMA standards, the National Electrical Code, and such local codes as may apply.
- Specify how unit heaters are to be suspended.
- Specify finish and color of the unit heater casing.
- Specify volume of discharge from each heater, _______ cfm, and at _______ rpm.

(k) *Radiator Air Valves, Automatic*

- Specify kind of radiator air valves, as adjustable port, simple venting, bottom or side inlet, etc.
- Specify that air vent valves, as selected, shall be placed on each radiator, cabinet convector, baseboard, and unit heater, in locations indicated by the radiation manufacturer.

(l) *Thermostatic Radiator Traps*

Note: These traps are not needed by one-pipe systems.

- For two-pipe radiators, convectors, and baseboard installations, specify thermostatic radiator traps, functioning by means of a nonferrous bellows and needle valve.
- Specify the capacity of the trap in equivalent feet of direct radiation.
- Specify the steam pressure, _______ psig, and the temperature, _______ F, on which the trap will operate.

Note: For steam pressures 25 psig or greater, it is customary to rate the trap capacity in condensate pounds per hour.

- Specify number of each size of thermostatic trap.

(m) *Steam Coils in Ducts*

- Specify that coils shall be capable of raising the temperature to _______ pounds of air an hour, _______ F, when the air is moving over the coils at a velocity of _______ fpm, and entering the coils at a temperature of _______ F.
- Specify the steam conditions entering the coils as

_______ psig pressure and _______ F total temperature.

- Specify construction and arrangement of coils, as seamless copper tubing, _______ in. OD, with (aluminum) (copper) fins mechanically bonded to the tubing. Specify fin spacing as _______ fins per lin ft, and give dimensions of fins as _______ in. by _______ in.
- Specify coil casing as steel, _______ gage, protected against corrosion by (galvanizing) (other).
- Specify test pressure as _______ psig and maximum working pressure as _______ psig at _______ F.
- Specify construction of tubes headers, as (cast iron) (steel) with the tubes rolled in.
- Specify rate of pitch to be given coils, so that condensate is sure to drain by gravity and will not lie in tubes to freeze.
- Specify that flow of steam and drainage of condensate are to be in the same direction.
- Require bidders to submit dimension and construction drawings.
- Require bidders to submit a table of conversion factors applying to the combinations of various entering air temperatures and various steam pressures.
- Require also a table or chart of face areas, sq ft, for tube lengths and tube arrangement in the casing. Bidders to furnish also data on air friction loss going through the coils, in in. of water.
- Require bidders to provide full information on the kind of air vent being supplied, the method of relieving a vacuum in the coil (as by a check valve designed to open at _______ in. of vacuum) and data on the kind and capacity of the condensate trap.

(n) *Condensate Receiver*

- Specify capacity of receiver, _______ gal. If filled by a gravity return, or otherwise subject to collections of air, specify an automatic vent valve.
- Provide drawing showing location of receiver in the return system, and relative to the condensate (boiler-feed) pump.
- Specify if pump is to be operated by means of a float in the receiver, permanent magnet controller, etc.
- Specify construction of receiver, as steel, _______ gage, welded, etc.

- Specify how makeup water is introduced into the receiver, and method of blending makeup with condensate to avoid cold spots, as perforated distributor pipe, etc.

(o) *Condensate Return Pump*

Note: Pump may be specified with the receiver as part of a proprietary return system, or it may be provided as an independent purchase and mounted where convenient to both the receiver and the boiler. Either way, the pump is specified as:

- Centrifugal pump with a capacity of _______ gpm when pumping water of _______ F against a head of _______ ft. Alternatively, specify a turbine pump.
- Specify that casing is to be of _____________ (material) with approved shaft seals of _____________.
- Bearings to be (ball) (roller) (babbitt), etc.
- Specify lubrication system desired, or alternatively, permanently sealed bearings.
- Specify speed of pump, _______ rpm.
- Specify driving motor, through a (flexible) coupling, motor and pump to be set on a common bedplate.
- Motor to operate on _______ volts, _______ phase, _______ cycle current.
- Require motor, starting equipment, and electrical wiring to meet requirements of National Electrical Code, the NEMA standards, and local codes.
- Require bidders to furnish drawings showing arrangement of impellers, shaft, thrust bearing, line bearings, seals, etc.

(p) *Condensate Return Vacuum Pump*

Note: The vacuum pump can be used only with two-pipe systems, and requires a thermostatic trap on each radiator. There must be no direct connection between the steam supply side of the system and the return piping. As a general rule, the condensate should flow back to the pump by gravity; in some instances where this is not possible, a lift fitting may be used, or a prereceiver tank may be placed in a pit, the vacuum pump taking its suction from this prereceiver. Whatever means of condensate elevation is chosen, the maximum practical lift is about 5 ft, and the reduction in the system vacuum will approximate 1 in. Hg for every foot of lift. In operation, the vacuum pump separates the air from the condensate and discharges this air to the ambient atmosphere, the remaining condensate then being pumped into the boiler.

- Specify capacity of pump in terms of vacuum, _______ in. Hg; or in terms of equivalent direct steam radiation.
- Specify how pump is to be coupled to its driving motor, as by a flexible coupling or by a belt. If the latter, specify if belt is to be flat, V, or other form.
- Specify the elevation above sea level, and the discharge pressure, _______ psig, of the condensate.
- State if this installation is to be a single pump or of the duplex type, with two separate pumps that can operate either singly or simultaneously.
- Specify if operation is to be continuous or intermittent. If the latter, state how operation is to be started and stopped, as by vacuum or by condensate level.
- Specify operating arrangement of the vacuum pump, as water and air impellers to have (common) (separate) motor drive.
- Specify materials of construction, as casing of _______ _______; shaft of _______________; seals (mechanical) (stuffing box); impellers, bronze or _______________, etc.
- Require bidders to submit detail drawings showing the internal arrangements and dimensions of the pumps; also weights, speed _______ rpm.
- Specify all motors and electrical materials must satisfy the National Electrical Code, the NEMA standards, and local codes.
- Specify the electric service available, _______ volts, _______ phase, _______ cycles.
- Require bidders to report motor horsepower, kind of motor, model, normal full-load amperage, locked rotor amperage, etc.
- Specify the range over which the pump is to operate, as starting from a minimum vacuum of _______ in. Hg and stopping at _______ in. Hg, maximum.
- Specify if a float is to be provided that will operate the pump from a high condensate level of _______ in. of condensate to a low of _______ in. of condensate, this float-initiated action to occur independently of the vacuum on the system.
- If it is desired to operate over the night hours strictly as a condensate pump without establishing any speci-

fied vacuum, require bidders to furnish a night switch with adjustment by hours, that will automatically keep the pump on float control during the night, and automatically return it to vacuum control during the day. Night switch to be time-clock directed, with settings for Sundays and holidays as well as for nights.

(q) *Boiler Return Trap or Alternating Receiver*

- Used with a high condensate return, this unit alternately accepts condensate from the heating system, and discharges the condensate into the boiler by means of the boiler steam pressure. The normal application is on a gravity return system where a low or floor-level return pipe is not practical.
- Specify the capacity of the trap as _______ pounds of condensate an hour, with steam boiler pressure at _______ psig.
- Specify the size of the trap, _______ in., the normal operating pressure, _______ psig, and the test pressure, _______ psig.
- Specify the kind and size of the automatic vent valve for discharging air to the atmosphere.
- Provide a drawing or sketch showing the location of this valve with reference to the boiler steam header and condensate return connections.

(r) *Hartford Connection*

- Specify the size of the equalizing line from the steam header on the boiler to the condensate return main, as _______ in., pipe.
- Specify the distance below the water level of the boiler at which the returning condensate is introduced into the boiler, _______ in. *Note:* 2 in. is common for a straight gravity return, and up to 4 in. for a pumped return.
- Specify that condensate return main must be increased in size to equal that of the equalizer pipe at the foot of the return riser which forms the return leg of the loop.

(s) *Piping Insulation*

- For steam heating piping, specify insulation shall be of _______________ (material) and _______ in. thick.
- Specify that insulation joints shall be (abutting) (lapped) (scarfed), etc.
- Specify that k factor shall not exceed _______ Btu per hour per degree F of temperature difference.

- Specify that all insulation joints are to be sealed with an approved mastic or cement, as recommended by the manufacturer of the insulation.
- Specify that all insulation for pipe, fittings, and valves shall be of the premolded type, dimensioned and shaped to fit the various sizes of pipe, valves, and fittings it is to cover.
- Specify painting, color, number of coats, and surface preparation.
- Specify that condensate piping is to be insulated similarly, but with the thickness to be ________ in. because of the lesser temperature difference across the condensate piping.
- If piping is to be jacketed over the insulation, specify kind of jacket, as ________ oz canvas jacket, cemented on, and using ________________ cement. Alternatively, specify an aluminum jacket, preformed to size, and snapped on around the insulation.
- Specify that all fittings and valves are to be insulated (and jacketed if desired) neatly. Careless, sloppy applications will not be accepted.

(t) *Pipe Sleeves*

- Specify that pipe sleeves, wherever the piping passes through the walls, shall be of standard (iron)(brass), etc., pipe, sized two sizes larger than the steam or condensate piping, or as necessary to accommodate the overall thickness of piping and insulation. Each sleeve shall be of the same length as the thickness of the finished wall.
- Specify that space around pipe inside the sleeve shall be packed with oakum, or with ________________ (material) prior to installing the plate covers.
- Specify that contractor must inform other trades of the exact location for each sleeve, and furnish the sleeves to these other trades as appropriate, for placement.
- Specify that this contractor is responsible for coordinating his schedule with that of the trades which will place the sleeves.

(u) *Plate Covers*

- Specify that plate covers shall be of the split type, stamped steel, chrome plate, and sized to fit the various sizes of pipes where these pass through a wall.

- Alternatively, specify other materials and other finishes if desired.

(v) *Comfort Controls of Steam Systems*

- For pressure gages and thermometers, refer to Sec. 11-1(j).
- For steam inlet control on district steam, see Sec. 11-1(k).
- For room thermostats, specify the thermostatic element, as bimetallic, liquid filled, etc.
- Specify dial range as from ______ F to ______ F.
- Specify degree of sensitivity as one-half of 1 degree F.
- Specify thermometer range on the thermostat.
- Specify what thermostat is to control, as a zone valve motor, radiator valves, condensate pumps, boiler firing equipment, steam admission valve on a heat exchanger type of steam generator, etc. Note if control of these components is to be singly, or if several are controlled simultaneously.
- Specify if locking cover is to be provided.
- Specify if thermostat is to provide both night and day setting, either manually set, or automatically controlled by electric, pneumatic, or electronic means.
- Specify operating and available voltage, ______ volts, ______ phase, ______ cycle, (two) (three) wire service.
- Specify kind of switch, as mercury, single pole, double throw, etc.
- Specify finish, as silver, bronze, etc.
- For remote bulb control, specify length of bulb, ______ ft, and length of tubing, ______ ft.
- Specify tubing and bulb material, as copper, monel, stainless steel, etc.
- If electric, specify number of potentiometers desired.
- For outdoor thermostats, specify sensitivity to sun and wind conditions as well as to temperature. This thermostat anticipates the need to turn heat on and off before the indoor thermostat acts, and thus prevents periods of overheating or underheating by decreasing or increasing the heat storage in the structure as required.
- For electronic thermostats with central panel control, specify resistance elements, indicators, lights, push-button switches, and adjustment potentiometers.

Note: While electronic installations may be used on single-room operations, the nature of electronic control permits it to be used effectively for multizone and remote-control problems. The ultimate moving unit, the zone valve for example, operates by means of a pneumatic or electric motor, the electronic control merely dispatching the signal to the motor. In this respect, the ultimate action is the same with all kinds of monitoring methods. The resistance element reflects the changes in temperature in place of the bimetallic or liquid-filled bulb used in more familiar arrangements.

(w) ***Time Controls***

- Specify the program on which the control is to work. Example: 24 hours, 7 days, with omission of any day desired, to function in reaction to the rising and setting of the sun, etc.
- Specify if control is to be mounted outdoors or indoors. For the former, specify weather protection to be provided in control or to be supplied by purchaser.
- Specify if spring control is to be provided to maintain operations automatically during periods of power failure.
- Specify voltage operation, rated amperes per pole, number of poles.
- Specify color painting of case, and require case to be fitted with an approved lock.
- Require bidders to furnish drawings and photographs as necessary to illustrate the design and construction of the controller.
- Require bidders for all forms of electrical or electronic controls to furnish complete wiring diagrams.
- Require all materials and installing practices to satisfy the National Electrical Code and local codes.

DIVISION 12

Premises Heating, Hydronic

12-1. LOW-PRESSURE CAST-IRON BOILERS

(a) *Design and Construction*

- Specify that boiler must be designed and constructed in accordance with the ASME "Code for Low Pressure Heating Boilers."
- Specify capacity of boiler as _______ Btu per hour, based on the I = B = R method for rating cast-iron boilers.
- Specify that boiler is to be delivered complete (if access to boiler room is large enough); otherwise specify that sections are to be brought in separately and assembled and jacketed on the site.
- Specify jacket to be of _______ gage steel, with _______ in. of _______________ (material) insulation, and finished with baked-on enamel.
- Specify firing equipment, as oil, coal, or gas.
- Specify the calorific value of the fuel as _______ Btu per pound (or gal) for oil, _______ Btu per pound for coal, and _______ Btu per cu ft for gas.
- Specify that fuel-oil pump shall have a capacity of _______ gal per hour, and shall be of the (single) (dual) stage design.
- Specify a solenoid-operated, delayed oil supply valve, stipulating time, _______ seconds, elapsing from energizing of oil burner until oil is admitted through this valve.
- Specify an automatic burner shutdown control, to function in the event of ignition failure.
- Specify if burner is to be gun type or rotary.
- State if gas fuel is natural or artificial; if natural, identify field.
- Specify the kind of gas burner desired.
- Specify gas pressure to burner, _______ psig.

- Specify if burner is to be of the atmospheric or power burner type.
- Require bidders on power burners to provide drawings and complete information on the control of these burners.
- For coal fuel, specify the kind of fuel, as (bituminous) (semibituminous) (anthracite). Specify the percentage of ash and sulfur.
- Specify the kind of coal-firing equipment to be provided, as hand grates, mechanical stoker, pneumatic stokers.
- If automatic ash removal is desired, so specify.
- Require bidders to furnish full information on the kind of stoker, model, and capacity in pounds of coal per hour.
- In asking for bids, specify the height and flue diameter of the stack or chimney.
- If either a forced or induced draft fan is to be used, so state. Specify the capacity of the fan, _______ cfm, against a friction loss of _______ in. W.G.
- Require bidder to state draft loss through the grate, stoker and boiler passes at full load, as _______ in. W.G.

(b) *Boiler Appurtenances*

- If local authorities or state authorities require the provision of a low-water cutoff, specify the cutoff.
- Specify the water column, size and length of glass.
- Specify low-water alarm (optional), indicating whether signal is to be visual, auditory, or both.
- Specify that safety relief valve is to comply with all requirements of the ASME low pressure heating boiler code. Carry discharge to safe place.
- Provide a drain valve for the boiler.
- On makeup water supply, specify the kind, type, and model of pressure-reducing valve on the supply to bring pressure below the 30 psig working pressure of the boiler.
- Specify the kind and material of the hand-operated makeup water inlet valve.
- Specify a stack switch control to shut down firing equipment in the occurrence of excessive stack temperatures. Specify switch setting, _______ F.
- Specify a submerged bulb thermostat to control the water temperature in the boiler. Set the _______ F tem-

perature at which the firing equipment starts and the temperature, _______ F, at which the firing equipment stops.

- Specify the kind and model of pressure or altitude gage to be furnished with the boiler. Also the kind and model of the water temperature thermometer. Specify the calibration of both gage and thermometer.
- Specify that boiler shall be lighted off and tested in strict accordance with the manufacturer's recommendations, and in the presence of the owner or his assigned deputy.
- Specify _______ gage steel in breeching, and painting.

12-2. LOW-PRESSURE STEEL BOILERS

(a) *Note:* The steel boiler for hot water is of approximately the same construction as the steel boiler for low-pressure steam heating.

For specifying this kind of boiler for a hydronic system, refer to Sec. 11-2(b), and make these changes:

- Maximum working pressure for water, 30 psig.
- Eliminate the surface blowdown connection, but include the washdown openings on each water leg.
- Specify a low-water cutoff control only if the area authorities require it.
- Eliminate the boiler-feed pump reference, since the circulator on the hydronic system performs this duty.
- Also specify that the rating of the steel boiler shall be in accordance with the Steel Boiler Institute Division of the I = B = R.
- Specify the boiler capacity, _______ Btu per hour.
- Specify the characteristics of the fuel in calorific value, Btu per pound, per gal, or per cu ft as appropriate.
- Specify firing equipment as detailed in Sec. 12-1(a).

(b) ***Steel Boiler Appurtenances***

- Specify as appropriate from Sec. 12-1(b), which lists appurtenances for a cast-iron low-pressure heating boiler.

12-3. HOT WATER FROM STEAM CONVERTERS

(a) ***Converter Operating Data***

- Specify the converter output, in gallons of water per hour at a temperature of _______ F.
- Specify the water-entering temperature, _______ F.

- Provide data on the range of water temperatures, from _______ F to _______ F inlet, and indicate expected output in _______ gpm at _______ F over this range.
- Specify the steam supply, as _______ psig and _______ F total temperature.

(b) *Converter Construction*

- Specify form of exchange surface, as ____________ (material) tubes arranged in a U-shaped tube bundle that may be withdrawn for inspection, cleaning, repair, etc.
- Specify material of tube sheet as steel, alloy, or other.
- Specify materials of shell and water box, as welded steel, _______ gage, and close-grained cast-iron, respectively. Other materials may be specified.
- Specify the working design pressure of the unit as _______ psig and the hydrostatic test pressure of _______ psig.
- Specify if converter is to be set vertically or horizontally, and if mounted on supports or suspended.
- Specify the number of passes, as 2, 4, 6.
- Specify design and construction to comply with ASME "Code for Unfired Pressure Vessels."
- Note if ASME inspector's stamp is to be placed on the converter. This usually involves an extra charge.
- Require bidders to furnish drawings showing the dimensions and arrangements of the converters, position of piping connections on water box, and the location and size of the condensate drain opening.
- Condensate trap may be specified from Sec. 11-1(g).

(c) *Control of Water Temperatures from Converters*

- Specify a steam regulator valve on the steam admission line to the converter, this valve to be actuated by a thermostatic bulb submerged in the hot water outlet piping. Give type and model.
- Specify size of steam valve, _______ in., and operating range up to _______ pounds of steam per hour with a steam pressure entering the valve of _______ psig.
- Provide drawing showing location of the steam regulator and submerged thermostatic bulb in relation to the converter.

(d) *Instruments*

- Specify the number of pressure gages. Give range and steps of calibration for each. Also kind of sensitive element. Stipulate the diameter and face of the dial, as silver finish, etc.

- Specify the number of thermometers. Give calibration from ________ F to ________ F. Specify whether straight or angular stem. Specify case and finish.
- If dial type is wanted, specify diameter of dial and length of bulb, ________ in.
- Specify accuracies to be guaranteed for both gages and thermometers, as ________ per cent plus or minus, deg, oz, etc.

(e) ***Compression Tank with Converter System***

- For specifying this tank, refer to Sec. 12-5(b).

12-4. GRAVITY SYSTEMS

Note: The circulation of hot water through pipes by gravity, or the variation in water densities as the result of differences in temperature, is presently obsolete as a primary heating system. It may still be used to advantage in special applications. For example, by mounting a water back or pipe coil in an open fireplace, and then connecting this to a combination storage and expansion tank in the attic, natural circulation will fill the tank with hot water. This hot water may be used for lavatories, for room heating, or otherwise as desired, to the extent of its availability.

(a) ***Piping Characteristics***

- Since the only circulating force is the difference in water temperature, very large pipes must be specified in order to reduce friction loss to the economic minimum.
- Specify the size of the pipe, ________ in., and require that its inside surfaces be clean and smooth.
- Specify the size of the hot water tank, in ________ gal capacity.
- Specify a pressure relief valve, to be set at ________ psig, and with a test lever for checking its operating condition.
- Specify discharge pipe from relief valve to be carried to a point of safety outside the building.
- Specify working pressure of the pipe, tank, and fittings as ________ psig.
- Specify the schedule number of the pipe and the material thereof, as wrought iron, black iron, brass, etc.
- Specify a manually operated vent valve to be mounted on the tank. Also a drain valve and an air charging valve.

- Specify insulation on the tank to be ______ in. of ____________ with joints sealed or cemented as recommended by the insulation manufacturer.
- Specify where tank is to be mounted, or suspended, to suit conditions.

12-5. ONE-PIPE FORCED CIRCULATION

(a) *Series Arrangement*

- Specify only for a small installation, rarely with more than three or four rooms to be heated. In this arrangement, all the hot water is forced through all the radiation, in series. Since the temperature of the circulating water is lowered in every radiator through which it passes, either more heating surface is needed toward the end of the loop, or some form of damper control must be provided over the first radiators on the loop to prevent excessive heat loss and overheating in the beginning of the circuit. Reduced heating surface at the beginning may be suitable as long as the water temperature is held at the peak level, but this reduced surface may prove inadequate at the end of the heating season, when tempering rather than full comfort heating is needed. Dampering therefore may be preferred.
- Specify the size and schedule number of the circulation piping.
- Specify the material, as (steel) (wrought iron) (brass) (copper tubing).
- Specify the working pressure, ______ psig, and the maximum water temperature, ______ F.
- Specify the radiation as follows:
  For cast-iron, column radiation, see Sec. 11-3(f).
  For cabinet convectors, see Sec. 11-3(g).
  For baseboard, cast iron, see Sec. 11-3(h).
  For baseboard, finned tubing, see Sec. 11-3(i).
  For unit heaters, see Sec. 11-3(j).
- Specify vent valves for each unit of radiation or high points in the circuit as indicated. Vent to be opened and closed manually.

(b) *Compression Tank*

- Specify the capacity as ______ gal.
- Require compression tank to meet specifications of the ASME "Code for Unfired Pressure Vessels."

- Specify that compression tank shall be fitted with an air charging valve and a drain valve.
- Specify that tank shall be mounted as shown on the drawing provided with the specification, in relation to the boiler and the circulating pump.
- Specify the provision of a flow-control valve, to be mounted in the circuit as shown, and to prevent natural circulation when the pump is not running.

(c) *Circulating Pump*

- Specify circulator to be of the ______________ type, driven by an electric motor of ________ hp, through a pin and disk, spring, or other type of coupling.
- Pump to have a capacity of ________ gpm when circulating water of ________ F temperature against a total head of ________ ft.
- Require bidders to submit a pump characteristic curve showing the head versus capacity and horsepower relationships.
- Specify that electrical material, design, and installation must meet the National Electrical Code, the NEMA standards, and local codes.
- Specify position of the pump, as on the return side of the boiler in (horizontal) (vertical) return piping; or on the boiler outlet side in (horizontal) (vertical) supply piping.
- Specify control for the circulator, as from a room thermostat, or from a temperature bulb submerged in the boiler.
- Specify room temperature, ________ F, or boiler water temperature, ________ F, at which pump starts, and similarly, temperatures at which it is stopped.
- For specifying the thermostat, refer to Sec. 11-3(v).
- Specify who provides foundation if floor mounted.

(d) *Valves and Fittings*

- Specify valves to be (gate) (globe), (rising) (nonrising) stem, with ______________ (material) body and kind of seat and disk.
- Specify ends, as (screwed) (soldered), etc.
- Specify working pressures, ________ psig.
- Specify fittings similarly.

(e) *Insulation*

- Specify that piping, valves, and fittings are to be heat insulated with ______________ (material), ________ in.

thick, with joints sealed or cemented as recommended by insulation manufacturer.

- Specify a jacket of _______ oz duck, pasted or sewed on. Alternatively, specify snap-on aluminum jacketing.
- Specify painting, color, coats, preparation of surface.

(f) ***Pipe Sleeves and Covers***

- Specify pipe sleeves where piping penetrates wall. Refer to Sec. 11-3(t).
- Specify plate covers as detailed in Sec. 11-3(u).

12-6. ONE-PIPE, FORCED CIRCULATION, DIVERSIONARY ARRANGEMENT

(a) *Note:* In this arrangement, the hot water main is of a uniform size from the point of its departure from the boiler to its return thereto.

Each radiator, convector, length of baseboard, or whatever unit of heat emitter is used, is supplied the amount of hot water for which it is designed, by means of a "diversionary fitting" usually of proprietary origin, and incorporated in the supply main.

This diversionary fitting may be of the fixed orifice type, or it may have an adjustable orifice that is readily set or reset from outside the fitting and without disturbing its position in the pipeline.

The water diverted to the radiator on leaving the radiator is returned to the supply main, sometimes through a conventional tee, sometimes through another diversionary fitting.

The principle of operation is to introduce an area of flow resistance in the main at the point of branch takeoff that is enough to force the desired flow from the main to the radiator. A fitting of this type is almost essential for inducing a flow downward from an overhead main to a radiator at a lower level.

The constant dilution of the original hot water temperature places a practical limitation on this arrangement; this limitation is not too severe to rule out the system in buildings of moderate size, and it is in good repute with designers.

(b) ***Diversionary Fittings***

- Specify if fittings are to be of the fixed orifice or adjustable orifice type.
- Provide a table of fitting sizes, capacities in _______

gph of hot water, and with the location of each fitting stated.

- Specify the materials of construction, as (iron) (bronze), etc.
- If a proprietary unit is desired, specify the name and model number.
- Specify if fitting ends are to be screw or soldering kind.

(c) ***Other Components***

- Since these are the same as for the one-pipe series system, each component for a diversionary system, excepting the fittings themselves as specified above, may be described in accordance with Sec. 12-5.

12-7. TWO-PIPE, REVERSED RETURN

(a) *Note:* The reversed-return arrangement of the piping was a common feature of the gravity hot water heating systems. It equalized the length of the supply and return to the boiler of every radiator with every other radiator. In theory, these equal circuits, in conjunction with the variations in pipe size appropriate to the load, presented equal resistances to the flow of the water to each radiator and thus avoided short circuiting. Performance generally was so satisfactory that the practice carried over to forced circulation and is still used by many system designers. The alternative is a direct return with a balancing cock in each radiator circuit, which will be examined in Sec. 12-7 as a subject for specifying.

(b) ***Hot Water Source***

- For a low-pressure cast-iron boiler, refer to Sec. 12-1.
- For a low-pressure steel boiler, refer to Sec. 12-2.
- For a steam converter, refer to Sec. 12-3.

(c) ***Piping, Valves, and Fittings***

- Specify material as steel, wrought iron, brass in iron pipe sizes, or copper tubing, Type K, L, or M, as appropriate.
- Specify working pressure, _______ psig, and give schedule number for the ferrous piping.
- Specify how piping joints are to be made, as screwed, flanged, soldered, welded, etc.
- Specify pipe "dope" for screwed joints, kind of flanges for flanged joints, together with kind of gasket, the kind of solder, as soft or hard, tin-antimony, tin-lead, etc.
- For fittings, specify kind, ends, joints, etc.

- For valves, specify the material as iron body, bronze body, screw ends, etc. Note if rising or nonrising stem.

(d) *Pipe Hangers*

- Specify kind of hangers, as strap, rod, or proprietary.
- Specify point of suspension and distance between hangers, _______ ft.

(e) *Insulation*

- See Sec. 12-5(e).

(f) *Pipe Sleeves and Covers*

- See Sec. 12-5(f).

(g) *System Compression Tank*

- Specify location, as basement, attic, etc.
- Refer to Sec. 12-5(b) for details.

(h) *Flow-control Valve*

- Specify a flow control on boiler hot water supply outlet or on return ahead of the boiler, to prevent natural circulation when pump is still.

(i) *Radiation*

- For cast-iron, column radiation, see Sec. 11-3(f).
- For cabinet convectors, see Sec. 11-3(g).
- For baseboard, cast-iron, see Sec. 11-3(h).
- For baseboard, finned tubing, see Sec. 11-3(i).
- For unit heaters, see Sec. 11-3(j).
- Specify vent valves for each unit of radiation and high points in the circuit. Specify manual vent valves or automatic vents to suit conditions.
- Specify that automatic vent discharges shall be carried outside building if there is danger from water.

(j) *Circulating Pumps*

- Specify number of pumps.
- Specify the capacity, _______ gpm, of each, for water of _______ F temperature, and for a total head of _______ ft.
- For construction details, location of pumps, and data to be required of bidders, see Sec. 12-5(c).

(k) *Pump and Firing Controls*

- For a single circulator, see Sec. 12-5(c).
- For multiple pumps and zones, specify that each zone pump shall be started and stopped at _______ F and _______ F respectively, as monitored by its zone thermostat.
- For specifying the thermostat, see Sec. 11-3(v).
- For a multizone installation, a master time clock set for

both night and holiday interruptions is generally justified economically.

- For time clock specifications, see Sec. 11-3(w).
- Specify an overriding thermostat, set above freezing, to prevent dangerous drops in inside temperatures over holidays when the normal controls are inoperative under time clock monitoring.
- Specify a thermostatic bulb submerged in the boiler water, to actuate the firing equipment as necessary to maintain the desired boiler water temperature independently of the running or nonrunning of the circulator pumps.

12-8. TWO-PIPE DIRECT RETURN

(a) *Note:* This arrangement, by the use of balancing cocks in each radiator or zone circuit, economizes on piping as compared with the reversed return. Another advantage is that the cocks permit ready adjustment of the hot water flow to the various circuits for correcting overheating or underheating therein.

Changes in the lengths of the pipe runs and pipe sizing and the addition of the balance cocks constitute the only significant variations from the two-pipe reversed return. Consequently, the specification for a direct, balanced-return layout can be taken from the data in Sec. 12-7, with the balance cocks to be added as new components.

(b) ***Hot Water Source***

- Same as for Sec. 12-7(b).

(c) ***Piping, Valves, and Fittings***

- Same as Sec. 12-7(c).

(d) ***Pipe Hangers***

- Same as Sec. 12-7(d).

(e) ***Insulation***

- Same as Sec. 12-5(e).

(f) ***System Compression Tank***

- See Sec. 12-5(b) for details.
- Specify position of tank, basement, attic, etc.

(g) ***Flow-control Valve***

- Same as Sec. 12-7(h).

(h) ***Radiation***

- Same as Sec. 12-7(i).

(i) *Circulating Pumps*

- Same as Sec. 12-7(j). For construction details, locations, data by bidders, etc., see Sec. 12-5(c).

(j) *Pump and Firing Controls*

- Same as Sec. 12-7(k).

(k) *Balancing Cocks*

- Specify the design and materials of the cocks, as (iron) (brass) (bronze), with (screw) (soldering) ends.
- Specify method of adjustment, whether by a square head or by a lever on the side of the cock, or other.
- If not already on the drawings, set up a table of sizes, capacities, ________ gph, and indicate where cocks are to be placed.

12-9. RADIANT PANELS, CEILING AND FLOOR

(a) *Ceiling Panels*

- Specify the kind of piping: (copper tubing) (steel pipe) (wrought iron pipe).
- For copper tubing, specify the type, Class L, M, or K, and give diameter, ________ in.
- For ferrous piping, specify the schedule number and diameter, in.
- Specify the joints, as silver soldered, screwed, welded.
- Specify piping arrangement, as return bends, tube and header grid.
- Specify desired cover depth of plaster over the piping, ________ in.
- Specify the center-to-center distance between the parallel runs of piping, ________ in.
- Specify maximum allowable surface temperature, ________ F.
- Specify maximum allowable water temperature, ________ F.
- Specify that plaster cover over piping must be allowed to air-dry ________ days before hot water is admitted to the coils.
- Specify kind of air vent valve for automatically venting air from the coils. Require vent discharge to be carried to a convenient drain or to the outside of the building, using ________ in. diam copper tubing.

(b) *Wall Panels*

- Specify in accordance with preceding Sec. 12-9(a).

(c) *Floor Panels*

- Specify similarly to ceilings, Sec. 12-9(a), with these exceptions:
  - Specify thickness of concrete on top of coils, ______ in.
  - Specify that the covering concrete must *not* be of the insulating kind.
  - Specify kind and thickness of bed on which the coils are laid, ______ in. of ____________ (material).
  - Specify that concrete floor over coils must not be painted or covered with any material until the entire mass of concrete has had time to dry out.

(d) *Panel Temperature Controls*

- Specify the furnishing and installation of an outdoor anticipating thermostat that responds to an automatic evaluation of the related factors of sun and shadow, wind and ambient temperature. This thermostat reduces hot water flow to the panels of ceiling or floor in time to prevent a temperature overrun in the mass of plaster or concrete which otherwise could follow a rising outdoor temperature, or equivalent temperature. Similarly the outdoor thermostat increases panel temperatures on a falling thermometer before the indoor thermostat senses the need, thus preventing a comfort lag.
- Specify a room thermostat to actuate the circulating pump in accordance with room needs.

Note: For specifying guidance concerning the pumps, boilers, heat exchangers, piping, zone valves, etc., used with hydronic panel installations, refer to the appropriate sections of Division 12, as they apply to these components.

DIVISION 13

Premises Heating, Warm Air

13-1. WARM AIR FURNACES, DIRECT FIRED

(a) *Capacity and Construction*

- Specify the capacity of each furnace, _______ Btu per hour, when fired with _______________ fuel, measured at bonnet.
- Specify the volume of air being passed around the heat exchanger as _______ cfm and the discharge temperature *at the registers* as _______ F.
- Specify the calorific value of the fuel as _______ Btu per gal for oil; _______ Btu per one cu ft of gas; _______ Btu per pound for coal.
- Provide proximate analysis to bidders of the fuel, including a note as to whether the gas fuel is natural, manufactured or liquefied petroleum gas.
- Specify the construction of the furnace, as the furnace is to be formed of (stainless steel of not less than _______ gage) or (refractory of the _______________ kind suitable for temperatures up to _______ F). Specify that the heat exchanger is to be fabricated of _______________ steel of not less than _______ gage, and with all-welded joints.
- Specify the casing to be steel, not less than _______ gage and insulated with _______ in. of _______________ (material). Finish of casing to be _______________ enamel, baked on or other, and _______________ in color.
- If firing by oil, specify kind of burner, as (gun) (rotary) (other).
- If firing by gas, specify kind of burner and draft, (natural) (forced).
- If firing by coal, specify hand grate, mechanically driven stoker, or pneumatic fuel injection system over the grate.

- Specify the minimum grate area, ______ sq ft, for selected method of firing.
- For motor-driven stoker, specify kind of motor, and give service available as ______ volts, ______ phase, ______ cycles. Specify kind of coupling, as (flexible mechanical) (hydraulic) (other).
- Specify maximum firing capacity of equipment, as ______ gal of oil per hour, ______ cu ft of gas per hour (or per minute for large units), ______ pounds of coal per hour.
- Specify the range of combustion air temperature, as from ______ F to ______ F, as taken from the ambient atmosphere.
- For pneumatically injected coal fuel, specify pressure of air available, ______ psig, ______ F temperature, and state source. If proprietary firing system, state name of manufacturer. At the option of the specification writer, add, "Or approved alternate."
- For mechanical stoker firing, likewise state name of manufacturer with reference to an approved alternate, at the option of the specification writer.
- Specify the furnace location. State access limitations, if any.
- Specify locations of duct connections, as bottom, side, or top of casing.

(b) *Furnace Control Equipment, Oil Fired*

- Specify ignition monitoring system, with sensors to act on loss of ignition or failure to ignite, as manufactured by ____________ or ____________.
- If burner fails to light on the first attempt, state if control will attempt again to light off after a delay of ______ minutes for venting the furnace, or if complete shutdown is to follow and the control must be reset by hand.
- Specify provision of a stack switch and necessary relays to govern operation of the oil burner.
- Specify a solenoid-operated oil valve to delay injection of oil to burner until burner motor and fan are up to speed.
- Specify the electric service available for the controls as ______ volts, ______ phase, ______ cycles.
- Require bidders to furnish full information on the char-

acteristics of their control system, together with identification of line voltage and reduced voltage components.

(c) *Furnace Control Equipment, Gas*

- Specify that control units are to be wired in series so that opening of the contacts of any one will shut down the gas supply valve.
- Specify that gas to pilot is supplied from a branch ahead of the main supply valve, or ahead of the pressure regulator and manual valve, in accordance with local regulations and codes, *provided* this gas fuel is either natural gas or artificial gas. For liquefied petroleum gases, specify that a failure of the pilot flame must result in a complete shutdown, including the gas supply to the pilot.
- Specify a thermostat for the warm air outlet of the heater, set to shut off all combustion whenever the temperature of the warm air reaches or exceeds _______ F. Specify if thermostatic limit switch is to be of the mercury type, or other.
- Specify the gas pressure drop across the pressure regulator, from _______ psig supply line, to _______ psig to the main burner.

(d) *Furnace Control Equipment, Coal Stoker*

- Specify stack switch to stop stoker and cutoff underfire draft when flue-gas temperature reaches or exceeds _______ F.
- Specify a bonnet-placed thermostat to start the warm air circulating fan at _______ F and to stop the fan at _______ F. Alternatively, the fan may be controlled as a subordinate to the stoker operation rather than as an independent function. Designate which form of control is desired.

(e) *Time Controls*

- The actuation of the firing equipment in a warm air furnace is subject to time adjustments for meeting the needs of nights, Sundays, and holidays.
- For specifying master time controls, with an overriding thermostat to prevent freeze-ups, see Sec. 12-7(k).
- For all forms of furnace, specify _______ (number) push-button stations for manual control and stipulate points of location.

(f) *Furnace Fans (or Blowers)*

- When integral with the furnace, specify the kind of fan or blower, as (squirrel cage) (centrifugal), etc.

- Specify capacity as _______ cfm against a total resistance of _______ in. W.G., when handling air at _______ F at _______ rpm.
- Specify fan or blower construction as nonoverloading, driven by electric motor through (direct) (belt) drive.
- Specify blower location, as in separate compartment of the furnace, carried on resilient mountings.
- Require fan to be both statically and dynamically balanced at factory.
- Require ratings to meet the standards of the Air Moving and Conditioning Association, or the test code of the ASME, PTC 11-1946, as applicable.

(g) *Filters*

- Specify the kind of filter, as (cleanable) (throwaway). Specify particle removal efficiency, _______ per cent with particles _______ micron in size.
- Specify maximum acceptable pressure drop, _______ in. W.G., when clean and operating at the rated capacity of _______ cfm.
- Specify filtering materials as _______________, and with a minimum surface area of _______ sq ft.
- Specify if material is to be flammable, fire-retardant, or nonflammable.
- Require filter to meet the standards of the Air Filter Institute as appropriate.
- Require bidders to submit complete data on the filter being offered.
- Specify location of filter in the furnace casing, and in relation to the blower or fan.

Note: In some large and important installations, the test specified can be in terms of dioctyl phthalate smoke, usually called the DOP test.

- Specify that blower motor satisfy the requirements of the National Electrical Code, the standards of NEMA, and such local codes as apply. This includes the wiring.

(h) *Fuel-oil Tank and Piping*

- Specify the capacity of the tank, _______ gal.
- Specify the construction, as all-welded, copper bearing steel, and specify that it shall carry the approved label of the Underwriters Laboratories.
- Specify that tank shall be thoroughly cleaned, then given two coats of red lead, followed by one cover coat of bitumastic or hot asphalt.
- If level of water table indicates that some floatation of

the tank may occur, specify that the tank be guyed down by means of deadman anchors, with connecting straps of steel across the top of the tank. Alternatively, the tank may be anchored to a slab of reinforced concrete, not less than 6 in. thick, and dimensioned to extend at least 1 ft beyond the tank in both diameter and length.

- Show location of the tank on the drawing.
- Specify that fill and vent pipes shall comply with the requirements of the National Board of Fire Underwriters, Pamphlet No. 31, or later editions.
  - Specify fill cap to be provided with a padlock and two (2) keys.
  - Specify material of fill and vent lines, as steel pipe, Schedule No. ________, zinc coated. Fittings to be of malleable iron, zinc coated, then threaded.
  - Oil supply and return lines to be of Type L, copper tubing, with wrought brass fittings. Specify a bronze or brass shutoff valve near the furnace.
- Specify minimum earth cover over tank to be not less than ________ ft.
- Specify that contractor is to place an initial supply of 50 gal of No. ________ fuel oil in the tank, to be used in testing the system.
- Specify that vent pipe is to end with a gooseneck and screened opening, as shown on the drawing.

(i) *Breeching or Smoke Pipe*

- Specify breeching or smoke pipe to be fabricated of ________ gage steel, of welded or riveted construction.
- Specify outside painting to be of a heat-resisting nature, as manufactured by ________________ or approved alternate.
- Specify the provision of a barometric damper for furnaces fired by oil or coal.

(j) *Draft Hood for Gas-fired Furnaces*

- Specify a draft hood in accordance with the requirements of the American Gas Association, the National Board of Fire Underwriters, or other proper authority.

(k) *Draft Inducers*

- Where natural draft is insufficient, specify a motor-driven draft inducer. Stipulate where mounted, as in breeching or smoke pipe, on top of chimney, etc.

- Specify capacity of inducer, ______ cfm, with flue gas of ______ F temperature.
- Specify that electrical motor, wiring, switches, and all other electrical components shall satisfy the National Electrical Code, the standards of the NEMA, and local codes.
- Specify operation of inducer, as to whether it shall run continuously or be incorporated into the control circuit of the firing equipment so that it starts and stops concurrently.

13-2. WARM AIR DUCTS

(a) *Design and Construction, Metal*

Note: Many warm air heating systems are expected to function as channels for chilled air as well, during periods of summer cooling. Since the volume of chilled air is substantially greater than for warm air, the decision to specify for heating alone, or for heating and cooling, must be made before the specification is written. Other than for greater size, the cooling cycle's main difference from the heating-function-only duct, is in the insulation. The chilled air duct requires a vapor barrier to prevent sweating, and the integrity of the insulation joints is far more critical, since a little air seepage from the atmosphere that makes contact with the duct may produce enough moisture to destroy the effectiveness of the insulation over a large area of duct surface. Duct specifications as given in the following sections may be used for all-year service, but insulation for winter and summer will be distinguished from each other.

- Specify the material of the ducts, as galvanized steel, aluminum, etc.
- Specify that the metal gage is to conform to the standard practices tabulated by the American Society of Heating, Refrigerating and Air-Conditioning Engineers (ASHRAE) and detailed in the ASHRAE *Guide and Data Book*, latest edition.
- Specify kind of joints, also in accordance with the latest edition of the ASHRAE *Guide and Data Book*.
- Specify kind and placement of turning vanes, kind and placement of dampers, balancing and manually operated.

- Specify if zone dampers are to be actuated by an electric motor or by a pneumatic operator. If the first, specify if damper is to be modulating, with _______ number of steps between wide-open and closed; or a simple two-position damper to be either open or closed. Specify pneumatic modulation; this kind of damper operator by its nature makes an infinite number of positions available.
  - Specify electric motor on damper, giving voltage available, _______ phase, _______ cycle current, and require compliance with National Electrical Code, etc.
  - Specify air pressure available for pneumatic damper operators, as _______ psig and state source.
  - Specify automatic fire dampers, wherever the duct passes through a fire wall, or connects two separate areas of the building. Specify that fire dampers shall be constructed and provided in accordance with the National Board of Fire Underwriters, Pamphlet 90A.
- Specify kind of duct hangers and supports, as suspension rods, spring hangers, adjustable hangers, etc.
- Specify minimum acceptable supporting strength of hangers, when placed as shown on the drawings.
- Specify that duct connections to warm air furnace shall be by means of heavy canvas joints, treated to make the material nonflamable.
- Specify if branch ducts are to be taken off main ducts through splitter dampers, or not. Refer to drawings for details.

(b) *Metal Duct Insulation, Heating Only*

- Specify that duct is to be insulated with _______ _______ (material), _______ in. thick.
- Specify that joints are to be sealed with _______ adhesive strips, not less than _______ in. wide. The joints in the insulation are to be (abutting) (lapped) (beveled) or other.
- Secure insulation to ducts by adhesives, or as recommended by the insulation manufacturer.
- Specify if insulation is to be slow-burning, fire-retardant, nonflammable, etc.

(c) *Metal Duct Insulation, Heating and Cooling*

- Specify as for heating service only as given above, plus the following:
- Specify that insulation shall be completely covered by

a vapor barrier material, as Kraft paper or other suitable finish.

- Specify that vapor barrier is to be tightly sealed against the insulation by adhesives manufactured for this purpose. *Note:* Specifiers may give proprietary names to the adhesive if desired.
- Specify that the vapor barrier is to be joined to the insulation at the factory.
- Specify that damper operators normally projecting to the outside of the duct must not be covered, in order to permit adjustments as necessary without disturbing insulation.

(d) ***Warm Air Ducts, Asbestos-Cement***

- Specify if ducts are to be used (aboveground) (perimeter in the slab), etc.
- Specify if duct is to be fireproof, waterproof, and rotproof.
- Specify kind of joints, as rubber connectors and stainless steel straps.
- If not already on the drawings, tabulate the sizes and quantities required in these specifications.
- Similarly tabulate the fittings by sizes, kind, and numbers.
- State if duct is to be used for heating only, or for seasonal cooling as well.
- Specify vapor barrier for cooling cycle.

(e) ***Warm Air Ducts, Fiber***

- Specify if duct is to be used for cooling as well as for heating.
- Specify construction, as aluminum lining, externally coated with moisture-resistant black asphalt that is soundly bonded to the duct.
- Specify kind of adhesive that shall be used in laminating the duct.
- Specify if duct is to be connected by metal fittings or by mitering and tape-sealing the joints.
- For metal connectors, specify ________ gage.
- For tape-sealed joints, specify the kind of tape, kind of adhesive, and minimum lap of tape on each side of joint as ________ in.
- Specify that installation must be in accordance with the manufacturer's instructions.

- If not already on drawings, tabulate the sizes and lengths of duct required.
- Specify kind and placement of vapor barrier if ducts are to be used for a cooling cycle.

(f) *Warm Air Ducts, Clay Pipe*

- Specify if pipe is to be laid in earth or buried in concrete.
- Specify kind of joints, as (mechanical compression) (plastic) (rubber) (hot pour compounds), etc.
- Specify that all joints must be made in accordance with the requirements of the research division of the National Clay Pipe Manufacturers, Inc.
- Specify that the installation of the pipe shall comply with the appropriate recommended practice, tentative or permanent, of the American Society for Testing and Materials, ASTM C12-58T.
- Specify if clay pipe is to be laid in standard strength or extra strength, and name ASTM standard applying.
- Specify kind, sizes, and numbers of fittings required.
- If not already on drawings, tabulate the lengths, sizes, weights, and quantities of clay pipe in each strength.
- Carefully specify the placement of the pipe, its leveling and supports, thickness of encasing concrete, depth of cover, ________ in., etc.

Note: Vitrified clay pipe, originally used for sewer service, has been found adaptable in some locations to warm air heating installations. For that reason, the items to be specified are given herein.

(g) *Warm Air Ducts, Fiber Glass*

Note: These ducts are formed in the shop and on the site from boards of fiber glass. They are rectangular in shape.

- If not on drawings, provide a table of duct sizes, giving dimensions, thickness of fiber glass, and quantities.
- Specify that duct is to be formed in accordance with the instructions of the manufacturer.
- Specify temperature of air, ________ F.
- Specify how joints are to be made, as with metal connectors, by lap and seal, stapling, etc.
- Specify type, supporting strength, and spacing of duct hangers. Unsupported spans must not exceed manufacturer's recommendations.

- Specify if ducts are to be painted or otherwise finished. Specify colors.
- Specify the coefficient of heat transmission through the duct wall, _______ Btu per deg F temperature difference per in. thickness of material.

(h) ***Warm Air Ducts, Flexible***

Note: This is a recently marketed product, formed of spiral wiring covered with fiber glass insulation covered with a factory-applied vapor barrier. It is lightweight and avoids the use of elbows.

- Specify length of sections needed as _______ ft.
- Specify diameters as _______ in.
- Specify capacity, _______ cfm, handling air at _______ F.
- Specify _______ wire gage; specify thickness of fiber glass insulating covering, _______ in.
- Require bidders to state heat transmission, _______ Btu per sq ft per deg F of temperature difference.
- Specify that maximum pressure duct must withstand _______ in. W.G.
- Specify Underwriters Rating for fire resistance; _______ outside surface; _______ inside surface.

13-3. DUCT HEATERS

(a) ***Duct Heaters, Gas Fired***

- Specify the performance of the duct heater on these items:
  - Thermal input, _______ Btu per hour.
  - Minimum air volume, _______ cfm.
  - Temperature rise, minimum air volume, _______ F.
  - Maximum air volume, _______ cfm.
  - Temperature rise, maximum air volume, _______ F.
- Specify desired dimensions of the duct heater, as _______ in. wide, _______ in. high; _______ in. deep.
- Specify vent arrangement as horizontal, requiring a vent of _______ in. diam; or vertical, with a vent diameter of _______ in.
- Specify if heater is to be suspended or base-supported.
- Specify kind of supports, bearing strength, _______ pounds, and by whom furnished.
- Specify thickness of steel in heater construction, as _______ gage.

- Specify that heater is to be tested for tightness under _______ in. W.G. pressure.
- Require heater to bear the Underwriters' seal of approval. Also require approval by American Gas Association for the gas or gases to be burned.
- Specify how heater is to be controlled, as by thermostat, thermostat to close main gas valve or open this valve in response to heat demands.
- Specify provision of a temperature limit control that automatically shuts off gas on the development of excessive temperatures. After such a shutoff, require that control must be reset by hand.
- Require bidders to submit drawings showing arrangement of heater, with particular attention to the methods recommended for cleaning.
- Specify gas pressure regulator to act on an incoming pressure of _______ psig (or in. W.G.) and to reduce this to a burner pressure of _______ psig (or in. W.G.).
- Specify if gas is to be natural or artificial and give a proximate analysis with its calorific value of _______ Btu per cu ft.
- Specify if draft hood is to be (external) (built in).
- Specify the finish of the heater; color, material, etc.
- Specify voltage for controls as (line) (reduced), giving _______________ voltage in each case.
- Specify that all wiring and electrical materials must comply with the National Electrical Code.
- Specify method of sealing the connections between the heater and the connecting duct.
- Specify maximum allowable pressure drop, _______ in. W.G., measured at maximum air flow through the heater.

(b) *Duct Heaters, Electric*

- Specify the required heating capacity of the duct heater as _______ kw, or _______ Btu per hour.
- Specify the air flow, _______ cfm.
- Specify the voltage, _______ volts, _______ phase, _______ cycles.
- Specify the number of circuits in the heater, _______.
- Specify heater dimensions, as _______ in. wide, _______ in. high, _______ in. deep.
- Specify number and kind of electric heating elements, _______ fintubes per heater.

- If not on drawings, tabulate the number, wattage, and cfm of air flow of the heaters required, this tabulation to be part of the specification.
- Specify type of switch and control components. Include a temperature limit switch to shut down elements on the occurrence of excessive temperatures.
- Specify type and placement of thermostat for energizing and deenergizing the heater in response to heating needs.
- Specify temperature range for fan control, so that fan is not started until air temperature reaches the set level.
- Specify the thickness of the frame, splice box, and other structural members, noting ______ gage steel as appropriate.
- Specify stability of wattage, as not to fluctuate more than ______ per cent.
- Require bidders to submit complete data on their units, including information on protection provided against electric leakage, progressive circuit failures, and the complete stoppage of air movement through the heater for any cause.
- Specify maximum pressure drop through the heater, ______ in. W.G.
- Specify if heater (or heaters) is to be controlled remotely or by local switch and thermostat. If remote, give distance.
- Specify that all electrical wiring and components must satisfy the National Electrical Code and local codes.
- Specify that heater must carry the Underwriters' label of approval.
- Require bidders to submit wiring diagram for fan, elements, and controls.
- If multiple duct heaters are to be controlled from a central control panel, so specify. Provide a sketch showing bidders what is wanted on the control panel in the way of switches, relays, push buttons, etc.

13-4. REGISTERS, SUPPLY

(a) *Wall Registers*

- Specify position on wall, as with horizontal center line ______ in. below ceiling.
- Specify deflection, as (single) (double).
- Specify volume control, using damper behind the outlet

vanes and controlled manually by settings made outside the register and in the room.

- Specify adequate seal between register and wall, as a rubber gasket to prevent smudging.
- Specify colors of finish.
- If not already tabulated on the drawings, prepare a table as part of this specification and showing the following data:
  Identifying symbol, if used; size: _______ in. × _______ in.; _______ cfm discharge; quantity: _______ units.
- Specify materials of construction as _______ gage steel.

(b) *Ceiling Registers*

- Specify materials of construction, as _______ gage steel.
- Specify directions of discharge, as 2-, 3-, or 4-way.
- Specify how volume is to be controlled, as by manually operated register damper or valve.
- Specify gasket seal to prevent smudging.
- Specify painting; coats and final finish.
- If not already on drawings, prepare a tabulation similar to that for wall registers, showing symbol; size: _______ in. × _______ in.; (2)(3)(4) way discharge; _______ cfm; _______ number of units of each kind.

(c) *Baseboard Outlets, Register Type*

- Specify locations, as in baseboard, or low wall.
- Specify construction, capacities, quantities, etc., as detailed in Sec. 13-4(a), Wall Registers.
- Specify the arc of area over which the discharge of warm air is diffused, as (180) deg.

(d) *Baseboard Outlets, Continuous, Perimeter*

- Specify dimensions as _______ in. high, _______ in. projecting from the wall.
- Specify if to be furnished in 2-ft lengths or 4-ft lengths, which will be used either as units, or put together as a continuous outlet.
- Specify capacity, _______ cfm, or _______ Btu per hour, per _______ ft length.
- Specify construction, as _______ gage steel.
- Specify finish; coats of paint and final color.
- Specify if volume control valves are to be furnished to function with the baseboard, or if entire areas will be activated and inactivated by means of a central control governing a fan or blower.

- Specify that connections to supply boot shall be made by adjustable panel rather than by cutting out sections of the metal.

(e) *Ceiling Diffusers*

- Specify shape of diffuser, as (round) (square) (half round).
- Specify pattern of air discharge, as (fixed) (adjustable) (combined with light fixture) (combination supply and return unit), etc.
- Specify if diffuser is to be mounted flush with ceiling, to be projecting, to be set on exposed duct.
- Specify output, _______ cfm, at a maximum velocity of _______ fpm.
- If discharge is segregated into 2, 3, or 4 ways, specify the _______ cfm of each direction; also the maximum velocity, _______ fpm.
- If not appearing on the drawing, prepare a tabulation of the diffusers, giving the following data:
  Identifying symbol, if used; dimensions, as _______ in. round; _______ in. × _______ in.; (square) (rectangular), as appropriate, capacity as _______ cfm; velocity, _______ fpm; _______________ pattern of discharge; throw _______ ft in _______________ direction; throw _______ ft in _______________ direction; ditto for throw in each direction.
- Specify materials of construction as _______ gage steel.
- Specify gasket seal to prevent smudging of ceiling.
- Specify painting and final color.

(f) *Slot-type Diffusers*

- Specify diffuser locations, as window sills, ceiling, high wall, etc.
- Specify whether each unit is to be of the 1-way or 2-way blow design. Give listed size, _______ in.
- Specify the following items:
  Output, _______ cfm per linear ft of unit; jet velocity of _______ fpm; throw, _______ ft; static pressure, _______ in. W.G. Also state projected mounting height, since some designs have a minimum.
- Specify maximum acceptable noise rating, _______ decibels.
- If not already shown on drawing, prepare a tabulation giving number of units of each size, and the characteristics of each as detailed above.

- Specify construction materials, as stainless steel, aluminum, etc.
- Specify location of volume control damper with respect to each linear unit.

13-5. RETURN GRILLES

(a) *Return Grilles*

- Specify the number, size, and capacity of return grilles.
- Prepare a tabulation identifying each by its symbol if used, and stating other controlling data, as _______ per cent free area, _______ cfm, etc.
- Specify if grille is to have volume damper.
- Specify if grille is to be equipped with a "no sight" core.
- Specify the maximum velocity acceptable, as _______ fpm.
- Specify grille locations.
- If bars are fixed, specify the angles, as _______ deg horizontal bars and _______ deg vertical bars.
- Specify how grilles are to be installed, as with molding strips, flanged frames, etc.
- Specify material of construction as _______ gage steel, aluminum, etc.
- Specify finish and color.
- If fire damper is desired in the return, specify both the damper and a fusible link designed to close the damper automatically.
- If grille opening is to be finished in an ornamental design, specify the kind of design desired. Refer to manufacturer's catalogue for this.

(b) *Slot-type Return Grilles*

Note: These may be specified the same as the supply slots, Sec. 13-4(f), as appropriate.

13-6. HUMIDIFIERS

(a) *Humidifiers, Steam*

- Specify location of humidifier, as (heater plenum) (supply ducts), etc.
- Specify volume of air passing over humidifier, _______ cfm, at a temperature of _______ F and with a relative humidity of _______ per cent.
- Specify the required relative humidity, _______ per cent, to be achieved by the humidifier.
- Specify the steam service available, _______ psig, and total temperature of _______ F.

- Require bidders to submit complete data on the equipment offered, including steam consumption for the condition specified, _______ pounds per hour.
- Specify control method, as by humidistat in duct, in area of controlled humidity, etc.
- Specify kind of humidistat desired, sensitive element to be (human hair) (salt film for electrical resistance), etc.
- Specify kind of transmission from humidistat to the humidifier, (pneumatic) (electric relay), etc.
- Specify steam admission valve to be provided by humidifier supplier, to suit the transmission system.
- Specify electric service available, _______ volts, _______ phases, _______ cycles. Specify if electric signals are to be transmitted at line or reduced voltage.
- Specify locations of control relays and push-button station for starting and stopping humidification.
- Specify the kind and schedule number of the steam piping and state where contractor's work stops. Also specify the piping, valves, and fittings, together with any insulation, this contractor is required to furnish.

(b) *Humidifiers, Evaporative*

- Specify location of evaporator, condition of air before humidification, and required relative humidity, _______ per cent.
- Specify details as appropriate from Sec 13-6(a).
- Specify the temperature range and pressure of the water available, as _______ F to _______ F, at _______ psig
- Specify control of makeup water, as by float and valve.
- Specify means for cleaning the water pan, and the removal of mud, scale, and lime.
- Specify where contractor's work stops, and state material he is required to furnish, including piping and valves.
- Require bidders to submit complete information on their equipment, including drawings and photographs. Also dimensions, weights, and rate of water consumption under specified conditions.
- Specify location of humidistat control, with sensitive element in duct or room (if used).

(c) *Humidifiers, Atomizing, Pump*

- Specify location, as plenum, supply duct, or return duct.
- Specify before and after conditions as detailed in Sec. 13-6(a).

- Specify the maximum volume of water to be atomized, _______ gph.
- Specify motor mounting on the centrifugal pump, as outside the humidifier housing, moisture-protected, etc.
- Specify electric service available, _______ volts, _______ phase, _______ cycles.
- Require bidders to furnish complete data, including dimensions; weight; motor horsepower; casing, as _______ gage stainless steel; finish.
- Specify kind and location of control, as humidistat in plenum, duct, room, other.
- Specify kind of humidistat and accuracy, as ±_______ per cent.
- Specify transmitting means, (pneumatic) (electric).
- If pneumatic, specify available air pressure, _______ psig.

(d) *Humidifier, Atomizing, Pneumatic*

- Specify before and after air conditions, as in Sec. 13-6(a).
- Specify atomizing head, as brass, stainless steel, etc.
- Specify capacity, _______ gph of water.
- If compressed air is available, state pressure, _______ psig.
- Specify components to be furnished by contractor. If compressed air is not available, specify that contractor shall furnish and install the compressor to suit the humidification schedule detailed.
- Specify water available, temperature range, _______ F to _______ F, and pressure, _______ psig.
- Specify control sequences, as a fall in humidity registered by the humidistat starts compressor (if compressor is for this purpose only) and opens solenoid valve in the air line to the atomizing heads. At the same time, the water supply valve is opened by the air pressure, and closes when the needs for humidification are ended and the air valve closes, dropping the air pressure.
- Specify the range of operation of the humidistat, from _______ per cent to _______ per cent relative humidity.
- Require bidders to submit complete data on their equipment, including dimensions, weights, and accuracy, ± per cent.
- Specify position of thermostat and sensitive element.

DIVISION 14

Premises Heating, Electric

14-1. RESISTANCE BASEBOARD

(a) *Construction and Arrangement*

- Specify the construction of the baseboard, as in sections ______ ft long, with backs and covers of ______ gage steel, finished in ____________ color, and designed for standard wattage or high-wattage operation. Specify maximum cover temperature, ______ F.
- If not already shown on the drawings, prepare a tabulation of the required baseboard, giving the following data: Identifying symbol for each section; length of each section, ______ ft; output, ______ watts, total; output, ______ Btu per hour; ______ volts; catalogue model number; with or without integral thermostat; number of sections for each symbol.
  For blank sections, tabulate the length of each and the total number of blank sections in each length. Also tabulate the catalogue numbers.
- Specify controls, whether by integral thermostat or by room thermostat.
- Specify that all equipment must comply with the Underwriters' requirements, as well as those of the National Electrical Code and applicable local codes.
- Specify that all wiring and connections must be made by qualified electricians.
- If catalogue number is not given, specify the height of the baseboard and its projection into the room, ______ in.
- Specify if baseboard is to be set against the wall or is to be semirecessed.

(b) ***Humidification***

- Specify if humidifier is to be self-contained, with a water reservoir that is filled manually, or if the reservoir level is made up automatically by a float valve.

- For either arrangement, specify the capacity of the reservoir, ________ gal.
- Specify if humidifier is to be portable or fixed.
- Specify capacity, ________ gph.
- Specify air filter, as (replaceable) (washable).
- Specify general construction of the humidifier and its method of protection against corrosion.
- Specify pump motor, as ________ hp, ________ volts, ________ phase, ________ cycles.
- Require bidders to furnish complete data, including dimensions, weights, current consumption, ________ watts per hour, ________________ manner of bearing lubrication.
- Specify control humidistat, giving location, kind of sensitive element, and method of transmission from humidistat to pump motor, as line or reduced voltage.
- If humidifier is to be set in a fixed position, state location.
- Specify desired operating range of humidistat, ________ per cent relative humidity to ________ per cent.

14-2. HYDRONIC-ELECTRIC BASEBOARD

(a) *Note:* This system has been developed recently and provides sections of baseboard that contain both water and an electric heating element. Standard ratings are available in both watts and Btu per hour. Control is by thermostat for each individual room.

- If not already shown on the drawings, prepare a tabulation of the various lengths of baseboard required, giving the following data:
  Identifying symbol for each length; list each length, ________ in.; length of active element, electric, ________ in.; rating, ________ watts; rating, ________ Btu per hour; rating, ________ amperes at ________ volts alternating current; weight, ________ pounds; quantity, ________ units of each symbol.
- Specify maximum permissible surface temperature, ________ F.
- If there is winter danger of a freeze, specify a nonfreeze liquid in the baseboard instead of water.
- Specify that all materials and components must comply with the National Electrical Code and local codes, and must carry the Underwriters Laboratories approval.

- Specify that all electrical connections and wiring must be accomplished by qualified electricians.
- Specify height, ________ in., and projection from wall, ________ in.
- Specify construction of the baseboard, as cast iron, steel, or other material.
- Specify painting and finished color.

(b) *Safety Provisions*

- Specify a high-limit control to cut off all electric supply to the baseboard if the temperature of the baseboard exceeds ________ F.
- Specify a water pressure relief valve, set to open automatically if the pressure in the baseboard exceeds ________ psig.
- Specify that the discharge of each relief valve is to be carried outside the building to a nonhazardous area.
- Specify that each independent unit is to be fused so that an overload results in a blown fuse and interruption of the current.

(c) *Water Fill Connections*

- Specify how units are to be refilled, if ever necessary, with connections on each unit.
- Specify that units shall be filled originally at the factory.
- Specify provision of pressure gage connection on each unit, so that pressure reading is possible if desired.

(d) *Humidification*

Note: This is an adjunct to the electrical heating system, and not incorporated as in a duct system.

- Specify humidification as detailed in Sec. 14-1(b).

14-3. ELECTRIC RADIANT PANELS

(a) *Ceiling Panels*

- Specify maximum allowable surface temperature, ________ F.
- Specify that cable lengths as received from the manufacturer, together with their nonheating leads, must not be altered in any way.
- Specify that cable not properly identified and approved will not be accepted.
- Specify electric service available, ________ volts, ________ phase, ________ cycles.
- Specify each length of cable as required for an area

ceiling, and stipulate its color coding, as white for 120-volt operation, orange for 240 volts, etc.

- Specify that cable must not be closed in until after approval by the local electrical inspector.
- Check cable for integrity and capacity after installation, using an ohmmeter or ammeter.
- Specify capacity of each length of cable, _______ watts.
- Specify that all cable and the installation thereof must comply with the National Electrical Code, the National Board of Fire Underwriters, and local codes.
- Specify that only qualified electricians shall be used in making connections and running wiring.
- Specify the thermostat range, from _______ F to _______ F. State where thermostat is to be positioned.
- Specify operating voltage of thermostat; line or reduced.
- Specify that cables must be run to clear ceiling lights and general wiring. Also stipulate adequate clearance, _______ in., from metal lath.
- Specify cable spacing in the ceiling, as _______ in. center to center.
- Specify location of control panel and kind of push-button station mounted thereon for manual control.
- If heated area is to be unoccupied over holidays, nights, and Sundays, specify a program timer as detailed in Sec. 13-1(e).
- Specify method of fixing cables to supports, as by staples.
- Specify rate of heating up new plaster, if cables are in a plaster ceiling, as _______ degrees per day to a maximum of _______ F.
- Require installation procedures to follow instructions of cable supplier.

(b) *Floor Panels*

- Follow the detailed specification for ceiling panels as given in Sec. 13-3(a) for spacing, color coding, cable lengths, testing, etc., as these details fit the floor cables.
- Also specify that concrete over floor cables must not be poured of insulating concrete.
- Specify thickness of concrete over the cables. An average of 1½ in. is often satisfactory.
- Specify the maximum surface temperature as _______ F.

  Note: A floor temperature much lower than that of a ceiling panel is required to avoid overheating of the feet. 84 F max. is recommended.

(c) *Humidification*

- Refer to Sec. 14-1(b). This applies to both ceiling and floor radiant panels.

(d) *Wall Panels*

- Refer to Sec. 14-3(a) Ceiling Panels. Construction and limitations are equivalent.

(e) *Glass Panels*

- Specify type of panel, as wall panel or baseboard form of panel.
- Specify capacity in watts.
- Specify construction, as tempered Pyrex glass, _______ in. thick, with a conductive coating fired on the inside surface of the glass. Also, reflector of (aluminum) (other) material.
- Specify outer frame of ______________ steel.
- Specify maximum dimensions.
- Require units to meet standards of Underwriters Laboratories, the National Electrical Code, and local codes.
- Specify the service available, as _______ volts, _______ phase, _______ cycles.
- Specify method of control, as room thermostat, manual push button, other.
- State maximum allowable surface temperature of glass and of frame.

(f) *Electric Composition Panels, Ceiling*

- Specify that heat-emitting surface of panel is to be composed of a semiconductor material, graphite-impregnated, and formed into radiant panels _______ ft wide × _______ ft long.
- Specify if panel is to be surface-mounted on the ceiling or shall be arranged for placement on T-bar suspended ceilings.
- Specify panel rating, as _______ watts per sq ft.
- Specify electric service available, _______ volts, _______ phase, _______ cycle.
- Specify insulation behind panel, as _______ in. of ______________ (material).
- Specify mounting height above floor.
- Require bidders to furnish full information regarding area of floor covered by the radiation from the panel, _______ ft. Also state wiring required.
- Specify thermostatic control, thermostat to be mounted (location) and to operate on (line)(reduced) voltage.

- Specify that these panels must be installed in strict compliance with the instructions of the manufacturer.
- Specify that panel shall bear Underwriters Laboratories approval and shall meet the requirements of the National Electrical Code and local codes.

Note: As with the glass panels, this panel is of proprietary design. In order to obtain exactly what is wanted, a specification by name and model number must be issued.

(g) ***Infrared Electric Heaters***

- Specify the kind of element, as (metal sheath) (quartz tube) (quartz lamp).
- Specify the rating of each unit, _______ watts per in. of length.
- Specify the diameter of the tube or lamp, _______ in. Also specify the active or heated length and the overall length, _______ in.
- Specify the operating temperature, _______ F.
- Specify position of the tube or lamp, as (vertical) (horizontal). *Note:* Some designs must be set horizontally.
- If unit is exposed to wind or weather, so state.
- If important, specify warm-up time, _______ seconds.
- Specify mounting height above the floor, _______ ft.
- Specify material of element housing, as _______ gage steel. Also reflector, as (aluminum) (polished stainless steel), etc.
- Specify controls, as thermostat for temperature, and an automatic timer for operations over nights, Sundays, and holidays. See Sec. 13-1(e).
- If not on drawings, prepare a tabulation of infrared units, specifying key data for each, including quantities.
- Specify that all equipment and wiring must comply with requirements of the National Electrical Code, the Underwriters Laboratories, and local regulations.

(h) ***Electric Unit Heaters***

- If not already on the drawings, prepare a tabulation of the unit heaters projected, giving the following data: Identification symbol _______; capacity, as _______ kw; capacity, as _______ Btu per hour; fan capacity, _______ cfm; final air temperature, _______ F, based on an entering temperature of 60 F; outlet velocity, _______ fpm; motor, _______ hp; _______ rpm.
  Also give mounting height, _______ ft, and desired hori-

zontal throw, ________ ft; desired electrical characteristics as ________ volts, ________ phase, ________ cycles; amperage, ________ per supply lead.

- Specify general construction materials, as housing of ________ gage steel, bonderized or otherwise made resistant to corrosion; adjustable louvers for directional discharge settings; fan guard with isolators between guard and housing to protect against vibration; kind of heating element, as sheathed fin tubing with overload protection.
- Specify kind of motor, as totally enclosed, semienclosed with built-in thermal overload protection.
- Specify that all electrical components must comply with National Electrical Code, carry the Underwriters Laboratories seal, and satisfy local codes.
- Require bidders to submit complete data of their offerings, including operating weight of each unit, overall dimensions, limitations on placement, and wiring diagrams.
- Specify that all electrical connections must be made by qualified electricians.
- Specify painting, color of finish.
- Make sure tabulation shows correct number of each size and capacity required.

(i) *Console-type Electric Heaters*

- If not shown on the drawings, tabulate the console units similarly to the unit heater tabulation, Sec. 14-3(h) above.
- Also specify if units are to be free-standing, semi-recessed, or fully recessed.
- Specify that openings in each unit shall be suitable for (top) (front) discharge.
- Specify the number and kind of fans in each unit mounted on a common shaft.
- Specify that rotating elements and the shaft shall be statically and dynamically balanced at the factory.
- Specify how fans are to be driven, by direct connection or through a belt. If by belt, specify kind of belt as flat, V, etc.
- Specify ________ rpm of motor; ________ rpm of fans.
- Specify whether air output, ________ cfm, is against free atmosphere or against a static pressure of ________ in. W.G.

- Specify if motors shall be of single-speed or two-speed design. Require overheat cutout.
- Specify number and kind of electric heating elements.
- Specify how unit is to be controlled, by room thermostat or by thermostat built into the cabinet.
- State electrical service available, as _______ volts, _______ phase, _______ cycles.
- Require all electrical equipment and wiring to comply with the National Electrical Code and the local codes. Also, unit shall carry the Underwriters' label of approval.
- Specify air filter (if desired), giving area of filtering element, sq ft, and material of filter element.
- Specify if filter is to be washable or a throwaway type.
- Specify inclusion of fan delay thermostat to prevent circulation of cold air.
- Specify cabinet construction, as _______ gage steel.
- Specify painting and color of finish.
- Require bidders to submit complete data regarding dimensions, weights, and noise level, _______ decibels.

(j) *Electric Wall Heaters, Fan Type*

- Specify capacity of heater, _______ watts, and capacity, _______ Btu per hour.
- Specify number of each size.
- Specify fan output, _______ cfm.
- Specify kind of heating element, as metal sheathed.
- State characteristics of electrical service available.
- Specify that unit must comply with National Electrical Code and bear the Underwriters' label of approval.
- Require bidders to furnish drawings and data showing dimensions of wall box, weight of unit, and all wiring connections.
- Specify that fan motor and heating element must not interfere with radio or television reception.
- Specify noise limit, _______ decibels.
- Specify type of control, as by wall thermostat or unit built-in thermostat adjustable by manual settings.

DIVISION 15

Air Conditioning, Comfort

15-1. AIR CONDITIONING COMPRESSORS, RECIPROCATING

(a) *Hermetic Compressors*

- Specify number and capacity, _______ Btu per hour, of the central station compressors, and _______ rpm.
- Specify general construction, as welded steel, _______ gage shell; crankshaft of _____________ steel; pistons and connecting rods of _____________ (material); kind of valves, as (flexible) (nonrotating) (springless), etc; kind of lubrication system; kind of bearings, etc.
- Specify kind of mufflers for both gas intake and discharge; specify number, placement, and kind of vibration isolators; kind of suction strainer.
- Specify built-in overload protection.
- Require bidders to furnish complete drawings, including sectional view of compressor, showing arrangement and dimensions.
- Specify motor type and cooling methods.
- State electric service available, _______ volts, _______ phase, _______ cycles.
- Require all electrical components to meet the standards of the National Electrical Code, NEMA, and local codes as appropriate.
- Specify the kind of refrigerant.
- Specify that units are to be dehydrated and sealed at the factory before delivery.
- For protection against excessive temperatures, specify a built-in thermostat for the motor designed to shut down the compressor. Restarting is to be manual.
- State if compressor is to be used with an air-cooled or water-cooled condenser. For the former, state maximum ambient air temperature; for the latter, maximum water temperature to the condenser, _______ F.
- Specify kind of starting and stopping equipment to be

furnished. Indicate if panel is to be mounted adjacent to the compressors or elsewhere. Supply bidders with location drawing if possible.

- Specify kind of compressor foundations, and state by whom furnished.

(b) *Open-type Compressors*

- Specify if compressor is to be of the radial type, in line, V, etc.
- Specify number of cylinders, ________; bore, ________ in.; stroke, ________ in.; ________ rpm.
- Specify capacity in ________ Btu (or tons) per hour when operating with condensing temperatures of ________ F ambient air or ________ F water.
- Specify cylinder unloading arrangements so that compressor starts without load.
- Specify materials of crankshaft and materials of pistons and connecting rods.
- Specify type and materials of valves.
- Specify that crankshaft is to be both statically and dynamically balanced.
- Specify type of lubricating system, with oil filter and pump for positive circulation.
- Require bidders to furnish drawings and sectional views as necessary to illustrate the internal arrangement of the compressor.
- Require bidders to furnish statement of compressor volumetric efficiency at ________ F suction temperature and ________ F condensing temperature.
- Require bidders to state how their equipment is protected from slugs of liquid refrigerant.
- Specify that cylinder liner is to be removable, if so desired.
- Specify kind of motor, arranged for ________ volt, ________ phase, ________ cycles current, and ________ rpm.
- Specify if motor is direct-connected to compressor, or otherwise. For direct connection, specify kind of coupling; if belt drive, specify kind of belt, as flat, V, multiple, etc.
- Specify that all electrical components must satisfy the National Electrical Code, NEMA, and local codes as appropriate.
- Specify kind of foundations and by whom furnished.

- Specify kind and location of starting and stopping facilities.
- Specify kind of safety controls, as high head pressure cutout; temperature cutout both high and low. Specify actuating pressures and temperatures.

(c) *Compressor Instruments*

- Specify pressure gages for both suction and discharge, giving range of calibration from ______ psig to ______ psig.
- Specify dial diameter as ______ in.
- Specify if gages are to be indicating or recording.
- For details of the gages, refer to Sec. 4-21, Power Plant Instruments.
- Specify thermometers for placement on both the suction and discharge connections of the compressor.
- Specify calibration range from ______ F to ______ F.
- Specify if thermometers are to be straight stem or angle stem, and state submergence of bulb, in.
- Specify if dial is to be column or round.
- For details of thermometer construction, refer to Sec. 4-21.

15-2. AIR CONDITIONING COMPRESSORS, CENTRIFUGAL

(a) *General Data*

- Specify capacity of compressor, as ______ tons per hour (or Btu per hour), when operating with a condensing temperature of ______ F water.
- Specify suction temperature, ______ F, and delivery temperature, ______ F.
- Specify number of stages, ______, and ______ rpm.
- Specify construction of compressor, as impellers to be of __________ (material); mounted on a ______ ______ (material) shaft; rotating element to be statically and dynamically balanced at the factory.
- Specify material of the casing.
- Specify the operating range of the compressor, as from full-rated load down to ______ per cent of full load.
- Require load control vanes or other automatic adjusting mechanism to suit.
- Specify electrical service to compressor motor as ______ volt, ______ phase, ______ cycle current.
- Specify kind of motor and method of cooling, as by refrigerant gas, or other.

- Specify lubricating system.
- Specify the control panel, listing the items to be mounted thereon, as the overload limit switch, load control adjuster, pilot lights, gages, and thermometers. Gages to indicate the suction pressure, _______ psig; oil pressure, _______ psig; any others desired or recommended by the manufacturer. For details of gages and thermometers required, see Sec. 4-21, Power Plant Instruments.
- Specify safety controls to stop compressor on loss of oil pressure, or excessive temperature, or on overload of the motor.
- Require bidders to furnish complete drawings of the equipment offered, with sectional views to show the internal arrangement of the compressor. Bidders also shall state the overall dimensions of the unit, and the weight, _______ pounds.
- Specify the refrigerant as _______________.
- Require bidders to provide performance curves, with the brake horsepower or kilowatt input as ordinates, and the percentage of full load as abscissa.

(b) *Compressor Testing*

- Specify that compressor is to be operated at _______ per cent overspeed for _______ minutes to determine correctness of balance and absence of vibration.
- Specify that casings shall be tested underwater with _______ psig air pressure, to prove absence of leaks.
- Specify that water-cooling coil in lubricating oil tank shall be tested to _______ psig air or water pressure and _______ psig with refrigerant.

15-3. AIR CONDITIONING CONDENSERS, REMOTE

(a) *Air Cooled, Remote Location*

- Specify the total cooling load, _______ Btu per hour.
- Specify operating conditions, as _______ F ambient dry bulb and _______ F wet bulb, at maximum load.
- Specify general arrangement, as mounted (roof) (slab), with (horizontal) (vertical) discharge of fans.
- Specify fan capacity as _______ cfm at _______ F air temperature, at a discharge pressure of _______ in. W.G., and a fan speed of _______ rpm.
- Specify drive, as (direct) (belt).
- Specify motor, as (totally enclosed) (splash-proof) or weather protected by _______________ (means).

- Specify electrical characteristics, _______ volts, _______ phase, _______ cycles.
- If direct-connected, specify kind of coupling, as (flexible) (rigid), etc.
- If belt-driven, specify sizes of pulleys and kind of belt, (flat) (V), etc.
- Specify maximum acceptable sound level, _______ decibels.
- Specify wind deflectors if considered necessary.
- Specify kind of fans, as (propeller) (squirrel cage) (other).
- Specify the condensing surface, _______ sq ft of area, also kind of tubing _______________ (material); number and kinds of fins, _______ in. centers, _______ _______ (material) and how attached to tubing.
- Specify the casing construction, as _______ gage steel; (zinc coated) (Bonderized), etc. Specify number, sizes, and locations of access doors to coil, motor, fans, and belts.
- Specify that condensing coil shall be tested to _______ psig at the factory.
- Specify kinds and locations of vibration isolators.
- Specify that all electrical components shall comply with the National Electrical Code, NEMA, and local codes as appropriate.
- Require bidders to submit drawings and data showing general arrangement of the equipment and giving weights and dimensions. Also motor horsepower, full-load amps and locked rotor amps.
- Require bidders to furnish information on controls provided to permit safe and satisfactory operation at low ambient air temperatures, as by diaphragm valve responding to variations in head pressure, by reheaters, etc.
- Specify that all wiring is to be installed by qualified electricians.
- Specify that receiver shall be fabricated in accordance with the ASME "Code for Unfired Pressure Vessels." See Sec. 15-3(b) below.
- Specify subcooling coil if considered desirable for the extra efficiency.
- Specify the kind of refrigerant to be used. Also how system is to be purged and charged after installation.

- Specify safety controls, as high head pressure cutout, fan motor overload cutout, etc.
- Specify gages and thermometers to be furnished, and give the calibration of each and locations for mounting. For details of instrument construction, refer to Sec. 4-21, Power Plant Instruments.
- For centrifugal fans, specify blading, as (forward curving) (nonoverloading).
- Require bidders to state critical speed of fans, as at ________ rpm.
- Require shafts to be statically and dynamically balanced.

(b) *Water Cooled, Shell and Tube*

- Specify capacity of condenser, ________ Btu per hour, when supplied with condensing water at ________ F in the volume of ________ gpm. Gas temperature, ________ F, entering the condenser.
- Specify general construction, as shell of ________________ (material); tubes of ________________ (material); arranged as U bundle or straight tubes, rolled into a tube sheet of ________________ (material).
- Specify condenser head design, as locations of water inlet and outlet, and gas and liquid refrigerant connections.
- If gaskets are used, specify kind.
- Specify test pressures for tubes and shell, ________ psig, respectively.
- Specify number and spacing of intermediate tube sheets to prevent tube sagging.
- Specify location of condenser with respect to compressors.
- Specify thermometers on water inlet and outlet, calibrated from ________ F to ________ F. Indicate if straight or angle stem, and whether scale is on a vertical column or on a dial.
- Specify pressure gage, ________ psig to ________ psig, on refrigerant gas line to condenser.
- Specify thermometer, ________ F to ________ F, on gas line to condenser. Also on the liquid line from the condenser, ________ F to ________ F.
- If receiver is provided separate from the condenser, specify that receiver must comply with ASME "Code for Unfired Pressure Vessels."

- Specify number and capacity of relief valves.
- Specify capacity of receiver, ______ pounds of refrigerant.
- Specify any auxiliaries desired, as gage glass connections.
- For safety requirements, as well as ASME, refer to the ASA B9.1, "Safety Code for Mechanical Refrigeration."
- Require bidders to submit complete drawings and data on condenser being offered, including dimensions and weights.

(c) ***Refrigerant Condensers, Evaporative***

- Specify capacity of unit, ______ tons per hour, or ______ Btu per hour.
- Specify general arrangement of condenser, as of (horizontal) (vertical) design.
- Specify if unit is to operate primarily as a wet coil condenser, with recirculated water sprayed over the coils, or as a dry coil with air circulated over the coils, or if operation will fluctuate from wet to dry depending on the season.
- Specify use of a subcooling coil if considered necessary or desirable to increase capacity.
- Specify kind of refrigerant ____________ (ammonia can be used with this condenser).
- Specify if coil is for the entire connected load, or if ______ circuits are to be used.
- Specify the operating conditions, as ______ F dry bulb and ______ F wet bulb.
- Specify how close to the wet bulb temperature the condenser shall approach in the final temperature of the refrigerant.
- Specify the location of the unit, and if it requires ductwork on the air intake, the air discharge, or both.
- Specify the fan performance as ______ cfm, and ______ hp (1) without ducts, (2) with ducts, (3) duct on one connection only.
- Specify fan, ______ rpm, for the appropriate duct connection; also the air delivery pressure, ______ in. W.G.

  Note: Coil performance not only varies with the number or absence of duct connections, but is affected by the nature of the surface, wet or dry, and by the basic horizontal or vertical design. Make clear in the specification each of these performance-affecting factors.

- Specify kind of water treatment to be provided to avoid growth of algae.
- Specify motor data for both fan and water pump, giving characteristics of electric service available, _______ volts, _______ phase, _______ cycles.
- Specify kind of belt for connecting motor and fan.
- Specify kind of coils, _______________ (material), how supported, and test pressure, _______ psig.
- Specify maximum acceptable windage loss, _______ per cent, at guarantee point of unit.
- Specify general construction, as casing of _______ _______ (material) and how protected against corrosion of a wet atmosphere; also _______________ (material) of trays, and method of water distribution over the coils.
- Specify a damper control arrangement, so head pressure can be maintained at most effective point over various ambient outdoor temperatures.
- Specify receiver in accordance with Sec. 15-3(b) preceding.
- Require bidders to furnish sectional drawings and photographs of equipment, stating all key dimensions and weights.
- Specify maximum acceptable noise level, _______ decibels.
- Specify if motors are to be totally enclosed or splashproof.

15-4. COOLING TOWERS

(a) *Forced Draft Towers*

- Specify capacity of tower in _______ tons of refrigeration when circulating _______ gpm of water entering the tower at _______ F and leaving at _______ F with a wet bulb of _______ F.
- Specify general construction of the tower, as casing of (redwood) _______ (gage steel) (asbestos board), etc. Filling to be of _______________.
- Specify kind of eliminators and state maximum acceptable windage loss, _______ per cent of water circulated.

- Specify if water distribution is by gravity or by pumped sprays.
- Specify if air discharge is vertical or horizontal.
- Specify that all nails, screws, bolts, nuts, and other fasteners must be of nonferrous material.
- Specify material of fans.
- Specify capacity of fans, as _______ cfm against a head of _______ in. W.G. when handling air at _______ F.
- Specify kind of fan drive, as (direct-connected) (belt drive). If direct, specify coupling as flexible or rigid; if belt drive, specify belt as flat, V, etc.
- Specify kind of motors (splash-proof) (drip-proof) (totally enclosed). Give electric service characteristics, _______ volts, _______ phases, _______ cycles.
- Specify finish of casing, color, paint, etc.
- Specify location of tower; roof, slab on the ground, etc.
- Specify the water pump circulating the water from the refrigeration condenser and to the distribution trough or nozzles, this pump to be of the (centrifugal) (rotary) (other) type. Capacity of pump to be _______ gpm against a total head of _______ ft when handling water of _______ F temperature.
- Specify materials of the pump, rotating parts and casing respectively. Specify pump rpm, _______.
- Specify the kind of pump drive, as (direct-connected), etc.
- Require bidders to furnish drawings and all details of the tower offered, giving the dimensions and weights.
- Require bidders to supply operating data at three points of operation, as _______ F dry bulb and _______ F wet bulb respectively, the flow of water to be constant at _______ gpm.
- Specify internal piping as (brass) (copper) (galvanized), etc.
- Specify that all electrical components must conform to the requirements of the National Electrical Code, NEMA, and local codes as appropriate.
- Specify the kind of foundations required, and by whom provided.

(b) *Induced Draft Towers*

Note: Specify as for a forced draft tower, Sec. 15-4(a), but note that fans operate on an induced draft arrange-

ment. This is the customary arrangement rather than forced draft, although both are encountered.
Induced draft allows an excellent appearance of the tower, especially when a ceramic casing or shell is provided as shown in Fig. 15-1.

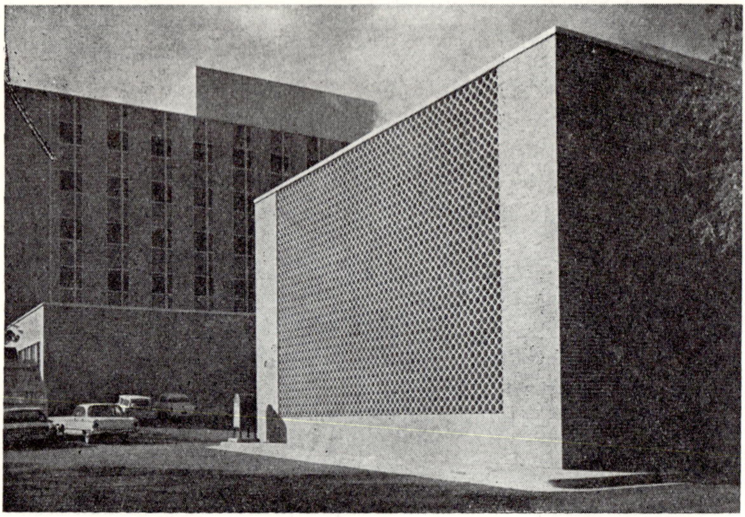

FIG. 15-1. Ceramic shell of 800-ton induced draft cooling tower. (*Courtesy Ceramic Cooling Tower Company.*)

(c) *Natural Draft Towers*

- Specify same as forced draft towers, omitting any reference to fans, fan motors, etc.
  - Construction of natural draft tower casing, filling, distribution system, basin, etc., can be of the identical material used with mechanical draft towers.
- Specify distribution; gravity or pump.

(d) *Hyperbolic Cooling Towers*

Note: Generally used for cooling circulating water for the condensation of steam. These are natural draft towers.

- Specify the capacity of the tower, to cool water from ________ F to ________ F in a volume of ________ gpm.
- Specify the atmospheric operating conditions, as ________ F dry bulb and ________ F wet bulb. (Alternatively, specify the dry bulb, ________ F, and relative humidity, ________ per cent.

- Specify the temperature of the leaving air, _______ F, and its relative humidity, _______ per cent.
- Specify that tower is to be constructed of _______ concrete, cast in place, or other construction.
- Specify limiting height, limiting overall diameter, or both, if applicable.
- Specify the normal cooling range, deg _______ F.
- Specify who clears and prepares ground site.
- Specify ground bearing capacity, _______ tons per sq ft, or require bidders to determine this and offer bids accordingly.
- Require bidders to furnish drawings and complete data, including all dimensions and weights.
- Specify the kind of internal piping.
- Specify the water pump; kind, capacity, _______ gpm, against a total head of _______ ft.
- Specify driving motor for pump, as kind; hp; speed, rpm; and connection to pump—whether direct, gears, or by belt.
- Specify the guaranteed approach to the wet bulb, _______ deg F.
- Specify electrical service characteristics for pump motor, _______ volts, _______ phase, _______ cycles.
- Specify maximum loss by windage, _______ per cent.

15-5. CHILLERS

(a) *Centrifugal Compressor Type*

- Specify the compressor in accordance with the details of Sec. 15-2(a) and (b).
- Specify the condenser in accordance with the details of Sec. 15-3(b).
- Specify the shell and tube type evaporator similarly to the condenser, which is also shell and tube, Sec. 15-3(b). In addition, specify the furnishing of a refrigerant float valve to function with the evaporator and control the flow of refrigerant.
- Specify the kind of liquid carryover eliminators desired for placement in the evaporator, naming _______ _______ material and effectiveness of liquid removal, _______ per cent.
- Specify the temperature drop in the water from _______ F incoming to _______ F leaving the evaporator.
- Specify the quantity of water circulated, _______ gpm.

- Specify the location and mounting of the control panel, as on the chiller, on the wall, etc.
- Specify all gages and thermometers to be mounted on the panel, or on the liquid and gas lines. Refer to Sec. 4-21 for details of the gages and thermometers.
- Specify the maximum acceptable pressure loss in the tubes of the evaporator and in the tubes of the condenser, respectively, ________ psig.
- Specify the safety controls of the chiller, as high and low limits on the compressor discharge, ________ psig.
- Specify finish of equipment; paint and colors.
- Describe foundation and state who furnishes.
- If vibration isolators are desired, specify type and placement.
- Require bidders to submit complete drawings and details of the equipment offered, showing internal arrangements of each component and the overall dimensions and weights.
- Specify the chilled-water circulating pump, stating type as centrifugal, rotary, reciprocating, other.
- Specify capacity, ________ gpm against a total head of ________ ft.
- Specify rpm ________ and drive as (direct) (belt) (gears); speed of driver, ________ rpm.
- For direct drive, specify kind of coupling, (flexible) (rigid) (resilient).
- For belt drive, specify kind of belt; flat, V, other.
- For rigid coupling or resilient coupling, specify material, (cast iron) (steel).
- For gear drive, specify the kind of gears and method of lubrication.
- For detailed specifying of the pumps, refer to Sec. 12-7(j), Circulating Pump.
- Require bidders to submit complete pump data, including ________ hp and performance curves showing relationships between delivery, gpm; head, ft; and hp or kw input.

(b) *Reciprocating Compressor Type*

- If compressor is hermetically sealed, specify in accordance with Sec. 15-1(a).
- If compressor is of the open type, specify in accordance with Sec. 15-1(b).
- Specify evaporator and condenser as in the preceding

Sec. 15-5(a), unless the evaporator and condenser are contained in a single housing. For single-housing design, specify how the evaporator and condenser tubes are separated.

- Specify controls, safety limit switches, gages, and thermometers as in Sec. 15-5(a).
- If receiver is to be used, specify as indicated in Sec. 15-3(b).
- Specify other data from Sec. 15-5(a) as may be appropriate to a reciprocating installation.

(c) *Chiller Miscellaneous Equipment*

- Specify as desired:
- Kind of insulation; thickness, ________ in.; and where to be used.
- Liquid line sight glasses.
- Refrigerant driers and strainers.
- Expansion valve. Specify kind, capacity, and with or without equalizing connection.
- Manual switches for starting and stopping. State locations.
- Subcooling coils.
- Any arrangement of superheating heat exchanger.
- Refrigerant piping. Specify material.
- Stop valves on chilled-water connections. Specify material; ________ psig test pressure; rising or nonrising stems; ends for soldering, welding, other.

(d) *Chiller–Gas-engine Package Units*

- For the gas engine, refer to Sec. 5-5(a)–(f) as appropriate.
- Specify if package unit is to operate air-cooled, water-cooled, or with an evaporative condenser.
- Specify clearly where the package unit supplier's work begins and ends. Example: All condensing equipment (or cooling equipment), together with interconnecting piping, shall be supplied by others.
- If water circulating pumps to tower are to be driven by engine power takeoff rather than by separate pumps, so state.
- If package supplier offers choice of compressor to be hooked up with the gas engine, specify what compressor, model and type, is desired.
- Specify compressor details as indicated in Sec. 15-1(a) and (b) as appropriate.

- Specify controls and safety switches and cutouts as indicated in Sec. 15-5(a).
- Specify miscellaneous equipment as in Sec. 15-5(c).
- Require bidders to submit complete data and drawings showing performance, dimensions, weights, materials, etc.
- Specify location of unit and give maximum acceptable sound level, in decibels.

DIVISION 16

Refrigerant Absorption Systems

16-1. HEAT SOURCES FOR ABSORPTION INSTALLATIONS

Note: In specifying an absorption refrigeration installation, the source of the heat must be identified. In most instances, the source of heat has been provided primarily for other purposes, such as power generation, industrial processing, and others. Heat diverted to absorption cooling is of an excess or waste nature, but the vehicle carrying the heat as steam or a hot gas basically affects the absorption design, and is therefore of importance to the specification writer.

(a) *Steam from a Low-pressure Boiler*

- Boiler specification data are given in Sec. 11-2(a) for cast-iron boilers, primarily used in premises heating.
- For steel boilers, similarly used, refer to Sec. 11-2(b). Specify delivered steam pressure and temperature.
- Specify how much of the connecting piping is to be provided and installed by the absorption cooling contractor. Define his work limits clearly.

(b) *Exhaust Steam*

- Refer to Sec. 11-2(c) and (d).
- Specify who furnishes and installs interconnecting piping. Define the absorption contractor's work limits clearly. Specify pressure, ________ psig.

(c) *Steam from a District Steam Supply*

- Refer to Sec. 11-1, all subheads.
- Specify steam pressure, ________ psig, that is to be delivered to the absorption generator.
- Specify steam temperature, ________ F, or quality, as delivered to absorption generator.
- Set work limits clearly for absorption contractor.
- Specify ASME safety valve. Discharge to be led outside of building.

(d) *Steam from a Hot Water Converter*

- Refer to Sec. 10-2(a).

- Specify clearly where absorption contractor's work starts and stops.
- Specify steam condition to absorption generator as ______ psig, pressure, and ______ F temperature.

(e) *Steam from a Waste-heat Boiler*

- Refer to Sec. 11-2(h).
- Specify steam conditions of pressure and temperature.
- Specify material and installation limitations as these affect the absorption contractor.

(f) *Heat from Hot Water*

- Specify volume of hot water supply, ______ gpm.
- Specify temperature of hot water supplied to generator, ______ F, and temperature of water leaving, ______ F.
- Specify pressure of water as delivered to generator, ______ psig.
- State source of hot water, as from a hot water boiler, from a steam-to-water converter, or other source.
- Specify who furnishes and installs the hot water system, pumps, piping, controls, etc.

(g) *Direct-fired Generators (Concentrators)*

- Specify fuel, as (natural) (artificial) (liquefied petroleum) gas; or fuel oil.
- Specify calorific value in ______ Btu per cu ft, or per pound, or per gal of oil.
- Specify the design and construction of the generator.
- Specify the kind of refrigerant to be used, as (a proprietary chemical) (lithium chloride) (lithium bromide) (ammonia) (other).
- For specifications of the boiler, refer to Sec. 11-2(a) for low-pressure cast-iron boilers and to Sec. 11-2(b) for low-pressure steel boilers.
- For specifications applicable to fuel handling, refer to Sec. 4-9, Fuel-oil Storage and Handling for Steam Plants; also to Sec. 4-10, Gas Fuel Handling for Steam Plants. Follow each section's subheadings as appropriate.
- Specify clearly how much of the direct-firing equipment and its connections shall be supplied and installed by the absorption machine contractor.
- Require bidders to furnish complete data on the direct-fired equipment, including arrangement drawings, dimensions, and weights.

16-2. ABSORPTION MACHINERY DETAILS

Note: The absorption cycle commences in the evaporator, where a partial vacuum above the surface of the water causes evaporation of the water and a consequent lowering of the temperature of the water remaining in the bottom of the evaporator. This chilled water is then pumped to the points of desired cooling.

The partial vacuum in the evaporator is produced by the affinity of a lithium bromide (or some other suitable salt) for water; consequently the water vapor in the evaporator is piped to another vessel called an absorber, and is there absorbed. This mixture of water and bromide is then pumped through a heat exchanger to the generator, a vessel in which steam coils or direct firing boils off the water and permits the concentrated solution to flow back to the absorber. Meanwhile the water vapor from the generator is led to a condenser, from which, as pure condensate, it proceeds to the evaporator to begin the whole cycle anew.

Figure 16-1 shows this cycle schematically.

(a) *Operating Conditions*

- Specify the capacity of the unit, ________ tons of refrigeration per hour.
- Specify the volume of chilled water to be circulated, ________ gpm, leaving the evaporator at a temperature of ________ F and returning at ________ F.
- Specify the amount of condensing water to the vapor condenser, ________ gpm.
- Specify the temperature of the water, ________ F, entering the condenser coil, and ________ F, leaving.
- Specify the amount of condensing water to the absorber coil, ________ gpm. Also specify the temperature of this water, ________ F, entering the coil, and ________ F, leaving.
- Specify maximum allowable pressure drop, ________ psig, for the condensing water in each coil, respectively.
- For hot water heat source, see Sec. 16-1(f).
- For steam as a heat source, refer to Sec. 16-1(a), (b), (c), (d), or (e), as appropriate.

(b) *General Construction*

- Specify shell material as ________ gage steel.
- Specify tubes and headers of ________________ mate-

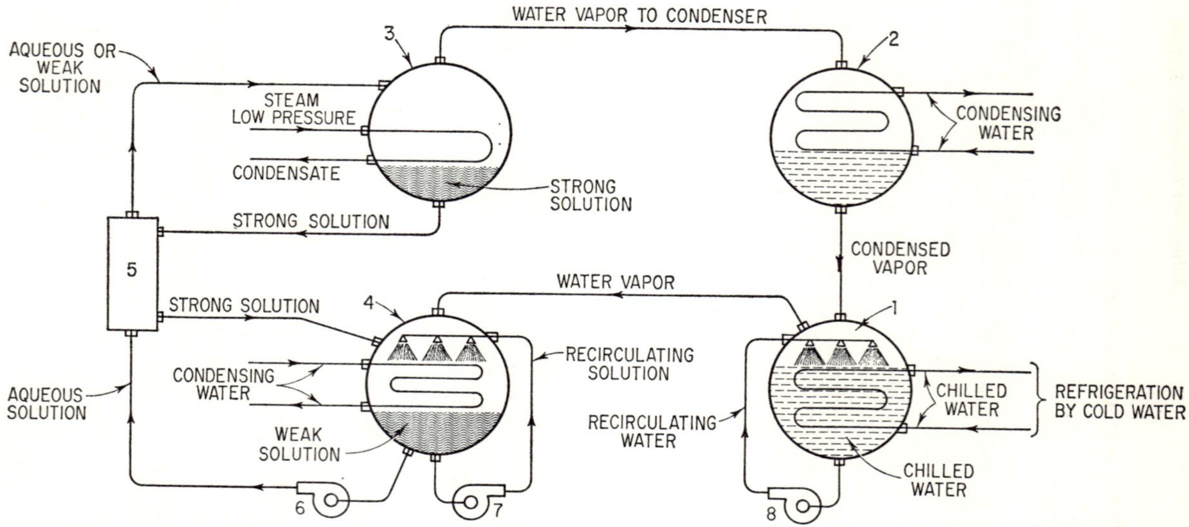

FIG. 16-1. Schematic diagram, lithium bromide system of absorption with low-pressure steam. (1) evaporator or chiller, (2) water vapor condenser, (3) generator where heat boils water out of solution, (4) absorber where water vapor enters solution, (5) heat exchanger, (6) weak solution pump, (7) absorber spray pump, (8) chiller pump.

rials, respectively. Specify design working pressure as ______ psig and test pressures as ______ psig.

- Specify how tubes are to be rolled into tube sheets or headers.
- For welded shells, specify kind of weld, as electric, gas. Require qualified welders only.

(c) *Pumps*

- List pumps to be furnished with equipment, as absorber pump; strong solution or generator pump; evaporator pump.

  Note: The generator is sometimes referred to as a "concentrator," since the lithium bromide is concentrated in this vessel by the driving off of the water.
- Since these pumps are an integral part of the machine, require bidders to submit complete drawings and data concerning these pumps and the locations with respect to the absorption machine.

  Data submitted shall include the pump capacity, ______ gpm, the discharge head, ______ ft; the construction of the pumps, ____________ materials; method of lubrication; how driven; if electric motor, bidders to state needed voltage ______, phase ______, cycle ______ current.

  If pump and motor are hermetically sealed, data should so state. How motors and bearings may be replaced shall be described.

(d) *Purge System*

- Require bidders to describe the purge system being incorporated in the machine, stating kind and model of vacuum pump being furnished.
- Bidders to state the maximum time needed to purge off all noncondensable gases, ______ minutes or hours per purge. Also state predicted frequency of needed purges, ______ days, ______ weeks, etc.
- Bidders also indicate how a power failure of the vacuum pump is guarded against in relation to the reentry of air into the system.
- Specify that purge pump shall not draw solution from the machine.

(e) *Decrystallization*

- Require bidders to describe methods or give recommendations for clearing up crystallization on and fouling of the heat exchange surfaces.

- Specify if action is to be automatic or manual.

(f) *Control Center*

- Specify location of central control panel and list items to be mounted on the panel, as time delays, for whatever purpose; motor starters; push-button stations; fuses; high-temperature and low-temperature cutouts; lights to indicate the operating or nonoperating status of the unit and of its individual pumps; a light to indicate the presence of excessive fouling of the condenser; any other items suggested by the particular installation and operating arrangements.

(g) *Instruments*

- Specify the details of the gages and thermometers as shown in Sec. 4-21.
- Specify the calibration range of each, as from ______ psig to ______ psig for pressure; from ______ F to ______ F for temperature.
- Provide a list of all the instruments to be provided by this system and to be furnished with the absorption machine. Indicate on this list where each is to be mounted.

(h) *Insulation*

- Specify where insulation shall be placed.
- Specify kind of insulation, as ____________ (name) and thickness, ______ in.
- If this contractor is to insulate the chilled-water piping for any distance adjacent to the evaporator, or over the entire piping system, so specify. Otherwise designate clearly where the work of others begins.

(i) *Valves*

- Require bidders to list all valves normally furnished with their equipment.
  - Giving kinds, sizes, and materials. Also design working pressure.

(j) *Performance*

- Specify maximum full-load steam rate, ______ pounds per ton.
- Require bidders to state coefficient of performance (COP).
- Require bidders to submit performance curves showing consumption of steam or hot water at the pressure, ______ psig, of the steam, or ______ F temperature of the water, over a chilled-water-leaving temperature

range of ______ F (usually 40 F) to ______ F (usually 50 F).

- For direct-fired absorption systems, bidders shall substitute for the steam or hot water, the fuel burned, as ______ cu ft of ____________ gas; ______ gal or pounds of oil. Specify the calorific value of the fuel, ______ Btu per unit of weight or volume.

16-3. AMMONIA SYSTEMS

Note: Although the thermodynamics of the ammonia system are identical with that of a lithium bromide arrangement, there are two notable differences in the cycles.

First, water in the latter functions as the refrigerant, and by evaporating into a high vacuum, lowers the temperature in the body of the evaporator water to something above freezing, as desired. By contrast, in an ammonia system the ammonia is the refrigerant and does the evaporating; and since it does not become solid until reaching subzero temperatures, it is useful in achieving very low temperatures where necessary, at positive pressures.

The absorbent for the water refrigerant is the lithium bromide; the absorbent for the ammonia is water. Separation of the ammonia from its water absorbent is more difficult than the separation of water from a lithium bromide absorbent. This means that for ammonia systems, as a rule, some auxiliary separating equipment, such as an analyzer or a rectifier, is needed; sometimes a system will include both. These auxiliary units must be specified as additions to the lithium bromide details of Sec. 16-2, Absorption Machinery Details, when using Sec. 16-2 as a basis for an ammonia system specification.

Sources of heat for ammonia systems are the same as for lithium bromide machines, as given in Sec. 16-1, all subheadings.

(a) ***Operating Conditions***

- Refer to Sec. 16-2(a).
- Specify maximum amount of water vapor in the mixture of ammonia and water vapors entering the condenser.

(b) ***General Construction***

- Specify the materials in each component, as the condenser, absorber, etc. Tubes and tube sheets likewise.
- Specify that all pressure vessels must satisfy the re-

quirements of the ASME "Code for Unfired Pressure Vessels" as appropriate.

- Specify how tubes are to be rolled into tube sheets for the condensing vessels.

(c) *Pumps*

- Refer to Sec. 16-2(c). List pumps needed.
- Specify what pumps shall be suitable for handling various concentrations of aqua ammonia and what pumps will handle water.
- If brine is involved, specify number, size, and kind of brine pumps to be furnished.

(d) *Purging*

- Require bidders to describe means or methods supplied or recommended for eliminating air and noncondensable gases from high side of the machine.
- Ditto for the low-pressure side. If foul-gas remover is recommended, so specify.

(e) *Analyzer or Bubble Column*

- Specify working pressure of column, ________ psig. Require compliance with ASME "Code for Unfired Pressure Vessels" as applicable.
- Specify number, placement, and materials of the baffles or plates in the column.
- Require bidders to submit drawings and complete data of this component as well as of the others, giving dimensions and weights.

(f) *Rectifier (Heat Exchanger)*

- Specify type of rectifier, as (double pipe)(multitube) (other).
- Specify operating conditions, as the temperature, ________ F, of the strong liquid entering the unit, and the temperature, ________ F, leaving. Ditto for the vapor from the generator, ________ F.
- Specify working pressures of tubes, ________ psig, and of the multitube shell.

(g) *Condensers*

- Specify type of condenser, as (shell and tube)(double pipe)(multitube single pass)(multitube multipass) (other).
- Specify materials of tubes and shells as appropriate. Give required working pressures, ________ psig.
- Specify cooling water for each condenser, ________ gpm at a head of ________ ft, and at an entering temperature

of ______ F. Stipulate temperature leaving, ______ F, also pressure drop, ______ psig.

(h) *Generator*

- Specify heat medium, as steam, hot water, or direct-fired. For steam, refer to Sec. 16-1(a)–(e).
- For hot water, refer to Sec. 16-1(f).
- For direct-fired, refer to Sec. 16-1(g).
- Specify type of generator; (coil)(multipass)(other).
- Specify working pressures, ______ psig.
- Specify any special features, as analyzer built into the generator.
- Specify insulation ____________ (name and kind); thickness, ______ in.

(i) *Absorber*

- Specify if horizontal or vertical, multitube type.
- Specify baffle plates, materials and placements.
- If weak liquid is sprayed into absorber, specify kind of spray heads.
- Specify liquid level controller.
- Specify liquid gage glasses.
- Require bidders to furnish drawings and complete data showing piping connections, dimensions, and weights.

(j) *Controls*

- Refer to Sec. 16-2(f).
- Expand, modify, or otherwise change to suit requirements of the ammonia system.

(k) *Instruments*

- Provide a list of all instruments and gages to be furnished with the system, indicating the location of each.
- Specify the range of calibration for each, from ______ psig to ______ psig on pressure gages, and from ______ F to ______ F on thermometers.
- For specifying the physical details of both gages and thermometers, refer to Sec. 4-21.

(l) *Valves and Fittings*

- Specify materials and working pressure of each valve and fitting. Prepare a list showing locations, quantities, and other pertinent data. Designate those valves and fittings to be designed for ammonia service.

(m) *Performance*

- Specify as shown in Sec. 16-2(j).
- Require bidders also to submit table, or curve, showing operating pressures in the condenser for various tem-

peratures of condenser circulating water from ______ F to ______ F.

- Require bidders to submit a recommended table of steam pressures, ______ psig, to be admitted to the generator in relation to the condensing water temperature, and to the strength of the strong liquid, degrees Baumé.
- Require bidders to state pressure range, from ______ psig to ______ psig, of the absorber, based on a cooling water flow through the absorber of ______ gpm at ______ F. These data may be furnished as a table or on a curve.

(n) *Operating and Maintenance Instructions*

- Specify this contractor must supply, in duplicate, a complete set of operating instructions, including such details as the packing of the ammonia pumps, the adjustments on the automatic regulator, and similar information. Also methods for testing for ammonia leaks.
- Require contractor to furnish a list of safety equipment recommended, as helmets or gas masks suitable for service in ammonia fumes. Specify if the equipment on this list is to be furnished by the absorption system contractor, or will be purchased separately.

(o) *Responsibility for Work*

- Specify clearly what the contractor is to supply and where his work begins and ends. If necessary for clarity, list every component of the system and indicate by whom furnished and by whom installed.

16-4. STEAM JET SYSTEMS

Note: As with the lithium bromide absorption system, the vaporizing of water in a high vacuum in a vessel known as the evaporator chills the remaining water in the evaporator down to a desired temperature—for air conditioning service, usually between 40 and 45 F.

The high vacuum in the evaporator is produced by means of high-pressure steam supplied to one or more steam jet exhausters (or ejectors) mounted on the evaporator. The air and water vapor in the evaporator are carried off by the steam jet to a primary condenser, where the vapors are condensed and removed by a condensate pump. Adequate vacuum on the primary condenser is maintained, often as high as 28 in., Hg column, by additional steam jet exhausters which, in turn, discharge into a second condenser.

Condensing water through both condensers flows through tubes which may be either straight or return bend, depending on the designer's judgment.

Condensing water may be obtained from a city supply and run to waste, or may be recirculated through a cooling tower to meet conditions.

The chilled water is circulated by pumps.

Steam pressure for efficient and economical performance should be 100 psig or higher.

(a) ***Capacity of System***

- Specify rating of system in _______ tons per hour, when supplied with steam at _______ psig and _______ F total temperature, with a condenser vacuum of _______ in. Hg.
- Specify chilled water to be circulated, as _______ gpm at a leaving temperature of _______ F and a return temperature of _______ F.
- Specify the number and arrangement of the steam jet exhausters mounted on the evaporator, noting if a single set of nozzles is incorporated in each exhauster, or if the total number of nozzles is divided into starting and running nozzles.
- Specify capacity of evaporator vessel, _______ gal or _______ cu ft.
- Specify if return water is to be sprayed into the evaporator from spray heads or from a perforated pipe.
- Specify source and temperature range of makeup water, from _______ F to _______ F.
- Specify the arrangement and materials of the primary condenser, its tubes and tube sheets, heads and water boxes.
- Ditto for the auxiliary condenser. Also specify how many stages shall be provided for this condenser.
- Specify the range of temperature in the condensing water, from _______ F to _______ F. Also specify the volume of condensing water to pass through each condenser, _______ gpm.
- Require bidders to submit a curve or tables showing relationship of vacuum in the condensers to the temperature of the condensing water at a fixed _______ gpm, as,

vacuum, _______ in. Hg versus temperature, _______ F.

(b) *Steam Jet Exhausters*

- Specify materials of construction, with particular attention to the nozzles, as (stainless steel) (molybdenum steel), etc.
- Require bidders to submit tables or curves showing steam consumption per exhauster at different vacuums in the condenser. Specify rating points on which a curve can be constructed by the bidders.
- Require bidders to show evacuating capacity of each jet exhauster in _______ cfm, as mounted on the evaporator.

(c) *Condensers*

- Specify type of condenser; shell and tube, straight tubes, return tubes, etc.
- Specify capacity, _______ pounds of steam and water vapor condensed per hour, when served with condensing water of _______ F at a rate of _______ gpm.
- Specify construction and materials of tubes, tube sheets, water boxes, shell, tube supports, etc.
- Require bidders to submit complete data and drawings showing arrangement of condenser interior and locations of air, water, and condensate connections, as well as steam opening. Also all dimensions and weights.
- Specify location of the steam jet complex, noting any restrictions of entry for large components.
- Indicate the kind of foundations to be provided and by whom provided.

(d) *Pumps*

- List the pumps required and give the capacity, _______ gpm, and total head, _______ ft, of each. As, chilled-water pumps; condenser condensate pumps; condenser circulating water pumps.
- Specify electrical service characteristics available for pump-driving motors, as _______ volt, _______ phase, _______ cycles.
- Require motors to comply with National Electrical Code, NEMA, and local codes as applicable.
- Specify coupling between pump and motor, as flexible, rigid, or resilient.
- Require bidders to furnish pump data, including speed,

________ rpm; ________ hp; materials of construction, and manufacturer's name and model number.

- Require bidders to furnish pump performance curves, showing hp versus head in ft and resultant capacities in gpm.

(e) *Cooling Towers*

- Specify kind of cooling tower, as forced draft, induced draft, or natural draft.
- Specify capacity in ________ gpm to be cooled from ________ F to ________ F, with an ambient dry bulb of ________ F and an ambient wet bulb of ________ F.
- For details of tower refer to Sec. 15-4(a), as these concern a forced draft design.
- For details of the induced draft design, refer to Sec. 15-4(b).
- Natural draft and hyperbolic towers are specified in Sec. 15-4(c) and (d).
- Specify that the operation of the tower fan or fans shall be started and stopped automatically as necessary to maintain the desired temperature of the condenser circulating water. Describe desired method, as by thermostat in the leaving water. Specify fan stopping and starting temperatures as ________ F and ________ F.
- If tower bypass is provided instead of fan control and is to be operated automatically, state this fact, and as above, give temperatures at which the bypass is to open and close.
- Specify the kind, size, and operating medium of the bypass valve, as (pneumatic) (electric) (other).
- Specify kind of water treatment to be provided, if any, to eliminate the growth of algae.

(f) *Steam Source*

Note: Applications of steam jet systems are met commercially more often on district steam supply lines than elsewhere.

As with all district steam connections, certain rules of installation must be complied with. If the specifications of a steam jet system are to include the initial entry of district steam to a building, refer to Sec. 11-1, all subheads, for information on the valves and piping to be provided.

Since high-temperature, high-pressure hot water can be used to provide steam from a converter at 100 psig or more, the possibility of a steam jet system being energized from such a source must be considered.

For information on how to specify such an arrangement, refer to Sec. 10-2(a).

If steam from a waste-heat boiler can be obtained at a high enough pressure, refer to Sec. 11-2(h).

- Specify the pressure, ______ psig, and temperature, ______ F, of the required steam.

(g) ***Instruments***

- Make a list of the gages and thermometers needed, specifying the calibration range of each.
- Specify in each case where the gage or thermometer is to be mounted.
- Specify if stems are to be straight or angle.
- For details in the construction of the instruments, refer to Sec. 4-21, subheadings as appropriate.

(h) ***Control Panel***

- Specify where panel is to be mounted, giving approximate dimensions. In addition to the gages and thermometers listed above, state what instruments and electrical or pneumatic devices are to be mounted on the panel.
- Specify material of panel and painting—color and finish.

(i) ***Safety Valves, Steam***

- Specify safety pressure relief valves, to be in accordance with the ASME Code.
- In all cases, lead the valve discharge to a point outside the building where it offers no hazard to persons or materials.

DIVISION 17

Heat Pump Air Conditioning

Notes. A refrigeration compressor properly sized to satisfy a building cooling load seldom is able to "pump up" the heat of evaporation of the refrigerant to a level adequate for meeting the same building's winter heating demands.

To compensate for this lack, a common practice is to furnish supplementary heat from electric strip heaters, arranged to be energized in steps of about 4 kw each, as the weather makes necessary.

Heat sources and heat sinks may be available in the atmospheric air, or in water circulated from wells through a heat exchanger. Various attempts to use earth coils have not been uniformly successful and in general are given little consideration by commercial designers.

Heat pump systems can be designed for the use of package units, or with inside-outside combinations, commonly termed "split systems."

On a split system installation, certain limitations must be recognized in the maximum practical distance, feet, between the inside and outside components, and in the differences of elevation. The specification writer therefore must call bidders' attention to proposed distances and to the relationship of one unit to the other in elevation.

17-1. GENERAL NOTES

(a) *Air-to-Air Heat Exchange Designs*

- Specify the cooling capacity required of _______ Btu per hour.
- Specify if package is to be located indoors or outdoors. Specify duct connections, whether for indoor coil, outdoor coil, or both.
- Specify finish of casing, both colors and manner of weatherproofing for an outdoor installation, as well as painting and colors for indoor unit.
- Specify compressor to be resiliently mounted.
- Specify maximum outdoor summer temperature, _______

F, and minimum outdoor winter temperature, _______ F.

- Specify defrosting methods for outdoor coil, as (electric) (hot gas bypass), etc.
- Provide drawing showing unit location, manner of support, and runout of the drain pipe from the catch basin under the cooling coils.
- Specify fan capacity for circulating the treated air through the indoor spaces, _______ cfm, when handling at a temperature of _______ F.
- Specify fan capacity for fan blowing outside air over the outside coil, _______ cfm, at _______ F.
- Specify form of supplementary heat, as _______ (number) of strip heaters, operating in increments of _______ kw.
- Require bidders to furnish drawings and complete operating data of the unit, giving dimensions, weights, and internal arrangement of the unit. Also require information by bidders on motor horsepowers for the compressor and for each of the fans, together with rpm of each.
- Specify the kind and effectiveness of the filters, whether throwaway type, washable, etc.
- Specify the kind and thickness of insulation inside the casing for deadening sound, as (fiber glass) _______ in. thick.
- Specify the coils material and construction, as (aluminum fins bonded to copper tubing) and tested to a pressure of _______ psig. If coil guard is desired for the outside coil, so specify.
- Specify methods of control, as heating-cooling thermostat, outdoor thermostat, automatic heating-cooling reversing valve directed by the heating-cooling thermostat.
- Specify if controls are to function at line voltage or reduced voltage. Give service available as _______ volts, _______ phase, _______ cycles.
- Specify that heat pump must carry the ARI certification of correct rating in the form of a seal. Also require unit to comply with the stipulations of the Underwriters Laboratories and the National Electrical Code as applicable.
- Require bidders to state area of each coil, _______ sq ft, and to give pressure loss through the coil of _______ in. W.G.
- Specify the heating load, _______ Btu per hour.

- Because the volume of the treated air is substantially less on the heating cycle, consider specifying a two-speed motor, the lower speed to circulate warm air.
- Require bidders to submit operating curves showing the changes in the unit's heat capacity caused by the outside temperatures varying from a minimum of _______ F to a winter high of 65 F.
- Curves should indicate temperatures at which the supplementary heaters are energized.

(b) ***Water-to-Water Design***

Note: Water from a convenient well, lake, river, or city supply becomes a heat sink on the cooling cycle and a heat source on the heating cycle. Equipment is identical on both cycles, the major components being the compressor, heat exchanger, and condenser, together with interconnecting piping, valves, and sometimes a raw water pump. The water, after giving up heat on the heating cycle, or absorbing heat on the cooling, should be disposed of advantageously if possible, as in the irrigation of gardens, farms, or to a dry well. Changeover, as with all heat pumps, is automatic, a changeover valve reversing the refrigerant flow through the evaporator and condenser as necessary.

- If the water supply is dirty, specify filters, giving the kind of filter, efficiency, and maximum allowable pressure drop when passing _______ gpm at an entering pressure of _______ psig.
- Specify maximum summer water temperature, _______ F, and minimum winter temperature, _______ F.
- Specify capacity on cooling cycle, as _______ tons or _______ Btu per hour.
- Specify heating capacity of the pump: _______ Btu per hour from compression and _______ Btu per hour supplementary heat under rated conditions.
- If raw water pump is also provided by this contractor, specify kind (centrifugal) (rotary), etc., and capacity of _______ gpm against a total head of _______ ft, when handling water at a temperature of _______ F.
- Specify controls, as by thermostat, and motor overload safety cutout devices.
- Require bidders to submit drawings and data providing all desired details of dimensions, weights, internal arrangements, and other physical data.
- Require performance curves over the operating tempera-

ture range of _______ F in winter to _______ 65 F, showing kw input and temperatures at which successive steps of the supplementary heating equipment are energized, and the additional kw at each step.

- Specify if unit is to be vertical or horizontal, and state the proposed location.
- Specify who drills the wells or pipes the river water, lake water, or city supply to the heat pump.
- Specify starting equipment, as push-button station or safety switch. State location.

(c) ***Water-to-Air Units***

Note: In this arrangement, air coils function alternatively as cooling coils or heating coils in accordance with the condition of the refrigerant entering them. For cooling, liquid refrigerant is admitted through an expansion valve, vaporizes, and thus absorbs heat from the passing air. It is drawn off by the compressor. On the heating cycle, refrigerant passes into the coils as a hot gas direct from the compressor, there giving up heat to the air. The gas is drawn through the heat exchanger by the compressor, completing the condensing action begun in the coil.

The direction of flow is reversed from cooling to heating, condenser (heat exchanger) to coil on cooling; compressor to coil on heating.

- For specification details, refer to Sec. 17-1(a) and (b) as applicable.
- Specify any particular conditions to be satisfied.

17-2. SPLIT SYSTEMS

- Specify the location of the outdoor unit, as (roof) (ground), and kind of foundation being provided.
- Specify painting, weatherproofing, and other treatment of the exposed metals.
- Specify kind of operation, as air-to-air, water-to-water, or water-to-air.
- Specify location of the indoor unit.
- Specify drip pan to be provided under cooling coils, with drainpipe to be led outdoors.
- Specify maximum distance, _______ ft, between the outdoor and indoor components of the system.
- Specify difference in elevation, _______ ft.

- Require ratings to carry the ARI certification.
- Require bidders to provide drawings and details of both outdoor and indoor sections, giving dimensions, weights, and full operating data.
- For specification details, refer to Sec. 17-1(a)–(c) as appropriate.
- Specify compliance with National Electrical Code, NEMA standards, and local codes as applicable.

DIVISION 18

Air Distribution, Central System

18-1. AIR-HANDLING UNITS

(a) *Capacity*

- Specify output of central air handler, as ______ cfm of air at ______ F temperature, and delivered against a system resistance of ______ in. W.G.
- Specify whether unit is to be of the blow-through or draw-through design, and state desired direction of discharge, as vertical or horizontal.

(b) *General Design and Arrangement*

- Specify form of cooling coil, as for the direct expansion of refrigerant or for the circulation of chilled water.
- Specify form of heating coil (if heating is to be included), as for steam or circulating hot water.
- Specify number and kind of filters on the airstream entering the coils; pressure drop, ______ in. W.G.
- If humidification is added, specify kind of humidifier, as water spray, steam jet, water evaporation, water troughs, etc. Specify source of water, and state hardness in terms of calcium carbonate. Give range of humidity desired, ______ per cent relative humidity.
- Specify removal system for condensate resulting from dehumidification of the air as it passes over the cooling coils. Drain line to be of copper tubing or other.
- Specify kind and thickness of insulation to be provided on drain pan.
- Specify casing insulation; kind and thickness.
- Specify the number and kind of dampers to be included, as bypass damper, mixing dampers, face dampers, etc.
- Specify if handler is to be dampered for multiple zones or for a double duct connection.
- If multizone, state number of zones.
- If double duct, state temperature of air for heating

duct and temperature of air for cooling, _______ F respectively.

- Specify the volume of air to be handled for each zone, _______ cfm at _______ F. If not shown on the drawings, prepare a table giving the number of each zone, cfm, temperature, maximum acceptable velocity _______ fpm.
- Specify that zone damper adjustments shall be conveniently provided outside of the handler casing.
- Specify color and finish of the handler casing.
- Specify material of casing, _______ gage steel.
- State location, and whether unit is supported on a foundation, structural steel, or is suspended. Give details of foundation or supports, if these are to be furnished and installed by the air handler contractor.
- Specify the maximum volumes of air to be handled in the mixing box, _______ cfm of return air and _______ cfm of fresh air. State temperatures of air, as _______ F on return during heating and _______ F during cooling. Outdoor fresh air temperatures, _______ F on heating cycle, _______ F during cooling at guarantees.
- For temperature reduction on cooling cycle in coils, specify _______ F. For temperature rise in coils during heating, specify _______ F.
- Require general arrangement drawings showing dimensions, weights, inlet and discharge openings, etc.

(c) *Coils*

- Specify construction of the coils, both cooling and heating, as aluminum fins on copper tubing or other, leak-tested at _______ psig.
- Specify that coils must be accessible by removal of a panel or the opening of a door in the casing.
- If coils are for the direct expansion of a refrigerant, specify the refrigerant, and require the coils to be factory-sealed with a holding charge of nitrogen or other gas acceptable for the purpose.
- If coils are for chilled water, specify that coils must be completely drainable and fitted with vents as necessary, these vents to open outside the casing. Also require the coil casings to be insulated with _______ in. of _______________ to avoid sweating.
- Heating coils for steam shall be pitched in the direction of condensate flow and provide complete gravity drain-

age of condensate. If tubes require flow-control orifices to provide uniform heating of the coil surface, so specify, giving material of orifices.

- Heating coils for hot water shall be pitched for complete gravity drainage, vented and orificed as may be necessary.
- Specify the temperature of the chilled water entering the coils as _______ F and in a volume of _______ gpm.
- Specify temperature rise in chilled water as _______ F.
- Specify maximum acceptable pressure drop in the coils as _______ psig, when passing the maximum flow of chilled water, _______ gpm.
- For steam heating, specify pressure of steam, _______ psig, and total temperature, _______ F.
- Specify kind and capacity of condensate removal trap, _______ pounds of condensate an hour.
- For hot water heating, specify the temperature drop in the coils, _______ F.
- Specify the entering temperature of the hot water, _______ F, and give the volume being pumped, _______ gpm.
- Specify maximum pressure drop in the coils, _______ psig.
- Require submission of coil details, areas, and pressure loss at rated air passage through the coils.

(d) *Fans*

- Specify capacity of each fan, _______ cfm, when handling air at _______ F against a discharge pressure of _______ in. W.G. Specify number of fans.
- Specify if the fan capacity is based on a blow-through or draw-through arrangement.
- Describe kind of fan wanted, as centrifugal, double width, double inlet, nonoverloading design.
- Specify kind of fan bearings and method of lubrication.
- Specify that fan shall not go through a critical speed in reaching its rated rpm.
- Require fans to be statically and dynamically balanced at the factory.
- Specify that rating of the fans must comply with the requirements of the Air Moving and Conditioning Asso-

ciation, Inc., and be entitled to the AMCA seal of certified ratings.

- Require bidders to submit drawings and data on their fans, including speed, _______ rpm; _______ hp; discharge in _______ cfm; pressure, _______ in. W.G.; diameters of fan, _______ in., and shaft, _______ in.; noise level, _______ decibels.

  Also electrical characteristics of driving motor (or motors), _______ volts, _______ phase, _______ cycles, full-load amperes and locked rotor amperes, respectively.
- Specify manner of drive, whether by (belt) (gears) (direct-connected) (flexible coupling) (rigid coupling) (resilient coupling).
- If drive is by belt, specify kind of belt, as (flat) (V) (round) (other).
- If two-speed fan motor (or motors) is required, specify the speeds, as _______ rpm and _______ rpm.

(e) *Electrical Work*

- All wiring, conduit, switches, relays, motors, and other electrical equipment shall conform to the requirements of the National Electrical Code, the NEMA standards, and local codes as applicable.
- Work shall be carried out only by qualified and licensed electricians.
- Controls shall operate on (reduced voltage) (full line voltage) (some of each).

(f) *Controls*

- Specify if control system is to be pneumatic or electrically operated. If the former, specify air pressure available, or require this contractor to furnish and install his own air compressor. State electric service available for driving the motor of the compressor.
- Specify what dampers are to be automatically controlled, what are to be manually controlled, and what are subject to automatic and manual control.
- Specify that temperature controls will be by (indoor thermostat) (outdoor thermostat) (both, with the indoor instrument overriding).
- Provide a sequence of operation for the guidance of bidders, showing what fans are in action, what coils are energized, the settings of the various dampers (similar to the chart, Table 18-1).

- Specify a general control panel to be mounted (in the machinery room) (other). Detail the items to be set on this panel, as pressure gages, thermometers, relays, switches, push-button stations, etc.
- Specify a linkage between the fresh air inlet damper and the return air damper so that the opening of either closes the other in proportion to the system temperature requirements.
- Specify a high-pressure cutout to stop the compressor under abnormal pressures, and require this cutout to be reset by hand.
- For outdoor thermostat, specify a unit that provides a composite action based on the combination of temperature, sun, shade, and wind.
- For details of gages and thermometers, refer to Sec. 4-21.
- Specify a temperature-sensitive bulb to be mounted in the return duct to stop fans at excessive temperature.

(g) *Piping*

- Specify refrigerant piping to be of ____________ (material), sizes as shown on the drawing, with _______ joints, and suitable for a working pressure of _______ psig.
- Specify the kind and type of vibration absorbers to be incorporated in the refrigerant pipelines, noting position of each.
- Specify the drain piping to be of ____________ (material) leading from the cooling-coil drip pan to a discharge to (a floor drain) (outside).
- Water piping, chilled, to be of (copper) (brass) (wrought iron) (steel). Specify classification, schedule number, working pressure, etc., as appropriate.
- Specify valves and fittings to suit the piping, giving working pressures in each case, and the material.
- Specify cold insulation, as _______ in. thick, _______ _______ (material).
- Specify steam piping insulation similarly to the water piping, recognizing the difference in temperatures and possibly in pressures.
- For condensate piping, specify material, working pressure, and indicate point of discharge.
- Indicate if piping is to be insulated or left bare.

- For all piping, designate the kind, placement, and spacing of pipe hangers and supports.
- Specify pipe sleeves and caulking where piping goes through walls or ceilings.

(h) *Fire Dampers*

- Refer to location of one or more fire dampers as shown on the drawings.
- Specify temperature, _______ F, at which the fire damper closes automatically.
- Specify construction of fire damper, as thickness of steel, _______ gage, actuating spring or weight, etc.
- If smoke signal is placed in any duct, specify where it is placed, and describe kind of warning desired, (visual) (audible) (both).

(i) *Evaporative Cooling in Handler*

- If owner's design contemplates evaporative cooling to be provided in the air-handling unit to delay the need for compressor operation as an item of economy, specify arrangement of sprays.
- Specify water conditions, as from city supply, well water, recirculated or piped to waste, etc.
- Specify range of spray water temperatures, from _______ F to _______ F. Also hardness as calcium carbonate.
- Specify desired conditions of leaving air, _______ F dry bulb and _______ F wet bulb.
- Specify volume of spray water required, _______ gpm.
- Specify carryover eliminators or other means for keeping the leaving air dry.
- For a recirculating pump, specify kind of pump, as centrifugal, rotary, etc. Give electric service characteristics, _______ volts, _______ phase, _______ cycles.
- Specify materials of spray nozzles and piping.
- Specify size of spray trough and means for maintaining water level in the trough, as float valve.
- Require bidders to furnish complete data and drawings covering the evaporative cooling features of the air handler.

(j) *Air Washers*

Note: This is a form of combined air cleaning and cooling by passing the air through water sprays, pads, or capil-

Table 18-1. Suggested Control Sequence for a Central Station Air Handler

Outside temperature, dry bulb	Inside temperature, dry bulb	Position fresh air damper	Position return air damper	Cooling coils	Heating coil
Design temperature as 95 F	Design temperature as 78 F	Open, admitting ______ cfm	Open for ______ cfm	Active	Off
Falling	Same	Gradually opening wider	Gradually closing in proportion	Active	Off
Falling, say to 72 F	Design temperature or a little below	Wide open; 100% fresh	Closed	Off	Off
Still falling	70 F	Gradually closing	Gradually opening	Off	Off
Still falling as 65 F	70 F	Closed except a little for ventilation	Open	Off	Active
Still falling as 45 F	72 F	Same as above	Wide open	Off	Active
Rising as 65 F	72 F	Same as above	Wide open	Off	Active
Still rising	72 F	Opening wide	Closing proportionally	Off	Off
Still rising	76 F	Open, admits ______ cfm	Open, admits ______ cfm	Active	Off

Note: Actual temperatures and conditions under which these operations occur will be determined only in the field by experiment.

lary cells and finally moisture eliminators at the leaving point of the machine.

- If reheat is to be used, so specify, giving information on kind and capacity of the reheater.
- Specify the condition of the air entering, as ______ F dry bulb and ______ F wet bulb. Also the volume, ______ cfm.
- Specify condition of air leaving, as ______ F dry bulb and ______ F wet bulb (usually close to saturation).
- Specify recirculating arrangement, as ______ (kind of pump), capacity, ______ gpm against a total head of ______ ft.

- Specify materials of spray nozzles and of distributing piping.
- Specify types of eliminators, number, materials, and arrangement.
- Specify approach to wet bulb in the cooling process, as _______ F.
- Specify construction of the washer, casing of _______ gage steel, corrosion-protected by _______________.
- Specify painting; color and finish.
- Describe source of water, hardness, impurities, organic material, sulfur if any, etc.
- Indicate where washer is to be located.
- Require bidders to provide drawings and data on the equipment they propose to furnish, giving dimensions, weights, and internal arrangements.

Note: If this washer is to be used in conjunction with a cooling tower, information on specifying the cooling tower is given in Sec. 15-4, appropriate subheadings.

(k) ***Air Washer Instruments***

- Specify pressure gages on the pump discharge, giving range of calibration from _______ psig to _______ psig.
- Specify thermometers (straight stem) (angle) for placement in the air entering, the air leaving, the water to the sprays, and the water in the suction basin. Specify calibrations from _______ F to _______ F.
- Specify wet bulb thermometers, in addition to the dry bulb instruments, for indicating conditions of the air entering and air leaving the washer.
- Specify which instruments are to be indicating and which recording. For the recording instruments, stipulate locations, as on panel in machinery room or elsewhere convenient.
- For details on all instruments, refer to Sec. 4-21, including charts, as appropriate.

(l) ***Air Washer Controls***

- Refer to Sec. 18-1(f).
- Specify a humidistat form of controller if this seems desirable for controlling the humidity of the air leaving the washer.
- Specify alarm signals, (audible) (visual) (both) to call attention to an abnormal rise in the leaving temperature. Designate the temperature, _______ F, at which the alarm functions.

(m) *Rotary-type Evaporative Coolers*

- Specify air conditions, as temperature, _______ F, and relative humidity, _______ per cent, of air entering the cooler, and specify similarly for temperature and relative humidity of the air leaving.
- Specify the volume of air handled, _______ cfm.
- If air is polluted, specify nature of pollution, as sulfur, smoke, acids, etc.
- State location of the cooler, as inside, outside on the roof, outside shelf-supported and discharging through a wall, in attic space, etc.
- Specify kind of foundations and supports, and by whom furnished and installed.
- Specify the construction of the casing, _______ gage steel, and finished with _____________ paint of a _____________ color.
- Specify the construction of the rotor, as alternate layers of flat and corrugated bronze screen carried on a nonferrous drum of _____________ (material).
- Specify construction of the water tank into which the rotor rotates, as _______ gage (brass) (stainless steel) (other).
- Specify how rotor is driven, as by (belt drive) (gears) (direct through coupling).
- Specify how drive is to be lubricated.
- Specify how water level is to be maintained in the tank, and how concentration of salts and dirt shall be removed automatically.
- Specify kind and capacity of fan or blower, as centrifugal, backward curved blades, nonoverloading type, with a capacity of _______ cfm of air at _______ F when discharging against a total head of _______ in. W.G.
- Specify motor drive, giving service characteristics of _______ volts, _______ phase, _______ cycles. Also state if drive is by belt, gears, coupling.
- Require bidders to submit complete information on the equipment offered, including dimensions, weight, speeds: _______ rpm of rotor, _______ rpm of fan, _______ rpm of rotor motor, _______ rpm of fan motor. Also horsepower of each motor, current in full-load amperes and with locked rotor.

- Require also information on the means provided for eliminating moisture carryover, and on the noise level, _______ decibels.
- Specify how evaporator is to be controlled, by thermostat, humidistat, or other, and state location of central control panel.
- Specify design and construction of control panel, and list switches, push buttons, and instruments to be mounted on the panel.
- For details of instrument construction, refer to Sec. 4-21.
- Specify the water conditions as fed to the rotor in make-up, as _______ F minimum temperature and _______ F maximum temperature. State source and degree of hardness, and identify any contamination present, organic or inorganic.

(n) ***Exhaust for Evaporative Coolers***

Note: Because of the continuous evaporation of water into the air passing through the evaporative cooler, the recirculation of this air raises the relative humidity to an uncomfortable level, unless provisions are made to exhaust from the conditioned spaces a volume of air equal to that entering through the cooler.

- Specify the number, size, and locations of exhaust air devices. State if these are to be gravity or mechanically actuated, as an exhaust fan.
- Specify the capacity of each, as _______ cfm.
- For motor-driven exhauster fans, specify kind and type of fan, also nature of drive, as by belt, gears, direct, etc.
- Require bidders to provide complete data on dimensions, construction, and weights of exhausters. For motor-driven units, require full data on electrical characteristics, _______ rpm, _______ hp; current draw in amperes, _______ full load and _______ locked rotor. For all units, require information on means for rejecting the entry of rain.
- Specify manner of exhauster control, as by interlocks with the main fan, cooling rotor, or other. Include manual starting and stopping stations, and indicate their locations.
- Specify that all electrical equipment must comply with National Electrical Code, NEMA, and local codes as appropriate.

- Specify materials of construction for the exhauster, as aluminum, plastic, other. Specify finish, as polished aluminum, anodized finish, color, etc.
- Specify height limitations, if any.
- Specify relief shutters, designating where these are to be placed, the number required, and capacity of each. Express capacity in _______ cfm, passing air at _______ F temperature, and with a velocity not to exceed _______ fpm.
- Specify the material and design of the relief shutters.
- Specify maximum opening pressure acceptable, as _______ in. W.G.
- Specify kind and material of bird screen.

(o) ***Filters for Evaporative Coolers***

- If filters are included, specify type, size, and filtering efficiency, stating size of particles to be removed, as _______ microns.
- Specify maximum acceptable loss of pressure through the filter, _______ in. W.G.
- Specify the nature of the air pollution entering the filter, if any exists, and indicate its concentration as _______ micrograms per cu cm for particulate matter, and the smoke density on a scale of 1 to 5.
- Specify filter material.
- Specify test by which efficiency and performance of the filter is to be measured.

18-2. DUCT SYSTEMS, LOW PRESSURE

(a) *Metal Ducts*

- Refer bidders to drawings for details of duct dimensions and general layout of main ducts and branches.
- Refer to Sec. 13-2(a) for details on the design and construction of these ducts and for their insulation. Additional data is given for the insulation in Sec. 13-2(c).
- Specify heavy canvas fire-resistant joint for connecting ducts to air handler.

(b) *Fiber Ducts*

- Refer to Sec. 13-2(e) for details of duct construction, vapor barrier, etc.

(c) *Fiber Glass Ducts, Rectangular*

- For specifying details, refer to Sec. 13-2(g).

Note: These ducts are formed from fiber glass boards in the contractor's shop.

(d) *Flexible Glass Ducts*

- For specifying details, refer to Sec. 13-2(h).
- These ducts are circular in cross section and are factory-formed.
- Specify type of connections and transitions, as (sheet metal) (other).

(e) *Wall Registers*

- Refer to Sec. 13-4(a).

(f) *Ceiling Registers*

- Refer to Sec. 13-4(b).

(g) *Baseboard Outlets*

- Refer to Sec. 13-4(c) and (d).

(h) *Ceiling Diffusers*

- Refer to Sec. 13-4(e).

(i) *Slot-type Diffusers*

- Refer to Sec. 13-4(f).

(j) *Return Grilles and Diffusers*

- Refer to Sec. 13-5(a) and (b).

(k) *Anodized Color Finishes for Outdoor Aluminum*

Note: This is a special finish for grilles, shutters, and other aluminum forms exposed to the weather, and possibly to a corrosive atmosphere as well.

- Specify the color of the finish, in satin smoothness.
- Specify the thickness of the coating, ________ thousandths of an inch.
- Specify corrosion resistance capable of withstanding the equivalent of a salt spray test continued for ________ hours.
- Specify fade resistance, ________ hours, as measured on a ________________ fadeometer.

(l) *Intake Grilles for Duct Systems*

- Refer bidders to drawings for locations of grilles and for dimensions.
- Specify provision of burglar bars, set on ________ in. centers, each bar to be ________ in. diam.
- Specify louvers to be (fixed) (adjustable) and formed of ________________ (material).
- Specify blade slope to prevent entry of rain, ________ degrees.
- Specify material of frames, as steel angle, aluminum, etc.
- Specify maximum rate of air entry, ________ cfm, and give maximum velocity, ________ fpm.

- Specify if louvers are to be closed manually or by motor.
- If a fresh air intake damper is provided in the duct connecting to the intake, so specify, indicating if this damper is modulating or a simple open-or-shut device.
- Require bidders to submit detailed drawings of their equipment and other data, as weight, finish, etc. If anodized color finish is required, specify the color, etc. For anodizing guide, see Sec. 18-2(k) immediately preceding this section.

18-3. DUAL-DUCT SYSTEMS, HIGHER PRESSURE

Note: Basic design incorporates two parallel supply ducts, one carrying warm air, the other chilled air. By means of a mixing box, warm air and chilled air are admitted in the volumes necessary to provide the desired temperature in the spaces to be conditioned, and then the mixture is discharged into the space.

Return ducts are designed conventionally, with no separation, and may or may not be served by a return fan. In order to absorb fresh air into the system, a proportionate amount of the return may be dumped, or exhausted, to the outdoors. This is particularly true when evaporative cooling is being practiced, in order to avoid excessive humidity in the conditioned spaces.

To conserve space, higher velocity is permitted in the two parallel ducts; this usually is not necessary for the return ducts, which operate at normal velocities. The use of dampers and splitters is not recommended because of the frequently changing relationships between the warm air and chilled air admission rates to the mixing box.

(a) ***General Conditions Of Operation***

- Specify range of outdoor temperatures in summer, from _______ F dry bulb and _______ F wet bulb, to _______ F dry bulb and _______ F wet bulb.
- Specify similarly for winter conditions.
- Specify maximum temperature, _______ F, in air leaving the heating coil, and minimum temperature, _______ F, of air leaving cooling coil.
- If evaporative cooler is incorporated in system for service under favorable weather conditions, refer to Sec. 18-1(i) and (m) for specifying data.
- If an air washer is incorporated in system, refer to Sec. 18-1(j)–(l) for specifying guidance.

- For either the evaporative cooler or the air washer, specify the temperature of the air leaving, ______ F, and the relative humidity, ______ per cent. (*Note:* Generally assumed to range between 99 and 100 per cent.)

(b) *Central Station Equipment*

- For mechanical compression equipment, refer to Secs. 15-1 through 15-5, all subheads as applicable.
- For nonammonia absorption systems and equipment, refer to Secs. 16-1 and 16-2, all subheads as applicable.
- For ammonia absorption systems and equipment, refer to Sec. 16-3, all subheads as applicable.
- For steam jet chillers, refer to Sec. 16-4, subheads as applicable.
- For heat pumps, refer to Secs. 17-1 and 17-2 as applicable.
- For air handling, package units, refer to Sec. 18-1 as appropriate.

(c) *Coils, Heating and Cooling*

- Refer to Sec. 18-1(c) for details of construction and suggested specified items.
- List coils required, as heating, cooling, reheating, the preheating and precooling of fresh air, etc.

(d) *Fans*

- Refer to Sec. 18-1(d) for supply, return, and fresh air specifying data as applicable. Note that the discharge pressure will vary considerably in reflection of the circuit in which it functions.

(e) *Electrical Work*

- Refer to Sec. 18-1(e) for specifying guidance.

(f) *Piping for Coils*

- Refer to Sec. 18-1(g) for guidance on specifying the piping required for refrigerants and for water and steam.

(g) *Duct Construction*

- Specify maximum velocity air in the cold supply duct as ______ fpm; in warm supply duct, ______ fpm; in return air duct, ______ fpm; in fresh air inlet duct, ______ fpm; in the return air dumping duct, ______ fpm.
- Specify the materials of ducts.
- Specify that ducts must comply with the recommendations of the ASHRAE Guide, latest edition, as to gage of metals for metal ducts and the kind of joints used.

- Specify method of duct support to be provided, and designate distance between supports, _______ ft and in. Specify support design and material.
- For details of duct specifications, refer to Sec. 18-2(a)–(d), as appropriate.
- For air intake grilles and shutters, also relief shutter or louvers, refer to Secs. 18-1(n) and 18-2(l), as appropriate. Also Sec. 18-2(k) for exposed finish.

(h) *Mixing Boxes*

- Specify volumes of air that mixing box is expected to handle, as _______ cfm warm air, _______ cfm chilled air, and _______ cfm discharge of mixed air to conditioned space.
- Specify materials of construction of the mixing box, as _______ gage steel.
- If pneumatic cylinders are used to actuate the warm air and chilled air admission valves, specify air pressure, _______ psig; also source of compressed air for operating the valves.
- If separate air compressor is required for service with the dual-duct system, specify the kind and characteristics of the compressor, method of driving, etc., as suggested in Sec. 8-1(a)–(j), as applicable.
- Specify sound-absorbing material to be placed in mixing box, giving kind of material and thickness, _______ in.
- Since boxes will be of different capacities, and if such a table is not provided to bidders on the drawings, prepare a table of mixing box units, identifying each by a symbol and showing dimensions, capacity, and location of each, plus the quantity of each symbol to be furnished under the contract.

(i) *Noise Control*

- If the high velocity of the air in the supply ducts is expected to create objectionable noise, specify the installation of one or more sound silencers to be placed as shown on the drawing. Since these silencers are proprietary devices, specify the make and model desired by name.
- Specify the sound reduction to be accomplished by the silencers, as _______ decibels with a discharge maximum noise of _______ decibels, with an air flow of _______ cfm, at a velocity of _______ fpm.
- Specify sound-absorbing insulation for the *inside* of the

ducts, if this is necessary, giving kind and thickness of insulation, _______ in., also the lin ft of duct to be so treated.

(j) *Dual-duct Insulation*

- Refer to Sec. 13-2(a) and (c) for guidance in specifying duct insulation, including vapor barriers.

(k) *Controls for Dual-duct Systems*

- Each conditioned space may control its own environment without affecting adjacent spaces. Temperature is controlled by thermostat, the thermostat shifting the proportions of warm and cold air to the mixing box as desired. The flow from the box to the space is constant, and is kept so by a volume control damper.
- Specify the thermostat for each space to be calibrated over a range of _______ F to _______ F in steps of _______ degrees each. Thermostat details of construction appear in Sec. 4-21.
- For controls of the central station equipment, refer to Sec. 18-1(f).
- If humidification is provided for the winter months, specify the kind and give details as suggested in Sec. 13-6(a)–(d) as applicable.
- Specify a humidistat to control the relative humidity of the warm air. Give operating range of humidistat from _______ per cent to _______ per cent relative humidity.
- Specify a humidistat to show the relative humidity of the cold air leaving the central station, and control dew-point temperature in the cooling coil accordingly.
- Specify electric service as used in control equipment, giving _______ volts, _______ phase, and _______ cycles.
- Specify compliance with National Electrical Code, NEMA, and local codes as appropriate.

18-4. CABINET FAN-COIL CONDITIONERS, CENTRAL SYSTEM

(a) *Units Required*

- If not already tabulated on the drawings, prepare a table of all fan-coil units required, giving these data for each unit:

  (1) Identifying symbol; (2) Capacity heating, _______ Btu per hour; capacity cooling, _______ Btu per hour; (3) amount of fresh air brought in, _______ cfm; (4)

hot water flow into unit, _______ gpm at a temperature of _______ F; (5) chilled-water flow into unit, _______ gpm at a temperature of _______ F; (6) volume of air discharged into room, heating _______ cfm; volume of air discharged into room, cooling _______ cfm; (7) whether cabinet is freestanding, semirecessed, wall hung; (8) number of units for each symbol.

- Specify cabinet construction and finish, as _______ gage steel, painted with _______ coats and finished in _______________ color.
- Specify kind of filter; (replaceable) (washable), etc.
- Specify construction of heat exchange coil, as _______ fins on _______________ tubing.
- Specify kind of drain pan, _______________ material; specify that drainpipe is to be of _______________ tubing and shall spill outside the building, or as shown on the drawings.
- Specify kind of fan desired, as centrifugal, propeller, etc. Fan to have a maximum capacity of _______ cfm when handling air at _______ F and against a total resistance of _______ in. W.G.
- Require bidders to submit drawings and data of equipment offered, providing full information on dimensions, weights, fan motor horsepower, _______ full-load amperes, _______ locked rotor amperes, _______ rpm of motor, _______ rpm of fan. Pressure loss in coil, _______ psig. Also manner of motor and fan connection: (direct-connected) (belt drive) (other).
- Specify number of fan speeds and rpm _______ at each.
- Specify where unit is to be insulated, giving kind of insulation and thickness, _______ in.
- Specify fresh air inlet grille, giving location; free area, _______ sq in.; desired method of weather exclusion, if any; and insect and bird screen.
- Specify materials of the grille and screen.
- Specify if fresh air inlet is to be closed manually or automatically.

(b) *Control*

- Specify room thermostat, indicating if it monitors the hot water and cold water supply valves, the fan, or both.
- For details of the thermostat construction, refer to Sec. 4-21.
- For details on the central station control of the circulat-

ing hot water or chilled water, refer to the sections on hydronic heating, Secs. 12-3 through 12-7 as appropriate; and for chilled water, Sec. 15-5.

- Specify time and program controls designed to provide reduced operation during off-hours, as nights, holidays, and Sundays. Stipulate the temperatures to be provided during the off-hours, _______ F, and set the time dials accordingly.
  - Provide a safety override inside thermostat to prevent freeze-ups.

18-5. ROOM CONDITIONERS, SELF-CONTAINED

Note: As it is customary to use these individual units in individual rooms, consider summarizing what is necessary in tabular form, similar to table suggested in Sec. 18-4(a) for cabinet units served from a central station supply of chilled or warm water.

This is only to be done if the summary does not already appear on the drawings.

(a) ***All-year, All-electric Cabinets***

- Specify location of each unit by room or space, referring to the drawings as necessary.
- Specify the capacity of each, _______ Btu per hour cooling, and _______ Btu per hour heating.
- Specify that ratings and installation techniques shall comply with the standards, latest edition, of the Air Conditioning and Refrigeration Institute. Ditto for refrigerant and condensate piping.
- Specify the volume of conditioned air to be discharged from each unit, _______ cfm.
- Specify volume of fresh air inducted, _______ cfm, when unit functions as a ventilator without either heating or cooling added.
- Specify seasonal outdoor conditions, as _______ F minimum or design temperature in winter, and _______ F maximum or design temperature in summer.
- Specify the electric service available, _______ volts, _______phase, _______ cycles.
- Require the bidders to furnish drawings and operating data for the conditioned-air fan, including discharge volumes, _______ cfm at fan speeds of _______ rpm, when handling air at _______ F temperature, at a discharge pressure of _______ in. W.G. One group of

data to represent cooling operation, the other heating cycle. Also required: Horsepower for each rating, and amperage load, _______ full load and _______ locked rotor.

- Require bidders to furnish drawings and data similarity for condensing coil fan.
- Require bidders to submit description and data concerning the refrigerant compressor and motor, giving, among other items, the rpm, motor horsepower, whether hermetically sealed in or an open compressor; also methods of lubrication.
- Specify that compressor motor must have overload protection.
- Specify that conditioned-air fan shall be designed to operate at three speeds—low, medium, and high—suitable for cooling, heating, and ventilation service as needed.
- Specify the kind and construction of the heat exchange coils, as aluminum fins on copper tubing.
- Require information on area of each coil, _______ sq ft for the cooling coil, _______ sq ft for the condensing coil.
- Specify kind and capacities of strip or resistance units of other forms for air heating, giving steps of energizing and the electric load at each step, _______ kw.
- Specify the fresh air inlet grille, material, free area, _______ sq in.; and require insect screening. Specify how weather is to be excluded from entry.
- Specify the cabinet construction, as _______ gage steel, protected against corrosion by (galvanizing) (other) and finished with _______ coats of enamel, _______________ in color.
- Specify desired inside temperatures, _______ F winter and _______ F summer.
- Specify number and kind of air filters.
- Require bidders to submit information on dimensions and weights.

(b) *Controls*

- Specify a room thermostat, calibrated from _______ F to _______ F for cooling, and _______ F to _______ F for heating.
- If outside anticipating control is desired also, call for an outside thermostat that reacts in a composite manner

to the effects of temperature, wind, shade, and sunlight. The inside thermostat remains the overriding control. Designate outside location.

- Specify push-button stations for manual control and indicate locations.
- For humidity control on the cooling cycle, specify a humidistat, calibrated to maintain relative humidity as set between ______ per cent and ______ per cent by varying the cooling coil temperatures as necessary.
- Specify the location of the control panel, as on top, on side, etc., of cabinet.
- Specify if switch from heating to cooling and vice versa shall be automatic or manual.

(c) ***Unit Heat Pumps***

Note: This is an all-electric design and may be specified from Sec. 17-1, all subheads as appropriate, with the following items emphasized:

- Specify how the condensing coil is to be defrosted, as by hot gas, auxiliary electric heaters, other.
- Specify a frosting alarm, either audible or visual, if a manual defrosting is contemplated, or if an interruption in heating from this cause is not to be tolerated.

(d) ***Window Units, Cooling Only***

- Specify size of each unit in terms of capacity, ______ tons of refrigeration or ______ Btu per hour.
- Specify electrical service available, ______ volts, ______ phases, ______ cycles.
- Specify that unit must qualify for the Symbol of Certification as issued by the Air Conditioning and Refrigeration Institute, and require bidders to state that their units are rated in accordance with ARI Standard 210-62, or Standard 250-62, as appropriate.
- Specify that the compressor and fans in each unit shall be accessible for inspection and repair.
- Specify if compressor is to be hermetically sealed with motor, or otherwise.
- Require bidders to submit drawings and data showing dimensions, weights, casing finish, etc.

(e) ***Through-the-wall Units, Cooling Only***

- Specify as for window units, as in the preceding Sec. 18-5(d).

DIVISION 19

Commercial Refrigeration

General Notes. All the refrigeration machinery already specified in the sections dealing with comfort air conditioning, Division 15, reappears in commercial refrigeration applications. Actually, air conditioning with the basic objective of cooling the atmosphere is a development of commercial refrigeration. For this reason, many of the detail specifications for commercial refrigeration equipment will be obtained by reference to preceding sections.

Cryogenics is the name given to the production and use of temperatures from −250 F right down to absolute zero. Much of the work is in the liquefaction of gases, with considerable research in the effects of low temperatures on metals, chemicals, and other special materials. As a result of cryogenic actions in laboratories, knowledge concerning superconductors and semiconductors has been broadened and has become useful. Since this is a highly specialized field, detailed specification guidance is not within the scope of this Handbook. Attainment of subzero temperatures down to about −100 F by means of series compression will be discussed, and information will be submitted.

19-1. COLD-STORAGE SYSTEMS

Note: Room or space temperatures for the storage of furs, foodstuffs, flowers, etc., can be provided either by direct expansion coils in the space to be kept cool, or by ceiling or wall coils through which a chilled brine is circulated. By either method, a central compressor station is necessary, with liquid refrigerant vaporizing in various air coolers or in a brine cooler, as appropriate to the design.

Air is circulated over the air cooler coils by a fan; a fan for general air movement may or may not be provided for a room that is cooled by the circulation of a cold brine, depending on the need for uniformity of temperature within the space.

(a) *Central Compressor Station*

- Refer bidders to the drawings for layout of the cold-storage plant, showing machinery room, various storage rooms, and corridors, and indicating the temperature, _______ F, planned for each space. Also details of building construction.
- Specify the total refrigeration load, _______ tons.
- Specify system to be of the (direct expansion) (brine circulating) type.
- Specify kind of compressors, (reciprocal) (centrifugal) (absorption), and number required.
- For specifications applying to the reciprocating machines, refer to Sec. 15-1(a)–(c).
- For centrifugal compressors, refer to Sec. 15-2(a) and (b).
- For absorption machines, nonammonia, refer to Secs. 16-1 and 16-2, all subheadings.
- For absorption machines, ammonia, refer to Sec. 16-3, all subheadings.
- For steam jet refrigeration machines, refer to Sec. 16-4(a)–(i).

  Note: For temperatures lower than 38 F, use of steam jet cooling is not practical, since the water in the evaporator freezes at 32 F.

(b) *Cooling Towers*

- For mechanical draft towers, refer to Sec. 15-4(a) and (b).
- For natural draft towers, including the hyperbolic, refer to Sec. 15-4(c) and (d).

(c) *Direct Expansion, Fin-Tube, Unit Coolers*

- If not already on the drawings, tabulate as part of the specification the following data about each individual unit:

  (1) Identification symbol; (2) coil surface, _______ sq ft; (3) number of expansion valves per coil, _______; (4) temperature drop in air passing across the coil, _______ F; (5) volume of air passing over coil, _______ cfm; (6) number of coils _______ in each size.
- Specify directional louvers on unit cooler discharge if considered desirable or necessary.
- Specify kind of fan, (propeller) (centrifugal), and capacity, _______ cfm.
- Specify kind of motor driving fan, and give electrical

service characteristics, _______ volts, _______ phases, _______ cycles.

- Require bidders to furnish complete data for unit coolers, including dimensions, weight. Also fan and motor _______ rpm; motor hp, _______; amperes, _______ full load, and amperes, _______________ locked rotor; motor make and model.
- Specify how each unit cooler is to be supported, as from overhead suspension, wall shelves, floor stands, etc.
- Specify how each unit cooler is to be controlled, as by wall-mounted thermostat, or other arrangements.
- Specify kind and location for back pressure valves on the suction line from each unit cooler.

(d) *Brine Coolers*

- Specify kind of brine, as _______ per cent by weight of sodium chloride or calcium chloride.
- Specify total capacity of brine systems, _______ gal, and capacity of the brine cooler tank, _______ gal, if less than this. Specify brine storage or surge tank, having _______ gal capacity, if necessary.
- Specify the kind of brine cooler, as shell and tube, trunk coils arranged in the brine tank, or other.
- Specify the tank construction, as steel plate _______ in. thick, with welded joints.
- Specify how tanks are to be protected against corrosion, as by chromate treatment, galvanizing, etc.
- Specify insulation on bottom and sides of tanks, _______ in. of _______________ (material). State how joints in the insulation, are to be made tight as by mopping with bitumastic, or other.
- If tank incorporates a brine agitator or circulator, so specify. If brine is pumped, specify kind and capacity of pump, _______ gpm; give materials required to resist corrosion of the pump parts. Specify corrosion-resistant material in agitator blades.
- Specify pump for circulating brine to the various cold-storage areas, as (centrifugal) (rotary) (other), having a capacity of _______ gpm against a total head of _______ ft, when handling (salt) (calcium) brine of _______ per cent by weight, at a temperature of _______ F. Specify pump to be corrosion-resistant.
- Specify electrical characteristics for motor driving brine

circulating pump, ______ volts, ______ phases, ______ cycles.

- Specify temperature of brine leaving tank, ______ F, and temperature of brine returning to the tank, ______ F. *Note:* this last will be a variable, owing to the variation of brine flow to the different spaces. Specify the return temperature expected with *all* refrigerated spaces operating at maximum load.
- Specify tank support method, as concrete, structural steel, etc. Also who provides the supports.

(e) ***Refrigerant Piping***

- Specify the kind of piping suitable for the refrigerant, as copper or brass for freons, ferrous piping for ammonia.
- Specify the working pressure, ______ psig.
- For ammonia, specify the schedule number, as No. 80 for pipe 1 in. or less in diameter; No. 40 butt welded, 1¼"–2"; and for 2½" up No. 40 seamless steel.
- Specify the piping joints, as hard soldered for nonferrous piping, welded for ferrous piping.
- Specify the stop valves used in the system, their kind, material, and connecting ends, as welding, screwed, etc.
- Specify fittings to suit the piping weight, except that screwed fittings if used on ammonia lines, shall be designed for 2,000 psig pressure.
- Specify kind and placement of pipe hangers. Center-to-center distance not to exceed ______ ft.
- Specify kind and thickness of piping insulation to be placed on suction piping.
- Specify likewise for the compressor discharge lines, if considered necessary (usually not provided).
- Specify the insulation for the brine piping, as ______ in. thick of ____________ (material) for the cold-brine supply lines, and ______ in. thick of ______ ______ (material) for the brine return lines.

(f) ***Brine Coils in Cold Rooms***

- Refer bidders to drawings for arrangement of coils and their locations.
- Specify total cooling surface of the coils, ______ sq ft.
- Specify the design of the coils as ____________ (material) designed for a working pressure of ______ psig and a test pressure of ______ psig.

- For each refrigerated space, specify the volume of brine to be circulated through the coils at full load, _______ gpm.
- Specify kind of brine admission valve to be provided, as modulating, motor-operated, manual valve, etc.

  Note: The temperature in the room can be controlled by reducing or increasing the brine flow through the coils, as desired.
- If air circulating fan is wanted, so specify, giving capacity, _______ cfm, kind of fan, motor characteristics, placement.

(g) *Direct Expansion Pipe Coils*

- Refer bidders to drawings for arrangement of the coils and their locations.
- Specify total cooling surface, _______ sq ft.
- Specify the size and number of expansion valves to be provided for each coil.
- Specify the design pressure of the coils, _______ psig, and the test pressure, _______ psig.
- Specify the refrigerant to be expanded in the coils.
- Specify the connection of the coil outlet to the expansion valve by way of the equalizing connection when appropriate.
- Specify control for admission of refrigerant, as a thermostat, wall mounted.
- Specify an air circulating fan, if considered necessary, for producing uniform temperatures in the refrigerated space.

(h) *Defrosting*

- Specify method of defrosting the fin-tube coils in the unit coolers, and the pipe coils for both brine cooling and direct expansion.
- If by water sprays, state where drain is located, and give temperature of spray water, _______ F. Ditto if brine sprays are used.
- If hot gas defrosting, specify hot gas temperature, _______ F, and set time limit.
- If by infrared or spot electric heaters, specify size of heaters and characteristics of electric service to the heaters.
- Specify if defrosting is to be accomplished on a time program, or is initiated by the thickness of frost tripping a depth gage.

(i) *General Controls and Instruments*

- Specify the usual controls for the compressors, condensers, and cooling towers, as referred to in Sec. 19-1(a) and Divisions 15 and 16.
- Specify safety controls as follows:
  - An alarm, visual and audible, signaling a failure of the refrigeration in any particular space and an abnormal rise in temperature. Specify the signaling temperature for each refrigerated space in the cold-storage building.
  - An alarm signal, visual and audible, for a significant leak in the refrigerant piping. This is particularly important when the refrigerant is ammonia, since ammonia is poisonous and can be lethal.
- Specify the kind, size, and location of the general control panel. List all instruments, switches, relays, signals, and other control items to be mounted on the panel.
- Specify the calibration for each gage, thermometer, humidistat, thermostat, and other instruments as appropriate. Refer to Sec. 4-21 for details of instruments design and construction.
- Specify where instruments are to be placed locally, as outside each refrigerated space, set a thermometer with the bulb in the space, so that temperature in the space may be observed without entering.
- Specify the kind of thermostat to be furnished and installed for monitoring the refrigerant admission to the individual room unit cooler or room coils, and indicate location for each room.
- Prepare a list of *all* instruments to be provided, giving number of each and service on which to function, as brine, supply or return, refrigerant suction and discharge. On this list indicate which ones are to have alarm contactors.

(j) *Brine in Fire System*

- Specify that sprinkler system piping inside the cold-storage building shall be charged with brine, either sodium or calcium, of sufficient density to avoid freezing at the lowest temperature on the system. Opening of a sprinkler head releases the brine first and thus makes safe the water that follows.
- Alternative: Specify a dry system, charged with compressed air at ________ psig.

- Release of the compressed air permits a valve to open and admit water from an outside source.

(k) *Larvae Killing in Cold Storage*

Note: Moths and moth larvae, moth eggs, and other forms of destructive infestation of materials can be killed by dropping the cold-storage temperature to 15 F and holding it there for 4 days. A repetition of this cycle may be necessary several times during the storage year. In some arrangements, a separate killing room is provided; in others, the storage room temperature is lowered, and sufficient refrigeration machinery for establishing the low temperatures is necessary. The possibility of lowering a normal 40 F storage temperature to 15 F must be considered when specifying the equipment for any given room.

- For killing facilities in a storage room, specify the usually maintained temperature, _______ F, and the killing temperature of _______ F.
- Specify the allowed time for dropping the temperature, _______ hours and minutes.
- Specify the length of time, _______ hours, during which the killing temperature is to be maintained.
- Specify the manner of control, as by thermostat, back pressure valve, other.

Note: An air circulating fan is essential. Specify, giving fan capacity, _______ cfm, motor characteristics, etc.

19-2. LOCKER PLANTS

Note: The usual product movement in a locker plant is in five steps: (1) receiving room to chill room; (2) to aging room; (3) to processing room; (4) to sharp freeze; (5) to locker storage. Room temperatures are attained by any of three cooling methods: forced circulation chilled air from a unit cooler having fin-tube heat exchange surface; plate-type evaporator surfaces; evaporator pipe coils.

Some designs provide a separate condensing unit for each room; others combine several rooms, or all rooms, on a single unit for the entire plant. Condensers may be either air- or water-cooled. Pickling and curing rooms also may be included in some large plants.

(a) *Operating Conditions*

- Refer bidders to the drawings for information on the plant construction, thickness of building insulation, arrangement of rooms.

- If not shown on the drawings, tabulate all rooms, including those not refrigerated, giving the purpose of each room, temperature to be maintained, dry bulb _______ F, and relative humidity, _______ per cent.
- Specify the maximum loading of each refrigerated room, _______ pounds of _______________ (material), giving time required in hours and minutes for the initial loading, and the time allowed, _______ hours and minutes, for reducing the temperature of the newly placed load to the established temperature of the room.
- Specify the temperature of the produce, meats, or other products when first placed in the refrigerated atmosphere.
- Specify maximum ambient temperature, outside, _______ F in summer, and minimum ambient, outside, _______ F in winter.

(b) *Refrigeration Equipment*

- Specify number, type, and capacity, _______ tons, of each condensing unit.
- Specify kind of compressors, as (reciprocating) (centrifugal) (absorption).
- Refer to Sec. 15-1(a)–(c) for reciprocating compressors; Sec. 15-2(a) and (b) for centrifugal compressors. Absorption machines for brine cooling are detailed in Sec. 16-1 through 16-3, all subheadings.
- Specify kind of condensers required, (air-cooled) (water-cooled) (evaporative). For specification details, see Sec. 15-3(a) for air-cooled units; Sec. 15-3(b) for water-cooled shell and tube units; Sec. 15-3(c) for evaporative condensers.
- For cooling towers, mechanical draft, refer to Sec. 15-4(a) and (b). Natural draft towers are discussed in Sec. 15-4(c) and (d).

(c) *Refrigerant Piping*

- Refer to Sec. 19-1(e).

(d) *Fin-Tube-type Unit Coolers, Direct Expansion*

- Refer to Sec. 19-1(c).

(e) *Direct Expansion Pipe Coils*

- Refer to Sec. 19-1(g).

(f) *Pipe Coils, Brine*

- Refer to Sec. 19-1(f).

(g) *Plate-type Direct Expansion Evaporators*

- Specify placements, as overhead, other.
- Specify the heat exchange surface, _______ sq ft.

- Specify the materials of construction: (steel) (aluminum) (copper).
- Specify suction temperature to be maintained in the plate, ________ F.
- Specify heat exchange capacity of the plate under conditions of maximum loading, ________ Btu per hour.
- Specify working pressure, ________ psig, and test pressure, ________ psig.
- Specify temperature control, as thermostat controlling a liquid admission valve.

(h) ***Defrosting***

- Refer to Sec. 19-1(h).

(i) ***Controls and Instruments***

- Refer to Sec. 19-1(i) for specified items.

19-3. ICE-MAKING PLANTS

(a) *Ice Cake Data*

- Specify the number and size, ________ pound cakes of ice to be produced every 24 hours.
- Specify that ice is to be clear and tasteless.
- Specify source of water. Refer to the chart, Table 19-1, Kinds of Water Contamination, and specify the kinds and degree of contamination, if any. Express degree in terms of (1) trace, (2) slight, (3) moderate, (4) heavy.
- Specify if contamination is permanent or seasonal.

(b) ***Brine Tank***

- Specify kind of brine, sodium chloride or calcium chloride.
- Specify construction of tank, as ________ gage steel, galvanized, chromate treated. All joints welded. Tank to be set on ________________ foundation provided by ________________. Depth of tank: ________ ft.
- Specify tank arrangement as brine cooler in one end with agitator at other; or the use of evaporator coils in brine raceways in the center and sides of the tank.
- Specify velocity of brine flow, ________ fps.
- Specify internal partitions to guide direction of brine flow, fabricated of ________ gage steel, galvanized and protected against corrosion by chromate treatment.
- Specify the hydraulic gradient of the tank, ________ in. for the full length of the tank.
- Specify the temperature of the brine circulating in the

tank, _______ F. Specify temperature variation tolerance, _______ F, plus or minus.

- Specify the size and construction of the brine cooler for operation with _______________ refrigerant. Specify working and test pressures, _______ psig, shell and tube.
- Specify capacity of cooler, _______ Btu per hour.
- Specify velocity of brine movement through cooler, _______ fps.
- Specify agitator as vertical or horizontal, diameter of propeller, and kind of drive, belt or direct-connected. Also, _______ rpm.

(c) *Ice Cake Handling*

- Specify means and manner of withdrawing ice can from the brine tank, as (lifting rods frozen in cake) (lifting lugs on edges of can) (monorail hoist with water sprays to loosen ice in cans).
- Specify how ice cakes are conveyed to storage room: by skidding, by suspension from a monorail, other.
- Specify method of air agitation for each can, as by compressed air released from a brass or other tube into the water while it is freezing. Specify if tube is fixed or oscillating.
- Specify air pressure, _______ psig.
- Specify air piping, as (steel) (wrought iron) (other), Schedule No. _______, and suitable for a working pressure of _______ psig.
- Specify source of compressed air. If project calls for its own compressor and tank, refer to Sec. 8-1(a)–(j), as appropriate.
- Specify that lifting rods, frozen in cake, shall be hollow to facilitate freeing them after use.

(d) *Ice Storage Room*

- Refer to drawings for size, construction, and location.
- Specify method for maintaining subfreezing temperature. Specify maximum required low, _______ F.
- Refer to Sec. 19-1(c) for fin-tube unit coolers, Sec. 19-1(f) for brine coils, and Sec. 19-1(g) for direct expansion pipe coils.

(e) *Refrigerant Compressors and Condensers*

- Refer to Sec. 19-1(a) and (b) for central compressor stations for cold-storage plants. Also to Secs. 15-1(a)–

(b) and 15-2(a)–(b) for compressor specifying details.

- For condenser details, refer to Sec. 15-3(a)–(c).
- Absorption systems are detailed in Secs. 16-1 through 16-3, all subheads.

(f) *Spray Ponds*

- Specify the amount of water to be cooled in the pond, ________ gpm, under conditions of ________ F dry bulb and ________ F wet bulb.
- Specify water temperature emerging from the spray nozzles, ________ F, and the temperature of the water, ________ F, leaving the pond.
- Specify required approach to the existing wet bulb, ________ F.
- Specify size, kind, and capacity of water circulating pump, as (centrifugal) (rotary) (other), delivering ________ gpm at a discharge head of ________ ft.
- Specify pump drive, as electric motor driving through (belt and pulleys) (direct-connected through flexible coupling) (other).
- Specify available electric service, ________ volts, ________ phase, ________ cycles. State if pump motor is to be squirrel cage, semienclosed, splash-proof, etc.
- Specify number, size, and material of spray nozzles.
- Specify water piping, (steel) (wrought iron) (other), of Schedule No. ________ for ________ psig working pressure.
- Specify valves and fittings to suit pressure conditions.
- Specify pump control, as by submerged thermostatic bulb in the water leaving the pond, designed to reduce the flow by changing the speed of the pump if a two-speed motor is being used. Alternatively, flow could be throttled by an automatically controlled valve on the pump discharge or on a bypass from pump discharge back to the suction.
- Refer to drawings for location and size of the spray pond and layout of the piping. Indicate where and how makeup water is introduced.

(g) *Control Panel*

- Specify placement, size, and design of control panel.
- Prepare a list of all instruments, switches, relays, indicators, and other devices that are to be mounted on the control panel. For details of instrument construction for specifying, refer to Sec. 4-21.

- List all gages, thermometers, and telltales or indicators to be used in the operation of the plant but mounted elsewhere than on the control panel, showing where each is to be placed. Give range and steps of calibration in each case.

(h) *Core Pump*

- Specify kind of pump, as (reciprocating) (rotary) (other).
- Specify capacity, _______ gpm, against a total discharge head of _______ ft.
- Specify drive: (belt) (direct-connected motor) (other).
- Specify motor characteristics: _______ volts, _______ phase, _______ cycles. Either specify or request bidders to furnish data on motor horsepower, speed, kind of motor, amperage, etc.

(i) *Water Piping for Can Filling*

- Specify steel or wrought iron, Schedule No. _______, with sizes and arrangement as shown on drawings.
- Specify valves and fittings to suit piping, specifying the working pressure, _______ psig.

(j) *Auxiliary Equipment*

- Ice plants are providing cube ice and packaged ice for both direct sales and for vending stations.
- Specify these devices from available catalogue data, keeping in mind the rate of production required.

Table 19-1. Kinds of Water Contamination

Contaminating material	*Usual corrective treatment*
Mud and silt	Filtration.
Odors	Active carbon purification; atmospheric aeration.
Sewage	Eliminate cross connections. Relocate septic tanks.
Algae	Remove and prevent by chemical additives.
Sulfur	Aeration. Boiling and aeration.
Hardness, calcium, and magnesium salts	Zeolite and/or other chemical treatment. See Sec. 4-12.
Silica	Filtration or coagulation. Chemical treatment.
Iron	Aeration and chemical treatment.
Oil	Both suspended and emulsified can be removed by filtration or coagulation.

Note: The design and specifying of water-treatment systems is a specialty, and except for zeolite and other boiler-feed treatments, details of the specifying procedures are outside the scope of this Handbook.

19-4. ICE SKATING RINKS

Note: Primary refrigerants are usually chosen from the various chemical groups, as freon 12, 22, and others, or ammonia. The secondary brine, circulated through the coils, is usually calcium chloride.

Direct expansion of ammonia into the coils is not permitted in some areas, owing to the possible hazards to personnel using the rink at the time of a leak. Since the other refrigerants are not poisonous or flammable, direct expansion is usually legally acceptable. However, for several reasons, such as the possibilities for storing cold brine over periods in which use of the rink is interrupted by other activities carried out on the floor, brine is preferred for most projects.

The temperature maintained at the ice surface varies somewhat with the particular kind of skating being done, as hockey, figure skating, curling, general public skating. The use of three compressors is frequently favored, with two in action while the rink is in service, and a third as a spare or to help on a fast-freezing program. During hours of rink shutdown, one compressor may serve to maintain the ice.

(a) *Operating Conditions*

- Refer to the drawings for the layout and construction of the rink, including the placement of the ice-freezing pipes, location of the compressors, condenser, etc.
- Specify the kind of skating planned, and the desired surface temperature of the ice, _______ F.
- Specify the ambient temperature, _______ F, and indicate if this is for an outdoor or indoor rink. Specify ambient wet bulb also _______ F.
- If rink is operated during the summer, state maximum outside air conditions, _______ F dry bulb and _______ F wet bulb.
- For winter use of the rink, state minimum and maximum wet bulb temperatures, _______ F.
- Specify temperature of brine, _______ F, entering ice-freezing coil, and state maximum temperature rise, _______ F, in brine leaving the coil.
- Specify primary refrigerant suction temperature, _______ F, in evaporator in brine cooler.
- For direct expansion in freezing coils, specify suction temperature, _______ F.
- Specify kind of primary refrigerant.

- State maximum number of persons expected to be on the ice at the same time, and time of year when this maximum may be reached.
- Specify the volume of brine to be circulated per ton of refrigeration, as _______ gpm.

(b) *Refrigerating Equipment*

- Specify the total installed tonnage, _______ tons.
- Specify the number of compressors, _______ and capacity of each, _______ tons.
- Specify the kind of compressor, reciprocating, centrifugal, absorption, etc., for service with _____________ (name) refrigerant.
- For detailed specifying data on the compressors, refer to Sec. 15-1(a) and (b) for reciprocating compressors; Sec. 15-2(a) and (b) for centrifugal; Secs. 16-2 and 16-3 for absorption cooling.
- For condenser specifying data, see Sec. 15-3(a)–(c).
- Cooling tower data is given in Sec. 15-4(a)–(d).
- For spray ponds, see Sec. 19-3(f).
- For brine coolers, refer to Sec. 19-1(d) for details as applicable.
- For refrigerant piping, materials, working pressures, etc., refer to Sec. 19-1(e).

(c) *Brine Pumps*

- Specify number and kind of brine pumps.
- Specify the kind and density of brine, _______ per cent of _____________ (kind).
- Specify construction of pump and use of suitable materials for handling brine.
- Specify capacity of each pump, _______ gpm, when pumping against a total discharge head of _______ ft.
- Specify pump drive, as electric motor for _______ volts, _______ phase, _______ cycle service. Driving through (flexible coupling) (belt) (other).
- Require bidders to furnish full details of pump construction speed, _______ rpm. Also kind of motor, (splash-proof) (enclosed); speed, _______ rpm; hp, _______; amperes, full load; amperes, _______ locked rotor; overload protection.
- Specify that all electrical material and workmanship must comply with National Electrical Code, NEMA standards, and local codes.

- Specify if pump and driver are mounted on a common bedplate and common foundation.

(d) *Piping for Freezing Coils*

- Refer to drawings for arrangement of coils and the details of the construction, as sand base, concrete base, center-to-center distance, ________ in., pipe diameter, etc.
- Specify the kind of pipe to be used: (steel) (wrought iron) (proprietary corrosion resisting) (other).
- Specify if piping is to be galvanized.
- Specify kind of joints, as welded, screwed, other.
- Specify working pressure, ________ psig, and test pressure, ________ psig.
- Specify test for tightness after pipe coil grid has been placed but before freezing is begun, giving test pressure, ________ psig, and time test pressure is to be held, ________ hours.
- Specify clips or other means for holding the pipes at the correct distance from each other.
- Specify the thickness of the ice sheet to be frozen on top of the coils, ________ in.
- For plastic brine piping, see Section 19-4(h).

(e) *Brine Expansion Tank*

- Specify a brine expansion tank sized to accommodate the total expanded volume of the brine in the system, from a minimum temperature of ________ F to a maximum shutdown temperature of ________ F. *Note:* For a rink in summer service, the maximum temperature may well on occasion equal the maximum outside ambient, or even exceed that if sunlight or other radiant heat is imposed on the tank.
- Specify that the tank shall comply with the ASME "Code for Unfired Pressure Vessels," including relief valve.
- Specify the location of the expansion tank.
- For additional information in specifying an expansion tank, see Sec. 12-5(b).

(f) *Instruments and Controls*

- Specify manner of control for both the compressors and the brine circulating pump, as from a thermostatic bulb embedded in the ice. Alternatively, the bulb may be submerged in the returning brine from the coils.
- Specify the calibration and operating temperatures of

the thermostats, giving the temperature, ______ F, at which the refrigeration action is stopped and the temperature, ______ F, at which action is resumed.

- Prepare a list of instruments required, as pressure gages, thermometers, humidistats, and others, identifying each by location, service, calibration, and range of calibration.
- Refer to Sec. 4-21 for details of instrument construction and materials.
- Specify a central station control panel, giving size and kind of panel and its location. List all the instruments, switches, relays, signal lights, and other appropriate items to be mounted on the control panel.
- Specify number and locations of push-button stations to be provided.

(g) *Ice Floor Melting System*

Note: In some auditoriums, the melting of the skating ice for other events may have a time deadline of four or five hours. This is met by pumping the brine through a heat exchanger, the brine going through the tubes and with low-pressure steam filling the shell.

- Specify capacity of the heat exchanger, ______ gpm of brine heated from ______ F to ______ F, when supplied with steam at ______ F total temperature and ______ psig pressure.
- Specify the construction of the heat exchanger, as ____________ (material) for the tubes, ______ ______ (material) for the shell, and ______ (material) for the tube sheet. Also specify number of passes in the exchanger.
- Specify condensate removal from the exchanger by ____________ (kind) of trap, having condensate handling capacity of ______ gph, with a pressure differential across the trap orifice of ______ pounds.
- Specify disposal of condensate, whether to waste or to a receiver for return to the steam boiler.

  Note: Query local authorities on maximum temperature, ______ F, of water that can be disposed of legally into a sewer system. If cooling is necessary for condensate from the exchanger, provide a cooling radiator as indicated in Sec. 11-1(a).

- Specify source of steam. If a steam generator, or boiler, must be part of the rink specification, refer to the appropriate section in this Handbook for details on the items to be specified. District steam appears in Sec. 11-1.

- Specify the steam piping, as suggested in Sec. 11-3(b).
- Specify manually operated valves, as in Sec. 11-3(c). Motor-operated valves in Sec. 11-3(d).
- In the list of instruments and controls, include the gages and thermometers necessary for the functioning and monitoring of the brine-heating steam circuit. Also the motor-operated or pneumatically operated valves, if these are used.
- State if the melting cycle is to be controlled from a separate panel or is to be incorporated with the refrigeration panel.
- Refer to drawing for arrangement of the steam and condensate piping and pipe sizes.
- Specify the kind of insulation and thickness, ________ in., for the steam lines. Ditto for condensate, if necessary.
- Specify the kind and center-to-center distance of all pipe hangers.
- If pneumatic controls are to be used, specify air pressure, ________ psig, and source, as from an existing compressed-air system, or from a compressor and tank that must be provided for this particular rink.
- For an individual compressed-air system, refer to Sec. 8-1(a)–(j) as applicable.

(h) *Plastic Piping*

- Specify if plastic pipe is to be embedded permanently in concrete or asphalt floors.
- Specify the brine temperature, ________ F. *Note:* this may be several degrees lower than for metal pipe because of the plastic's lower coefficient of heat transfer.
- Specify the kind of plastic, as (polyethylene) (polyvinyl chloride) (acrylonitrile-butadiene-styrene).
- Specify standard to which plastic piping shall conform, as (CS 255-63 for PE pipe) (CS 254-63 for ABS) (CS 256-63 for PVC).
- For wall thickness of the plastic piping, refer to the Standard Thermoplastic Pipe Dimension Ratio (SDR) as it applies to the particular kind of plastic being specified.
- Specify how joints are to be made, as (threaded) (inserts and clamps) (solvent cementing) (other).
- Refer to drawings for general arrangement of the piping.

- Specify if particular attention must be given to thermal expansion and contraction, and describe desired method of compensation.
- Specify that inserts, if used for joints, shall be of (stainless steel) (brass) (other material).
- Specify use of (plastic) (metal) fittings. Specify working pressure, ______ psig, and design pressure, ______ psig, for all fittings.
- Specify pressure under which system shall be tested after installation but prior to freezing, as ______ psig. Also stipulate how long test pressure shall be held without drop in order to demonstrate the tightness of the system.

DIVISION 20

Ventilating Equipment

20-1. GRAVITY-TYPE ROOF EXHAUSTERS AND VENTILATORS

Note: Refer to drawing for locations of the roof ventilators, all types, and heights above ground.

(a) ***Ridge- or Box-type Ventilators***

- Specify the nature of the air that will vent through the device, noting contaminants, if any, as (process gases or vapors) (fumes) (dusts), etc.
- Specify the conditions on which performance is predicated, as (1) wind, _______ mph; (2) temperature of the effluent, _______ F; (3) ambient outside temperature, _______ F; (4) where makeup air enters, as at ground level perimeter; (5) height of ventilator.
- Specify the maximum wind velocity, _______ mph, which a roof ventilator must be structurally able to withstand in this location.
- Specify materials of construction, as (galvanized steel) (aluminum) (copper) (plastic; name kind).
- Specify provision of a bird screen if considered necessary. Ditto for insect screen and damper.
- Specify if damper is to be manually operated or motor-operated. Specify point of damper control, as floor chain for the manual damper, floor-level push-button station for a motor-driven damper.
- For motor operation, specify electric service available. Require data from bidders on horsepower of motor, kind of motor, and speed, _______ rpm. Also designate if motor is a simple two-position unit or modulating.
- Specify wiring and other electrical components to comply with National Electrical Code, NEMA standards, and local codes as applicable.
- Require bidders to submit drawings, photographs, and other information on dimensions, weights, and arrangement of weather-excluding baffles.

- Specify exhaust capacity, ______ cfm, at three wind speeds, four temperature differentials, and at the specified height.

(b) *Rotary Turbine Roof Ventilators, Wind Actuated*

- Specify as appropriate as indicated in Sec. 20-1(a) preceding, including conditions, dampers, windage, etc.
- Specify that rotor shall be fabricated of light metal, as aluminum or plastic. Give kind of plastic.
- Specify rotor diameter, ______ in., and throat diameter, ______ in.
- Specify the gages of the structural materials as ______ gage galvanized steel and ______ in. thickness of aluminum.
- Specify lubrication system for rotating element.

(c) *Stack-type Stationary Ventilator*

Note: This is the conventional round-throated ventilator with the wide weather band, designed to exclude rain.

- Specify in accordance with Sec. 20-1(a) as applicable.

20-2. MOTOR-DRIVEN ROOF EXHAUSTERS

(a) *Centrifugal and Propeller Fans*

- Specify number of units in each capacity, and refer to drawings for locations.
- Specify if exhauster is to be direct-connected to its motor, or shall be driven by a belt.
- Specify nature of the effluent handled by the exhauster, identifying any contaminants, particularly those which are corrosive or of a nuisance nature, such as smoke, offensive odors, or other unpleasant characteristics.
- Specify unit mounting on vibration isolators, and state maximum acceptable noise level, ______ decibels. Specify kind of vibration isolators desired if manufacturer's standard design is not approved.
- Specify duct connections to the exhauster, if any, on both the inlet side and discharge. If discharge duct is carried straight up, specify height, ______ ft, above exhauster.
- Specify materials of ducts, as (galvanized steel) (aluminum) (plastic) (other).
- Specify kind of fan wheel, as centrifugal, propeller, nonoverloading, etc.
- Specify the conditions on which exhauster performance shall be based. Give temperature of effluent, ______

F, and ambient outside temperature, _______ F, into which the discharge is released.

- Specify the electrical service available, _______ volts, _______ phase, _______ cycles.
- Specify compliance with the National Electrical Code, NEMA standards, and local codes as applicable.
- Require bidders to furnish drawings and complete data, including the following performance items:
  (1) kind of motor and motor horsepower; (2) motor _______ rpm; (3) fan _______ rpm; fan tip speed, _______ fpm; (4) a table of capacities, _______ cfm, at various static pressures, _______ in. W.G.
- Also require dimensions and weights of each kind of unit.
- Specify general construction, as _______________ (material) of ventilator body, and fan. Specify gages and thicknesses.
- Specify lubricating system for bearings.
- For belt-driven units, specify possible changes of speeds by adjustments on pulleys.
- Specify any form of damper desired, as butterfly or disk, and method of operation, whether manual or electrical.
- Specify that fan output rating must be supported by the AMCA test procedures, as followed in an AMCA-approved laboratory.
- Specify if fan air supply enters fan from a free space, or through the louvers of a relief shutter or damper.
- If through a relief shutter or damper, give resistance of louvers, _______ in. W.G. Also state the net free area through the louvers as _______ per cent.
- Specify diameter of fan, _______ in.
- Since each manufacturer has particular features of design, specify a type and model as made by _______________ or an approved alternate. In this way the distinction of appearance, as mushroom, cylindrical stacks, or other, is made clear to bidders.
- Specify bird and insect screens of _______________ (material).
- Specify if louvers are to operate by gravity, or by electric motor wired into the fan circuit so that louvers are opened or closed in consonance with the operation of the fan.
- Specify how each ventilator is to be controlled, as by

simple temperature thermostat reacting from inside temperatures, by humidistats, by fume or smoke detectors and photoelectric cells, or other methods.

- For each system of control, specify the operating range of the controller, as deg F, per cent relative humidity, smoke or fume concentration on the Ringelmann scale, or other.
- If several exhausters are connected in battery and are to operate simultaneously, so state, identify the units so connected.

- Specify panel control board, giving location and materials; and list all instruments, switches, controllers, relays, signals, and other items to be mounted on the board.
- Specify locations and kind of push-button stations for starting and stopping individual units.
- Specify overload protection for all motors.

(b) *Vaneaxial and Tubeaxial Fans*

Note: With the tubeaxial design, the fan and motor are centered in a tube or duct. The addition of guide vanes in the fan discharge changes the classification from tubeaxial to vaneaxial.

With both tubeaxial and vaneaxial, a higher velocity of air discharge is practical than is acceptable for centrifugal and propeller design.

- Specify the amount of step-down from duct size, both intake and discharge, to the fan openings, expressed in comparative diameters, ________ in.
- Specify the velocity of discharge as ________ fpm.
- Specify total volume of discharge, ________ cfm, when handling air at a temperature of ________ F against a total discharge head of ________ in. W.G., ________ rpm.
- Specify other aspects of tubeaxial and vaneaxial fans from Sec. 20-2(a), as applicable, including direct-connected and belt drives.

(c) *Wall Ventilators, Motor-driven*

- Wall ventilators, the same as roof ventilators, may be of centrifugal, propeller, tubeaxial, or vaneaxial design.
- Specify the kind of fan required, as centrifugal, vaneaxial, etc.
- Specify the capacity in ________ cfm at ________ rpm fan speed at a static pressure of ________ in. W.G., handling air at a temperature of ________ F.

- Specify size of unit, _______ in. diam.
- Refer to drawings for locations of wall ventilators.
- Specify other items as appropriate from Sec. 20-2(a), Centrifugal and Propeller Fans.
- If considering axial fans versus centrifugal fans, require bidders to furnish comparative operating curves showing per cent rated pressure and horsepower as ordinates and per cent rated capacity as abscissas, for both forms of fan.

(d) *Attic Fan Ventilators*

- Refer to drawings for locations of the fans.
- Specify fan capacity, _______ cfm, when handling air at a temperature of _______ F against a discharge pressure of _______ in. W.G.
- Specify fan blading design, as conventionally pitched propeller, or airfoil with _______ (number) of blades.
- Specify electrical service characteristics available.
- Require bidders to submit complete data including dimensions; weights; materials of construction, (aluminum) (steel); speed, _______ rpm; drive direct-connected or by belt; motor characteristics, _______ hp; speed, _______ rpm; whether totally enclosed, single phase, one or two speeds, etc.
- Specify placement, as horizontal in the floor or vertical in an end wall.
- Specify shutter size as _______ in. × _______ in. Also materials of the shutter, and free area, _______ sq in.

20-3. INDUSTRIAL EXHAUSTERS

(a) *Particulate and Corrosive Gas Exhausters*

- Specify nature of materials entrained with the air, as dust from various industrial operations, fumes from chemical processes or welding, vapors, and smoke.
- Specify volumetric capacity of the exhauster, _______ cfm, when handling the specified effluent at _______ F and with entrained particulates amounting to _______ pounds per hour.
- Specify nature of fumes, vapors, and corrosive gases, also particulates.
- Specify construction of exhauster intended to counteract and resist the effects of dust and corrosive gases, as with

rubber vulcanized to the metal, or by all parts of the exhauster being fabricated of polyvinyl chloride or other specified plastic.

- Specify exhauster drive, as by direct-connected motor or by belt.
- Specify driving motor for severe service and contaminated atmospheres, as totally enclosed, and designed to operate at a temperature of _______ F. Specify ambient temperature, _______ F.
- Specify the kind and location of the subbase to which the fan and motor are bolted. Specify vibration isolaters for reduction of noise, if necessary, to a maximum allowable of _______ decibels.
- Specify the number and location of cleanout doors to be provided in the casing.
- Specify kind of bearings, as self-aligning ball-bearings, roller bearings. Require information on kind of thrust compensation provided for the fan.
- Specify bearing lubrication system, grease, oil, forced circulation or manual.
- For rubber coating protection, specify kind of rubber, thickness, and method of application.
- For plastic exhausters, specify kind of plastic and its physical properties in terms of withstanding abrasion, temperature, corrosive fumes, and impact.
- Specify that shaft and all rotating elements shall be balanced statically and dynamically at the factory.
- Require bidders to submit drawings and complete data including motor hp, _______; full-load amperes _______; locked rotor amperes, _______; speed, _______ rpm. Also physical information on dimensions, weights, and relative arrangements of wheel and bearings.
- If exhauster is mounted in a duct system, specify dimensions of duct and materials.
- Specify starting equipment for exhauster motor, kind and location. Specify overload protection.
- Require all electrical components to comply with the National Electrical Code, the NEMA standards, and local codes as applicable.
- Specify control panel and location. List all switches, relays, push buttons, signal lights, and other items to be mounted on the control panel.

(b) *Wet Scrubber–Exhauster Combinations*

Note: May be used for the removal and sludging of dust, and for the reclamation of various chemicals that are air-borne. Design takes the form of an air washer, with either static or rotary separation, and single- or multiple-stage exhauster fans.

- For items to include in the air washer section of the specification, refer to Sec. 18-1(j)–(l).
- In addition, specify the sludge removal system, stating if sludge receiver tank is to be integral with the scrubber or to be located elsewhere.
- Specify construction of sludge receiver tank, as ________ gage steel, lined with ________________ (material).
- Specify if sludge is pumped or is to drain by gravity. If the latter, specify provision of a rotary-type mud valve.
- Specify if mud valve is to be manually or motor operated.
- If sludge is pumped, specify kind and material of pump, as (centrifugal) (rotary) (screw) (other), of (cast iron) (steel) (other casing). Rotor to be of (steel) (bronze or brass) (other). Specify single or double suction.
- Specify electrical characteristics for pump motor, ________ volts, ________ phase, ________ cycles.
- Specify percentage of the dust or particulate matter to be removed in the scrubber and converted into sludge, ________.
- Specify ultimate disposal of sludge, as to a settling basin with overflow across a weir to a sewer, or directly to a sewer.
- Specify if pump is to recirculate water in the hopper or sludge tank to the scrubber sprays to conserve water, or is to discharge directly to disposal.
- Specify pump capacity either way, ________ gpm when handling sludge at ________ F temperature and with a density of ________ at a discharge head of ________ ft.
- For the exhauster components of the machine, refer to Sec. 20-3(a), as appropriate.
- Specify the number of stages in the exhauster.
- Specify discharge pressure of scrubbed air, ________ in. W.G.
- Specify the final condition of the discharged air, as ________ grains of ________________ (material) per cu ft.
- Require bidders to submit complete drawings and data on the equipment being offered, showing dimensions,

weights, arrangement of scrubber nozzles, and sludge tank connections; also performance of pumps and fans.

- Require fans to carry certification of test ratings under the AMCA Test Code.
- For wet scrubbers in fly-ash removal, refer to Sec. 4-13(b).

(c) ***Exhauster Controls and Instruments***

- List all instruments to be used with the equipment, stating kind of instrument, as gage, thermometer, humidistat, globe thermometer for effective temperature, etc. Also state location, as control panel, on the machine, remote reading, etc. For details of the instrument construction, refer to Sec. 4-21.
- Specify the calibration for each instrument, giving total range and step markings.
- Specify operation control, whether automatic, as monitored by the actions of a process or by the development of a condition of the atmosphere. For the first, the starting and stopping of a particular unit of process machinery may suffice; for the latter, a photoelectric smoke detector has been used. Alternatively, a simple manual switch may be all the control needed.
- Specify alarm system to be visual, audible, or both, to signal failure of some exhauster component, as the sludge pump, the interruption of water sprays, the stoppage of the exhauster fan, the development of a motor overload, or other service failure.
- Specify the kind and location of the control panel.

(d) ***Bag-type Dust Separators***

- Specify the maximum air flow that must be cleaned, _______ cfm.
- Specify the desired ratio of cfm to sq ft of free cloth area, as (2 to 1)(3 to 1).
- Specify the range of sizes of the dust to be separated, _______ microns.
- Specify separation efficiency, _______ per cent.
- Specify the material in the bag, as (cotton)(nylon)(wool)(orlon)(teflon)(dacron)(glass)(other).
- Specify temperature of dust-laden air entering the separator, _______ F. *Note:* for temperatures over 500 F, only siliconized glass or teflon are recommended.
- Specify the diameter, _______ in., of the bag and the length, _______ ft and in.

- Specify if bags are to be shaken or cleaned by reverse air flow. If the latter, specify automatic air reversal valve.
- Specify air pressure, _______ psig, as made available for operating the reversing valve and any other control devices.
- Specify a timer, set to initiate cleaning over a period of _______ minutes on a _______ hour cycle.
- Specify construction of the unit, as casing or housing of _______ gage steel; bag holder frames to be of (bolted) (welded) fabrication. Specify painting finish, _______ coats of _____________ (color) paint.
- Specify capacity of hopper, _______ cu ft of material.
- Specify hopper emptying methods, discharge valves, etc.
- Require bidders to submit drawings and complete data including free cloth area of bags, _______ sq ft; ratio of air movement, cfm, to free cloth area, _______; number of compartments, _______; total number of bags, _______; weight, _______ pounds of the complete unit; loss of air pressure through bags, _______ in. W.G., at full rated load and with bags clean.
- Specify that all electrical materials and design must comply with the National Electrical Code, NEMA standards, and local codes as applicable.
- Refer to drawings for location and installation limitations, if any.

(e) *Electrostatic Precipitators*

- Specify material to be precipitated, giving nature, sizes where appropriate, etc.
- Specify efficiency of operation, _______ per cent.
- Refer to drawings for location of precipitator.
- Refer to Sec. 4-13(a) for specifying items guidance.
- Specify temperature of air or gas entering precipitator, _______ F.

(f) *Cyclone Separators*

- Specify the material to be handled, giving particle sizes, microns. Also temperature of air entering, _______ F.
- Specify efficiency of removal, _______ per cent.
- Specify construction, as _______ gage steel. Or (aluminum) (stainless steel) (plastic) (other).
- For some applications, a refractory lining may be desirable. If this is the case, specify the refractory by tem-

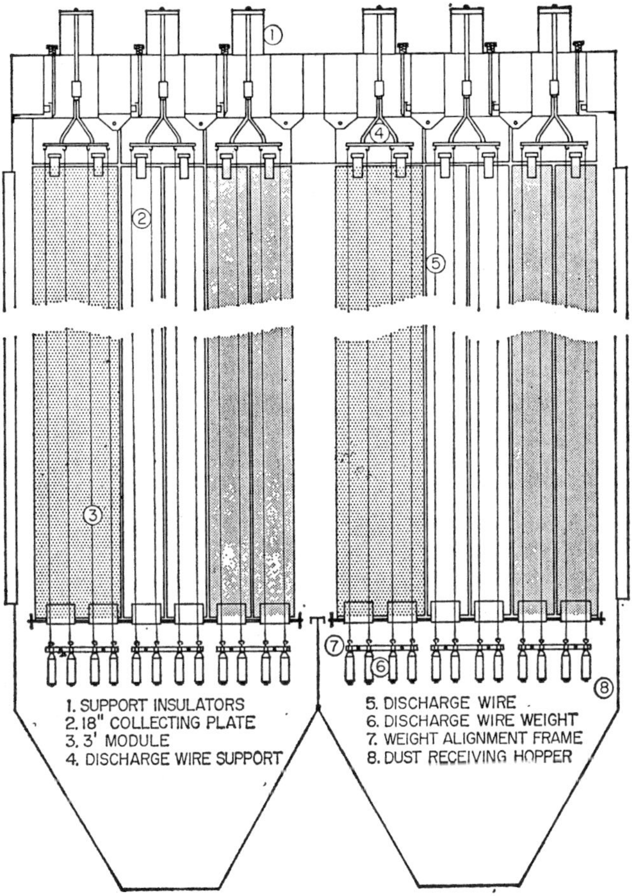

FIG. 20-1. Electric precipitator. Built in modules allowing any desired capacity. (*Courtesy Buell Engineering Company, Inc.*)

perature ______ F, and material, (fire clay) (high alumina) (silica) (acid-proof) (other).

- Specify analysis of mortar suitable for the refractory.
- Refer to the drawings for location and arrangement.
- Specify number of cyclones.
- Require bidders to furnish drawings and plans as necessary to indicate dimensions, weights, and performance of the units. Also to designate the materials used in construction.

(g) *Machinery Hoods*

- Refer to drawings for shape of hood and dimensions.
- Specify the kind of machine the hood is serving, as woodworking equipment, sander, planer, metalworking machinery, grinders, etc.
- Specify material of hood, as _______ gage steel; plastic, etc.
- Specify velocity of air across face of hood (face velocity), _______ fpm.
- Specify painting and finish, as _______ coats of _______ _______ (color) paint.

(h) *Kitchen Hoods*

- Specify equipment to be served, as stove, fryer, kettles, etc.
- Specify if hood is to be of the canopy or ventilator type.
- Specify cooking by gas or by electric burners.
- Specify height above floor of the hood opening edge, _______ ft, _______ in.
- Specify the number and kind of grease filters.
- Specify filter construction of (stainless steel) (bronze) (brass) (wire).
- Specify volume of air passing through filter, _______ cfm, at a temperature of _______ F.
- Specify face velocity of air entering hood, _______ fpm.
- Specify that the clearance between the lower edge of the filter and the cooking surface shall be _______ ft, _______ in., in accordance with Standard No. 96 of the National Fire Protection Association.
- If negative air pressure is maintained in the kitchen, specify that it shall not exceed (0.02 in. maximum) W.G.

(i) *Kitchen Makeup Air*

Note: The makeup air must be supplied both summer and winter to replace that lost through the hood ventilation.

- Refer to drawings for location of inlets and sizes.
- Specify size of fan, _______ cfm, when handling makeup air at a temperature of _______ F.
- If large volumes of makeup air are required, heating in the winter may be essential. If this is the case, specify the heating method to suit, as steam coils, Sec. 11-3(m); hot water coils, Sec. 18-1(c); duct furnaces, Sec. 13-3(a) and (b).

(j) *Ventilation Ducts*

- Refer to Sec. 13-2(a)–(c).
- For air outlet diffusers and grilles, see Sec. 13-4(a)–(f). Specify to suit.

(k) *Fire Detection in Kitchen Vent Hoods and Ducts*

- Locate a temperature-sensitive bulb as close to the stove as practical, designed to give an alarm if fire lights off in the grease filter or in the vent duct.
- Specify nature of alarm; visible, audible, or both.
- Specify temperature at which the alarm is energized.
- Specify the electric service available, ________ volts, ________ phase, ________ cycle.
- If alarm energizes any fire extinguishing equipment, as a water spray in the duct, steam jet to the grease filter, or any other method, so specify.
- Specify kind of hand fire extinguisher to station in the kitchen, as dry chemical or dry chemical multipurpose type, effective on grease fires.
- Specify number and size of each unit extinguisher.
- Require each unit to bear the approval of the Underwriters Laboratories, Factory Mutual Insurance Association, or other authorized testing laboratory.
- Specify locations of extinguishers in the kitchen.
- If alarm is connected to a municipal fire station, state identification name and location of station. Give approximate distance from kitchen to fire station.

DIVISION 21

Hot Water, Nonspace Heating

Note. Hot water other than for space heating is required for domestic use, restaurants, laundries, showers, etc. Systems will be classified as to source, distribution, and special features.

21-1. SOURCES OF HOT WATER

(a) *Coils in a Space Heating Boiler*

- Refer to the boiler specification, and include therein the provision of a domestic hot water coil submerged in the boiler water, designed to produce _______ gal of _______ F temperature hot water per hour, when supplied with cold water at a temperature of _______ F. Steam heating boilers are detailed in Sec. 11-2 for space heating, and hydronic space heating boilers in Secs. 12-1 and 12-2.
- Specify cold water pressure, _______ psig.

(b) *Hot Water from a Steam Converter*

- Refer to Sec. 12-3(a)–(d).
- Specify output in _______ gph of hot water at _______ F when cold water enters at _______ F and steam supply condition is _______ psig pressure and _______ F total temperature.

(c) *Instantaneous Hot Water Generator with Steam Boiler*

- Specify output of generator as in Sec. 21-1(b) preceding.
- Specify construction of the instantaneous generators, as shell and tube type, set horizontally, with water boxes of (cast iron) (other material), shell of (steel) (other), and tube sheets of (steel) (other). Also specify tube arrangement, as U bundle of _______________ (material) tubes, _______ in. diam supported by plate or bracket of _________________ (material).
- Specify working pressures of shell, _______ psig, and tubes, _______ psig, and also test pressures, _______ psig, for these parts.

- Specify if heater is to be furnished with suspension hangers or support saddles.
- For vertical mounting, specify supporting lugs of ________ (material) and indicate positions of lugs on heater.
- Require bidders to submit sectional drawings showing dimensions; weights; number of passes, (2) (4) (6); locations of air vent and vacuum breaker connections.
- Specify make and model of automatic air vent valve and vacuum breaker, if these are to be supplied by the heater manufacturer.
- Specify kind and capacity of condensate trap, ________ pounds per hour, with a back pressure across the trap orifice of ________ psia.
- Specify kind of steam admission control valve, as diaphragm, spring loaded, capillary to hot water outlet and bulb. See Sec. 21-2(g).

(d) *Tank Storage Water Heaters*

- Specify capacity of storage tank, ________ gal.
- Specify capacity of submerged heating coil, ________ gph of water raised to ________ F from incoming cold temperature of ________ F, when coil is furnished with steam at a pressure of ________ psig and temperature of ________ F.
- Specify tubes as a U bundle of ________ (material) tubes, ________ in. diam and ________ in. long, expanded into a tube sheet of ________ (material).
- Specify that tank construction must comply with the requirements of the ASME "Code for Unfired Pressure Vessels." Specify that all manholes must be reinforced.
- Specify arrangement of tank; vertical or horizontal.
- Specify vacuum breaker and automatic air vent valves.
- Specify pressures, working, ________ psig, and test, ________ psig.
- Specify condensate removal system, giving make and model of trap, capacity in ________ pounds per hour with a back pressure on the trap of ________ psia.
- If copper lining is desired, specify weight of copper to be used, ________ psf. (*Note:* ASME calls for 3-pound copper.)
- Specify that copper lining is to be welded at all seams and spun into the tube holes in the tube sheet.

- Specify that copper lining must be tested to a pressure of _______ psig, hydrostatic.
- If cement lining is desired, specify manner of cement placement, as thorough cleaning of the inside surfaces of the shell, then a build-up of cement to _______ in. thickness. Cement is to be cured over a period of _______ days, or _______ hours.
- State source of steam, as boiler, district steam main, etc.
- Specify a combination pressure-temperature relief valve to be mounted with bulb in area of hottest water.

(e) *Storage Tanks without Heating Elements*

- Specify capacity of tank, _______ gal.
- Specify tank design and construction to comply with ASME "Code for Unfired Pressure Vessels."
- If copper or cement lined, see Sec. 21-1(d) preceding.
- Specify tank supports, as suspension rods, saddles, or structural steel. Provide sketch showing dimensions of supporting equipment.
- If tank is in a natural, or gravity, circulation arrangement between the tank and the heater, specify that tank must be located at least 5 ft higher than hot water outlet on heater.
- If cold water makeup is introduced into the tank, specify this shall be accomplished by a dip tube, carried to within 3 or _______ in. of the bottom of the tank.
- Specify a pressure-temperature relief valve, with the bulb placed in the area of hottest water in the tank.

(f) *Gas-fired Storage Water Heaters*

- Specify capacity of heater, _______ gph of _______ F water, with a cold water supply at _______ F.

  Note: Alternatively, capacity specification may be expressed in the usual _______ gph raised in temperature from 40 to 140 F.
- Specify number of these heaters.
- If one heater is to be used as a temperature booster, commonly seen in restaurants where 180 F water is necessary for sterilizing, specify the temperature rise from _______ F to _______ F. Specify the capacity under these conditions, _______ gph.
- Specify that heater shall be as approved by the American Gas Association and shall comply with ASME Code.
- Specify if heater is to operate on upward or downward flow.
- Gas controls to include a high-limit temperature control

that will shut off the gas on development of excessive temperature, to be specified as _______ F.

Note: For liquefied petroleum gas, automatic shutoff shall apply to both the main gas valve and the pilot gas valve, with manual relighting required. (Pressure regulator not needed on L.P. gas.)

- Specify a gas pressure regulator set to maintain a gas pressure to the burners of _______ in. W.G., when supplied with natural or artificial gas at _______ in. W.G. or _______ psig, as applicable.
- Specify construction of heating element, as circular copper tubes, or other arrangement.
- Specify insulation as _______ in. of _____________ (material) and finish of jacket as (enamel baked on) (other).
- If glass-lined tank is specified, require glass to be fused to steel. Also require a magnesium rod for corrosion protection.
- Specify storage capacity of tank, _______ gal.
- Specify kind of gas to be burned, as (natural) (L.P. gas) (artificial) (other). Give calorific value, _______ Btu per cu ft.
- Specify a draft hood to be provided with the heater.
- Specify water pressure, _______ psig; also test pressure.
- Specify temperature-pressure relief valve to be mounted on heater.
- Require bidders to furnish drawings and full details of equipment being offered, as dimensions, weights, internal arrangement of heating coils, gas burner, pilot, and other components.
- If peak draw of hot water will continue for more than 1 hour, specify the peak term, as 2 hours, 3 hours, etc.
- Specify the water conditions, particularly the degree of hardness, in ppm.
- If two-temperature water is required from the same heater, so specify, giving the two temperatures desired, as 140 and 180 F.

(g) ***Storage Water Heaters, Electric***

- Specify storage capacity of tank, _______ gal.
- Specify recovery rate, _______ gal in _______ hours, when heating water from _______ F to _______ F.
- Specify construction of tank, as (galvanized) (glass lined) (other).

- Specify installation of magnesium rod to combat corrosion.
- Specify working pressure, _______ psig, and test pressure, _______ psig.
- Specify insulation, as _______ in. of _____________ (material).
- If booster heater is to be used, so specify, giving peak, _______ gph for _______ hours at a final temperature of _______ F.
- Specify number of electric heating elements, upper and lower, and give wattage of each, _______ watts.
- Specify a temperature-pressure relief valve, and give settings as _______ F and _______ psig.
- Require units to be approved by the Underwriters Laboratories.
- Specify electric service available, _______ volts, _______ phase, _______ cycle.
- Specify that contactors in thermostat shall operate on an off and on cycle, without arcing.
- Require all electrical components to meet National Electrical Code, NEMA standards, and local codes as applicable.
- Require bidders to submit drawings and complete data showing dimensions, weights, connection locations, internal arrangements, etc.
- Describe water conditions, hardness in ppm of calcium carbonate, presence of any organic material, iron, sulfur, etc.

21-2. HOT WATER DISTRIBUTION

(a) ***Tank Circulators***

Note: These pumps are installed to maintain a positive circulation between the storage tank and heater. They become necessary when natural or gravity circulation is inadequate or impossible because of the relative positions of tank and heater, and because of high friction in some forms of heater and interconnecting piping.

- Specify capacity of pump, as _______ gpm when handling water at _______ F suction temperature, when taking suction from bottom of storage tank. Specify the discharge head, _______ ft.
- Specify kind of pump, (centrifugal) (rotary) (other), and materials of construction, as _____________

rotor, ______________ casing, ______________ shaft, ______________ bearings.

- Specify electric service available for pump motor, _______ volts, _______ phase, _______ cycle.
- Specify connecting coupling; flexible, rigid, etc.
- Require bidders to furnish information including pump speed, _______ rpm; motor horsepower, _______; kind of motor, as splash-proof, totally enclosed, etc. Also dimensions and weights of pump-motor unit.

(b) *System Circulators*

Note: These pumps deliver water to the point of use and are mounted either on the hot water connections leaving the heater or storage tank, or alternatively, with a recirculating system, on the return.

- Specify position of pump, on supply or return.
- Specify temperature of water going through the pump, _______ F.
- Refer to Sec. 21-2(a) preceding for items to specify for this pump.

(c) ***Piping, Valves, and Fittings***

- Specify materials of piping: (copper tubing) (brass) (wrought iron) (steel) (other).
- For copper tubing, specify the classification, Type L, Type M, or Type K.
- For copper and brass, regular weight, specify percentage of copper: (67 per cent) (85 per cent) (100 per cent).
- For steel and wrought iron piping, specify the schedule number, as Schedule 40. (*Note:* Wall thickness is slightly greater in wrought iron welded pipe than in welded or seamless steel pipe.)
- Specify the kind of joints to be used, as brazed joints for nonferrous piping, socket type (no fillet brazing). Also compression fittings or flared fittings with hard solder as suited to the pressure and temperature. Specify that threaded joints shall conform to American Standard B2 for taper pipe threads.

  For ferrous piping, joints may be threaded or welded. Specify which shall be used.
- Specify valves to be of ______________ (material) body, with ______________ (material) seat and (rising) (nonrising) stem. Stipulate working pressure, _______ psig, and test pressure, _______ psig.
- Specify valve ends, threaded, welding, flanged, flared.

- Specify materials of fittings, as (cast iron) (steel) (malleable iron) (brass) (bronze) (other).
- Specify working pressure, ________ psig, and test pressure, ________ psig.
- Specify kind of flanges, as companion, Van Stone, tongue and groove, etc. If raised face, specify raise, ________ in.
- Specify flange material: (cast iron) (steel) (malleable iron) (other).
- For flanged joints, specify kind of gaskets, as fiber, metallic, flexible, etc., for ________ psig working pressure. Refer to the ASME Boiler Code for guidance on gasket selection.
- Specify the kind and spacing of pipe hangers, ________ ft on centers.
- To reduce expansion problems in a pipe run, it is possible to specify erection by "cold strain." If cold strain is planned, so specify and stipulate the amount of linear compression to be applied, ________ in. (usually one-half of expansion).

(d) *Insulation*

- Specify piping insulation, as ________ in. thickness of ________________ (material).
- Specify outer jacket ________________ (material), (sewed on) (pasted on) (banded) (other).
- Specify painting, if any, giving number of coats and color.
- Specify how valves and fittings are to be covered, as by preformed insulation, by applications of plastic insulation in the field, other means.

(e) *Mixing Valves*

- Specify temperature of valve discharge, ________ F.
- Specify temperature of hot water entering valve, ________ F, and temperature, ________ F, of diluting water.
- Specify temperature adjustment range, whether by removable key, screw, or otherwise.
- Specify number and size of mixing valves, and indicate the location of each.

(f) *Water Hammer Arrester*

- Specify make and model of arrester.
- Specify number, sizes, and locations of the arresters.
- Specify that arresters shall be provided with means for

relieving conditions of waterlogging and shall permit recharging with air unless of diaphragm type.

(g) *Hot Water Controls by Temperature*

- For control of temperatures leaving steam to water heat exchangers, specify a steam regulating valve that opens and closes as monitored by a thermostatic bulb immersed in the hot water discharge.
- Specify the temperature range of this valve, from _______ F to a maximum of _______ F.
- Specify the valve size, _______ in. diam. Also stipulate the capacity of the valve to pass steam into the exchangers, as _______ pounds of steam an hour, when the pressure of the steam to the valve is _______ psig.
- Specify capillary tubing from bulb to valve to be of _____________ (material). If to be armored, so state.
- For direct-fired water heaters, specify a thermostatic bulb that lights off and cuts out the flame in response to the water temperature around the bulb, as it leaves the heater.
- For storage, unfired hot water heaters, control temperatures by a thermostatic bulb that opens and closes the steam admission valve in response to the temperature of the water leaving the tank.
- For electric storage heaters, temperature in the tank is to be controlled by a thermostatic bulb that energizes first one, then the second, of the electric elements submerged in the tank.
- For system circulating pumps, specify a thermostatic bulb in the return or recirculating piping, to start and stop the pump as necessary to maintain a temperature of _______ F in the return water.
- In all cases, specify a manually operated push-button station that can be used in place of the automatic shutoff for the burners in a direct-fired unit, or for starting and stopping the system circulator.
- Manually operated stop valves shall be specified for all steam supply lines that serve steam to water heat exchangers. For details of valve specifications, see Sec. 11-3(c) and (d).
- Proprietary temperature control stations are available through which very close temperature stabilizing is possible. For this kind of installation:

- Specify the temperature, ________ F, to be maintained. Also specify the temperature range (as from 65 to 115 F) over which the control may be adjusted.
- Specify that the controller shall shut off the hot water supply if the cold water flow to the mixer is interrupted at any time.
- Specify the maximum temperature that must not be exceeded, ________ F.
- Specify the quantity of water to be delivered at the controlled temperature, ________ gpm.
- Specify a central control panel if system is large and varied in demands and performance.

(h) *Instruments*

- List all thermometers and gages to be installed throughout the hot water system, giving for each the calibration desired and range of operation, together with the location in which each is to be stationed.
- Specify the details of construction as suggested in Sec. 4-21.

DIVISION 22

Miscellaneous Equipment

22-1. INCINERATORS, DOMESTIC

- Provide location drawing or specify exact location of incinerator.
- Specify form, as package, portable, masonry, etc.
- Specify type of waste to be burned, as grouped by the standards of the Incinerator Institute of America. For example, Type 1 is combustible rubbish in the form of paper, cartons, floor sweepings, etc., but excluding material with more than 25 per cent moisture by weight. Type 2 is a mixture of combustible and garbage. Other types recognized are 3, 4, 5, and 6.
- Specify the incinerator classification (I through VII) as set by the Incinerator Institute of America.
- Specify storage capacity of unit, ________ cu ft, and burning rate, ________ pounds per hour.
- State if incinerator is to be direct fed or flue fed.
- Specify general construction, as casing of ________ gage steel, lined with refractory for temperatures up to ______ F, adequately insulated by ________ in. of ________________ (material). (*Note:* Neither refractory lining nor insulation is required for small units set outdoors.)
- Specify kind of fuel, as (natural gas) (L.P. gas) (artificial gas) (oil) (anthracite coal) (other).
- State calorific value of fuel, ________ Btu per cu ft, per pound, or as appropriate.
- For gas fuel, specify that installation must comply with requirements of National Fire Protection Association Standard No. 54, "Standard of Gas Piping and Gas Appliances in Buildings."
- Specify that incinerator meet the standards of the Underwriters Laboratories, American Gas Association, or Factory Mutual Laboratories, as appropriate.
- Call attention to local air pollution codes. Specify a design

and arrangement that will perform acceptably under the pollution code. Specify the concentration of particulate matter in the incinerator's discharge to atmosphere to be less than ________ grains per cu ft of the flue gas.

- For incinerator flues, breechings, and chimneys, refer to Sec. 4-13(a) and 4-13(c)–(e).
- For spark arresters, see Sec. 4-13(f).
- For any electrical devices and materials functioning with the incinerator, require compliance with the National Electrical Code, the NEMA standards, and local codes as applicable.
- Specify that in general, the incinerator shall meet the requirements of NFPA Standard No. 82, latest edition.
- Require bidder to submit drawings, photographs, or both, and complete data for his equipment. Information needed is size and weight, methods for restricting emission of particulates, fumes, and smoke. Construction of grates, kind of refractory, and residue removal facilities are also wanted.

22-2. INCINERATORS, COMMERCIAL AND INDUSTRIAL

Note: Under this heading are included apartment house and hotel incinerators involving waste chutes from upper floors to basement or ground-level incinerating equipment. Most of these incinerators are indoors.

Commercial and factory incinerators generally are set outdoors, although in some office or department store applications, basement locations are not unusual.

Particular attention must be given to the prevention or control of fire in waste chutes and the stopping of sparks at roof levels.

(a) *Classification and Capacity*

- Specify the incinerator classification in accordance with the IIA standard (I through VII).
- Specify the type of waste to be burned, as indicated by the IIA type standard (1 through 6).
- Specify the storage capacity of the unit ________ cu ft, and the burning rate, ________ pounds per hour.
- State if incinerator is direct fed or through a chute.

(b) *Form and Construction*

- Specify a package type of unit or masonry construction, as appropriate.
- Specify the number of chambers required for minimizing discharge of fly ash, smoke, fumes, etc.

- Refer to drawing for general location and arrangement for charging, withdrawal of residue, etc. Also for details of size, kind, and locations of flue charging doors at various levels.
- Require compliance with NFPA Standard No. 82, latest edition, as appropriate and applicable to this installation.
- Specify kind of fuel, as (gas) (oil) (coal) (other).
- Specify calorific value of fuel, _______ Btu.
- For gas firing, require gas connection to comply with NFPA Standard No. 54, applicable to gas piping in buildings, if this installation is inside.
- Specify refractory lining to extend upward in flue at least _______ ft above the incinerator. Specify refractory to withstand temperature of _______ F without spalling, cracking, crumbling, or otherwise deteriorating.
- Specify adequate insulation for inside units, _______ in. of _______________ (material).
- Specify maximum concentration of emitted particulate matter as _______ grains per cu ft of flue gas.
- Specify smoke limitation as No. _______ Ringelmann.
- For construction of flues and chimneys, refer to Sec. 4-13(a)–(e).
- For spark arresters, specify material of screen, as (stainless steel) (steel) (copper) (other). Refer to Sec. 4-13(f) for specifying particulars.
- Require bidders to submit drawings and complete data.

(c) *Fire Protection*

- Specify the placement of automatic sprinklers in flues to extinguish fires occurring in the waste that reaches the flues through the charging doors.
- Specify number and size of shower heads and temperature, _______ F, at which they shall open.
- Specify automatic fire alarm installation, with sensitive bulbs or wires located in the flues. Alarm signal to appear on a central control board, also, if considered desirable, in the nearest fire station. This will be a proprietary system, so specify by name and model, or approved alternate. *Note:* This kind of fire protection is not required for an outside incinerator with its own chimney.
- Specify electrical service available for alarm system and other auxiliary actions of the incinerator installation, as _______ volts, _______ phase, _______ cycles.
- Require all electrical equipment and materials to comply

with National Electrical Code, NEMA standards, and local codes as applicable.

- If rubbish bins and storage rooms are provided for the accumulation of waste between firing periods, these also must be protected from fire. Consequently specify automatic sprinklers and fire-sensitive alarm systems in these compartments.
- In areas having possible freezing, specify heating equipment adequate for maintaining above-freezing temperatures. For steam, hot water, infrared, and resistance electrical heating, refer to Divisions 11–14 as applicable.

(d) *Hospital Incinerators*

- In addition to the guidance suggested in Sec. 22-2(a)–(c), refer to Bulletin H as published by the Incinerator Institute of America. Title of the bulletin is "Selection of Incinerators for Hospital Use."
- This applies to veterinary hospitals and contagion wards, as well as to hospitals in general.

Note: Most important problem is to make sure that all matter, fumes, and smoke emitted to the atmosphere have been processed through the incinerator at temperatures so high that absolute sterilization has been accomplished.

- Specify this sterilization temperature: ________ F.

22-3. INCINERATORS, MUNICIPAL

(a) *General Notes*

Municipal incineration involves so many types of materials, and such a fluctuation in charging rates from irregular delivery of rubbish and garbage, that each incineration plant must be designed as an individual problem. To illustrate, in some plants and on some days, delivery directly into the furnace is possible and is practiced, since this reduces handling to a minimum. In other plants, and in the same plant on some days, delivery may be more rapid than the furnaces can accept; consequently a temporary storage bin becomes necessary. Transferring the waste from the storage bin to the charging chute is accomplished by clamshell bucket operated from the cab of an overhead bridge crane. This introduces an item of extra capital investment and extra labor each day of operation, so many designers avoid this arrangement if they can.

Fluctuations of furnace temperatures resulting from the irregular deliveries of material, together with a usual weekend and sometimes overnight shutdown, impose temperature stresses on the refractory. For this reason, some designs separate the fire side refractory from the casing by insulating brick and air spaces, the latter helping to keep down refractory temperatures. Other control methods involve outlet dampers, which may be used to prevent too-rapid furnace cooling by restricting the outflow of hot combustion gases.

For the reasons discussed, the specification guidance items given in this section cover several arrangements, so that selection is necessary to fit the kind of design adopted.

(b) *Capacity Specifications*

- Specify weight of material to be handled, _______ tons per day.
- Specify average percentage of water, _______ per cent, by weight.
- Specify number of furnace cells, _______, and burning rate of each, _______ tons per day.
- Specify kind of stoking to be used, as (manual) (traveling grate) (shaking grate) (rotary) (other).
- Specify quantity of material to be burned on the grate as _______ pounds per sq ft of grate per hour.
- Specify temperature range to be maintained, from _______ F to a maximum of _______ F.
- Specify the amount of forced draft, under-the-grate combustion air, _______ cfm. Also specify temperature of this air, stating if preheated, or from the area ambient, _______ F.
- Specify amount of over-the-fire air, _______ cfm, if introduced by forced draft fan. If induced by furnace draft, state size of opening, _______ sq ft, and if known or calculated, give draft over fire in the furnace, _______ in. W.G.
- Referring to the six types of waste as standardized by the Incinerator Institute of America, specify the percentage of the total represented by each type.

(c) *Construction*

- Specify casing of steel, or brick masonry, or other finish. For steel, specify the thickness as _______ gage; for brick, specify the kind of brick desired, color, crushing strength, tensile strength, etc. Specify the expected brick temperature at the outside surface, _______ F, with an atmospheric ambient of _______ F.

- Specify the number of furnace cells and their relationship to each other and to the stack or stacks.
- Specify general construction of each cell, as a suspended ceiling, air-cooled walls or noncooled walls, and specify the kind of refractory to be used.
- Specify area of drying hearth in each cell.
- Specify slope of grate as _______ per cent to the horizontal.
- Specify kind of outlet damper from each combustion chamber as (steel) (alloy) (other), and design as tilting, butterfly, etc.
- Specify control of grates and dampers, as by hydraulic cylinders, pneumatic cylinders, or electric motors. Specify operating pressure for water and air, _______ psig. For electric operation, specify the service available, as _______ volts, _______ phase, _______ cycle.
- Specify location and capacity of ash hopper, as _______ tons, or _______ hours of full incinerator operation between dumps.
- Specify size and capacity of receiving hopper over charging holes. Specify material, weight, and size of covers for charging holes. Specify if manually or hydraulically manipulated.
- Where applicable, specify number and operation of guillotine doors.
- As applicable, specify thickness of fire brick, _______ in.; thickness of insulating brick, _______ in.; thickness of _______________ (material) insulation, _______ in.
- Specify the construction of both receiving and ash hoppers, as _______ gage steel, (bolted) (welded). Refer to drawings for dimensions, slope of sides, supports, and other pertinent data.
- Specify the material and construction of the air ducts, as (galvanized steel) (aluminum) (copper) (other). Must comply with thicknesses, cross breaking, jointing, and other aspects of duct fabrication as given in the ASHRAE Guide, latest edition.
- If separate storage bin must be provided, refer to drawing for dimensions and construction of the bin. *Note:* Bin capacity should not be less than 2½ days delivery of waste.

- Specify the bridge crane for transferring the waste from storage to the furnaces, as suggested in Sec. 4-20, Power Plant Cranes.
  - Specify the capacity of the clamshell bucket as ________ cu ft.
- Specify the masonry chimney or chimneys as given in Sec. 4-13(d).
- Specify the steel stack or stacks from Sec. 4-13(c).
- For various ways to control fly-ash emissions, refer to Sec. 4-13(b).

(d) *Fans and Blowers*

- For induced draft fans, refer to drawings for locations and general arrangement with the connecting ducts.
- Specify the temperature of the combustion gases the fan must handle, ________ F, maximum.
- Specify the volume of the combustion gases the fan must handle, ________ cfm.
- Specify discharge pressure of the fan, ________ in. W.G.
- Specify the kind of fan wanted, as (centrifugal) (propeller) (axial) (other).
- Specify fan drive; electric motor through a (belt) (flexible coupling) (rigid coupling) (other).
- Specify kind of bearings and method of lubrication.
- Require bidders to submit drawings and complete data, including (1) material of fan blades; (2) material of fan casing; (3) thickness of material, ________ gage; (4) shaft material, ________________, thickness, ________ in.; (5) fan speed, ________ rpm; motor speed, ________ rpm; (6) kind of motor, totally enclosed splash-proof and with a temperature rise of ________ C; (7) full-load hp, ________; full-load amperes, ________.
- Specify the expected ambient maximum temperature in the area where the fan is located, ________ F.
- For the forced draft fan, specify similarly, making the changes for air temperature, location, etc.
- For temporary storage bins, specify a grid type of water spray to allay dust, and the intermittent use of a chemical spray to allay odors and insects.
- A wall type of ventilating fan, or exhauster, may be specified to service the storage bin under some conditions. Specify this fan from Sec. 20-2(c).
- Specify an adequate drain in the floor of the storage bin

or pit, to remove excess water from the materials and from the sprays.

- Specify the number of sprays and the material of construction, as brass; specify piping material and working pressure, _______ psig.

(e) *Hydraulic Pumps*

- For service with hydraulic control systems, or for water sprays, specify kind of pump, as centrifugal, rotary, other.
- Specify materials of construction, as _______________ for rotor and _______________ for casing.
- Specify capacity as _______ gpm at a total discharge head of _______ ft, when handling water at a temperature of _______ F.
- Specify details of motor drive, coupling, etc.
- Require bidders to furnish complete data on pump and motor, speed _______ rpm; hp, _______; voltage, phases, amperes, etc. Also method of starting, as a manual _______________ pole switch, push-button station, etc. Specify location of starting switch or panel.
- Specify piping valves and fittings to suit working pressures, giving materials of each.
- Since the cylinders are proprietary, specifications are simplified by naming a system manufacturer, following this with "or approved alternate."

(f) *Pneumatic Controls*

- Specify the compressed air available, _______ psig working pressure. Give source, as available, or if compressor must be furnished, to serve the control system.
- As in the preceding section, name a manufacturer of pneumatic control systems, followed by "or approved alternate."

(g) *Electric Operators*

- Specify kind of motors desired, as two-position, or modulating, multipositional. Also give temperature rise.
- Specify electric service available, _______ volts, _______ phase, _______ cycle.
- Specify kind and location of motor controls, as switches on a central control panel, push-button stations adjacent to the motor being controlled, etc.
- Specify time of operation, _______ seconds between full opening and tight closing. Alternatively, time over travel distance, if appropriate.

(h) *Incinerator Housing*

- Not within the scope of this Handbook.
- Refer to architect's drawings where necessary in issuing mechanical specifications.

22-4. SWIMMING POOL EQUIPMENT

(a) *General Notes*

Water circulating rates currently in use provide a complete change of water in the pool every 8 hours or less. The circuit may be expected to include the hair strainer, pumps, slurry feeder, filters, chlorinator, rate of flow indicator, and the return piping to the pool.

Disinfection by chlorine gas in water solution or by one of the hypochlorates is common, and although other chemicals may serve, their use may require approval by a local or state board of health.

Pools may be classified as follows:

Type A. Municipal, community, school, and country club.

Type B. Large hotel pools, with a water surface in excess of 1,600 sq ft.

Type C. Motels, apartments, small hotels with a water surface of less than 1,600 sq ft.

Type D. Therapeutic and treatment pools.

Type E. Indoor pools.

Filters may be of either sand or diatomaceous earth type.

The disinfecting of pool water with chlorine gas, which is common in large pools, requires provision of a special compartment for storage of the gas cylinders, proportioning equipment, and a ventilating fan for exhausting the compartment to atmosphere at a rate of one complete air change a minute.

If ammonia is added to the chlorine to maintain residual conditions of pool water with a fluctuating load of bathers, it must be handled with the same care prescribed for chlorine gas.

(b) *Hair Strainers, or Interceptors*

Note: Removal of hair and lint is necessary to prevent these materials from wrapping around the pump shafting and clogging the filters.

- Specify if strainer is to be of the single basket or duplex type.
- Specify materials of construction, as (stainless steel) (bronze) (copper) (aluminum) (other).

- Specify maximum diameter of strainer openings, _______ in. (⅛ in. is common).
- Specify that total areas of all perforations in the basket shall equal approximately 10 times the cross-sectional area of the strainer inlet piping.
- Specify the volume of water to be handled by the strainer, as _______ gpm.
- Require bidders to stipulate pressure loss through strainer immediately prior to cleaning, as _______ in. W.G. Also to provide drawings and information regarding dimensions, weights, method of cleaning, etc.
- Refer bidders to general layout drawings for strainer location, piping connections, and other details.

(c) *Recirculating Pumps*

- Specify kind of pump, as single-stage, double-suction, centrifugal, motor-driven.
- Specify construction of pump, as bronze impeller, steel shaft, cast-iron casing, etc. Direct-connected on a common bedplate and concrete foundation, by a (flexible) (rigid) (other) coupling, to an electric motor operating on _______ volt, _______ phase, _______ cycle service.
  - If noise is a factor, specify that a vibration absorption pad of _______________ (material) be placed under the foundation.
- Specify the pump capacity, _______ gpm, against a discharge head of _______ ft.
- Require bidders to submit drawings and data for the pumps offered, giving speed, _______ rpm, and speed of motor, _______ rpm; also motor hp, _______, kind of motor, and starting equipment.
- Specify that pump motor must have built-in overload protection.
- State number of pumps to be furnished under this specification.

(d) *Vacuum Cleaning Pump*

- If recirculating pump is to be used for suction or vacuum cleaning, so state, and require pumps to be furnished with open-type propellers.
- For a vacuum cleaning pump which is used for that purpose only, specify as for the recirculating pumps, but with open impellers.
- Also specify the suction lift on the pump, _______ ft.

(e) *Pressure Filter, Sand Type*

- Specify arrangement of filter, as horizontal tank or vertical tank.
- Refer to drawings for location of filters, number of units, and information on piping connections.
- Specify the thickness of the filter bed, _______ ft.
- Specify the size of the sand particles, _______ mm. Also state kind, as quartz, gravel, mixture, etc.
- Specify the kind of coagulant to be added, as aluminum sulfate or other.
- Specify the size and capacity of the coagulant tank, _______ gal.
- Specify the maximum rate of flow through the filter as _______ gpm per sq ft of filter.
- Specify the tank construction, as welded steel, _______ in. thick for filter tank, and similarly specify for the coagulant tank.
- Specify the anticorrosion treatment to be applied to the filter and coagulant tanks, as (galvanizing) (bitumastic coatings) (polyvinyl plastic) (other). Also specify installation of magnesium rod if electrolysis hazard exists.
- Specify operating pressures through filters and other tanks as _______ psig, and test pressures of _______ psig.
- Specify capacity of filters in terms of _______ hours required to filter entire contents of the swimming pool.
- Specify rate of backwash, as _______ gpm, at a flow rate of not over _______ gpm per sq ft of filter area. Water pressure for backwashing, _______ psig.
- Specify control of backwash, if it is to take place automatically when resistance of filter from clinging material actuates the backwash valve, or if this is to be undertaken manually. Also length of backwash, _______ minutes.
- Specify electric service available for control operations, _______ volts, _______ phase, _______ cycle.
- Require backwash drains beneath the filter bed to be uniformly spaced, so that channeling through the filter bed is not promoted.
- Require bidders to submit complete data and drawings covering general arrangement and construction, and

showing dimensions, weights, and connections, as these are standard to his particular design.

(f) ***Pressure Filter, Diatomaceous Earth Type***

- Specify materials of the filtering screens, as (monel) (stainless steel) (aluminum) (nylon) (other).
- Refer to the drawings for locations, number of units, and piping connections.
- Specify the rate of flow through the filter as _______ gpm per sq ft of filter area.
- Specify the anticorrosion treatment to be given to the tanks, inside and outside, prior to installation. See Sec. 22-4(e).
- Specify operating pressures, _______ psig, and test pressure, _______ psig.
- Specify capacity of filters in terms of _______ hours required to filter entire contents of swimming pool.
- Specify rate of backwash, _______ gpm, at a flow rate of not over _______ gpm per sq ft of filter area.

  Note: Additional water is not necessary when backwashing a diatomaceous filter; the direction of normal flow is simply reversed.
- Specify control of backwash as suggested in Sec. 22-4(e).
- Specify characteristics of electrical service available.
- Require bidders to submit complete data and drawings covering general arrangement and construction, showing dimensions, weights, connections, etc.
- If a slurry tank is used in conjunction with the filters, specify capacity of tank, _______ gal, and give rate of flow of slurry to filters, as _______ gal per day.
- Specify the construction of all tanks, filters, slurry, and precoat pots, as of steel, _______ gage, or _______ in. thickness, welded. Require manholes and removable filter head as necessary to permit inspection and access to tank interior.
- Specify the minimum size of impurity to be removed by the filter as _______ microns.

(g) *Gas Chlorination Compartment*

- Provide bidders with detail drawing of compartment, showing dimensions, windows, if any, doors, and arrangement of chlorine gas cylinders, proportioners, exhaust ventilating fan, and motor.
- Specify rate of chlorine consumption as _______ pounds per day. *Note:* Usually for requirements in excess of 1

pound per day, the gas into water solution is likely to be the more economical method of addition.

- Specify method of introducing the chlorine solution into the pool, as manually, or automatically by injection into the pool influent water.
- Specify the desired residual concentration in the pool as ________ ppm. Also specify an expected maximum requirement of ________ ppm when a heavy influx of bathers produces a peak hour.
- Specify exhauster fan kind, (propeller) (centrifugal) (axial) (other); driven by an electric motor operating on ________________ (characteristics) electric service.
- Specify fan capacity, ________ cfm, against a discharge pressure of ________ in. W.G.
- Specify the fan delivery duct, as (sheet steel) (aluminum) (galvanized) (other), and refer to drawing for data on dimensions of duct, its length to point of safe discharge, support, etc.
- Specify that duct construction shall comply with the requirements of the ASHRAE Guide, latest edition, in gage of metal, kind of joints, etc.
- Specify a makeup air louver to be placed high in the compartment wall and sized to provide ________ sq ft of free area when the louvers are open.
- Specify louvers to be of aluminum, of light weight, and designed to open automatically by outside air pressure when the fan begins exhausting air from the compartment.
- Specify explosion-proof light bulbs in the compartment, to be operated by a manual switch outside the door.
- Specify placement of a gas mask outside the compartment, mask to be of a type and style approved by the U.S. Bureau of Mines.

  Note: Construction of the compartment is not within the scope of this Handbook, since it may be of masonry, frame, or steel materials, involving structural problems.

(h) ***Rate of Flow Indicator***

- Specify that indicator shall be calibrated in terms of gpm, to a maximum range of 10 per cent excess above the designed maximum flow rate.
- Specify that indicator shall be a ________________ (name) type and model ________________, or an approved alternate.

(i) *Instruments*

- For automatic backwash and other operations of the system, flow registers may be specified, with totalizing dial and counters, and incorporating electrical contactors that initiate system action as desired.
- Specify the diameter of the dial, ________ in. Also the number of digits in the totalizer, (4) (6) (other).
- Specify operating voltage and amperes.
- Specify make and model, ________________, or approved alternate.
- List gages, thermometers, and sight glasses, giving sizes and calibrations for each to suit the particular service on which it functions. Designate location of each. For details of instrument construction, refer to Sec. 4-21.

(j) *Piping, Valves, and Fittings*

- Specify materials of piping, as (copper tubing) (brass pipe) (wrought iron) (steel) (other).
- Specify kind of joints, as threaded, welded, soldered.
- Specify working pressure, ________ psig, and the schedule number of the ferrous piping, as applicable.
- Specify materials and working pressures for all valves and fittings.
- Specify joints of all valves and fittings.
- If not already shown on the drawings, prepare a list of all valves required on the system, showing the sizes, number of each size, location in the system, materials, and ends for in-piping connections.

(k) *Swimming Pool Water Heaters*

Note: Pool water may be heated by circulation through a shell and tube heat exchanger, with steam or hot water in the shell and pool water passing through the tubes. Alternatively, the pool water may be pumped through the coils of a direct-fired water heater burning gas or oil.

- Specify heater capacity on basis of raising pool water temperature at rate of 1 deg F per hour.
- For suggestions in specifying shell and tube heat exchangers suitable for this service, see Sec. 12-3(a)–(c). Specifications for a boiler if used, are given in Sec. 11-2 (a) and (b) for steam, and Secs. 12-1 and 12-2 for hot water boilers.
- For direct-fired heaters, refer to boiler sections as above, and specify fuel, as gas or oil. For a gas-fired water

heater particularly suitable for heating pool water, refer to Sec. 21-1(f), Gas-fired Storage Water Heaters.

- Refer to drawings to show how water heater is incorporated in the swimming pool piping and equipment arrangement, with particular attention to the valves that either direct the pump discharge through the heater, or bypass the heater as desired.

(l) *Swimming Pool Covers, Metal*

Note: Mechanically operated aluminum covers are available that retract into accordion folds during pool use. Electrically powered, either by available service or by storage batteries, pushing a button extends the cover over the entire pool in a period of 1 to 2 minutes.

- Specify size of pool; length and width.
- Specify the kind of drive wanted, as wired or storage battery. Specify the characteristics of the service available from existing circuits, _______ volts, _______ phase, _______ cycle. For storage batteries, state operating voltage and length of time, _______ hours, between rechargings.
- Require bidders to furnish drawings, photographs, and complete data, including weight of aluminum sheets, floor area required in folded position, maximum height of folds, _______ ft.
- Specify color wanted.
- Specify that all electric components must comply with the National Electrical Code, the NEMA standards, and local codes as applicable.
- Specify that device must be approved by local safety and health authorities.

(m) *Instructions for Use*

- Require the successful contractor for the swimming pool equipment to furnish in duplicate complete instructions for operating all components of the installation, including chlorination, pool covers, backwash control, etc.

22-5. CENTRAL VACUUM CLEANING STATIONS, INDUSTRIAL

(a) *Operating Conditions*

- Describe the nature of the dust to be collected, as metallic, coal dust, rock dust, talc, cement, color pigments, pharmaceuticals, etc.
- Specify the temperature range of the air and dust drawn into the tools, from _______ F to _______ F.

- State the number of operators who will be using the system continuously, and the number of part-time users.
- Specify the number of operators who will be using the system simultaneously.
- Specify the desired rate of cleaning, _______ sq ft per hour.
- Describe the surface conditions where cleaning is to be accomplished, as polished, smooth, rough, etc.
- Specify maximum acceptable velocity in piping, _______ fpm.
- Specify maximum total friction loss _______ in. W.G.
- Specify total weight, _______ pounds per hour of dust to be collected under full-load conditions.

(b) *Exhauster*

- Specify capacity, _______ cfm of air at _______ F temperature, and discharged at a pressure of _______ in. W.G.
- Specify vibration isolators, as pads under the exhauster bedplate or foundation, and flexible or bellows form absorbers connecting the piping to the machine.
- For details of exhauster items to be specified, refer to Sec. 20-3(a) and (c).

(c) *Separators*

- Specify number of separators arranged in series, Fig. 22-1.
- Specify kinds of separators, as primary centrifugal or baffle design, and secondary of bag types.
- For bag-type separators, specify kind of material in bags, as cotton, nylon, etc. See Sec. 20-3(d).
- Specify kind of bag shaker, manual or mechanical.
- Specify minimum head room under the dump valves in the bottom of each separator hopper, _______ ft.
- Specify capacity of each separator, _______ cu ft, or in terms of continual collection, _______ hours between dumps.
- Specify maximum velocity of air through the bag filters, fpm.
- Require bidders to submit drawings of their separators, including dimensions, and showing construction.

(d) *Piping, Valves, and Fittings*

- Refer to drawings for layout of piping, sizes, and locations of tool connections.
- Specify piping material (wrought iron) (steel) (other).

Note: Some manufacturers recommend the use of cast-iron drain piping.

- Specify Schedule No. ______ and working pressure, ______ psig, for ferrous piping as applicable.
- For cast-iron piping, specify class, ______, ends, etc.
- Specify that fittings are to be long sweep.
- Specify kinds of valves, as (blast gates) (plug and lever valves), etc.

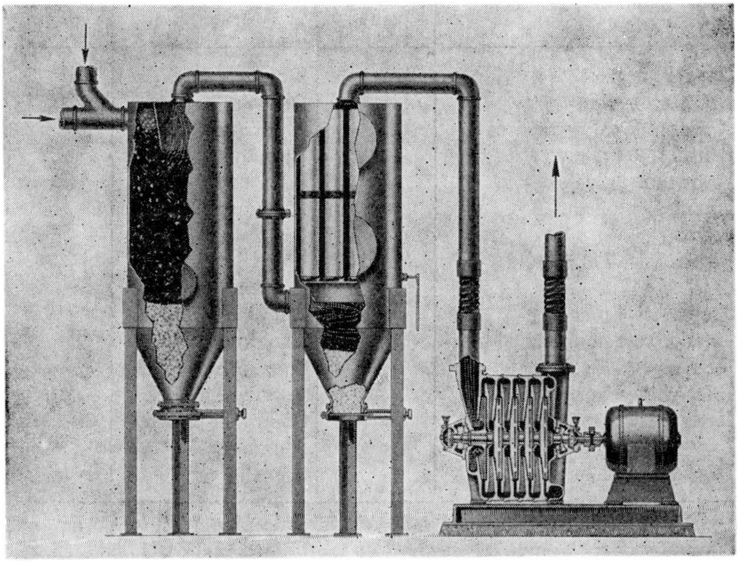

FIG. 22-1. Dust collection and separating system. (*Courtesy Lamson Corporation.*)

- List all valves by kind, locations, quantities, and working pressures.
- Specify kind of air hose, size, and number of pieces of each standard length. Example: 4 lengths of 25 ft each; or 50 ft each, metal wound.
- Specify kind of tool connector for use with the hose, as snap type, screw connection, etc.
- Specify kind of piping joints: (threaded) (welded) (flanged) (soldered) (other).

(e) ***Tools***

- Prepare and submit to bidders as part of the specification a list of tools required. Table 22-1 suggests the form of such a list.

- Require bidders to submit unit prices for each.

(f) *Scrubbers and Other Separators*

- Refer to Sec. 22-3(b) for wet scrubbers; Sec. 20-3(e) for electrostatic; Sec. 20-3(f) for cyclones.

Table 22-1. Vacuum Cleaning Tools

Tool	Quantity	Material*	Unit price
Nozzles, crevice		Fiber	
Nozzles, flexible		Rubber	
Nozzles, metal; give sizes†		Aluminum	
Brushes, floor; give sizes		Hair, nylon, bristles	
Floor tools; give sizes		Fiber or rubber	
Handles, wall and floor; give sizes		Aluminum	
Brushes, wall and hand		Wood backed hair, nylon bristles	
Angle attachment		Aluminum, other	
Handle hose coupling, shutoff		Aluminum, other	
Hose and tool cabinet		Steel or other	
Inlet valves, spring cover		Aluminum	
Recessed wall plate		Cadmium plated	
Inlet valve, floor type		Aluminum	
Inlet valve, floor key		Alloy	

* Materials will vary somewhat with different manufacturers. Evaluate materials after the bids are received, in light of original costs and estimated life.

† Specify capacity of each tool for air handling, ______ cfm.

(g) *Instruments*

- Specify pressure and vacuum gages in accordance with Sec. 4-21.
- Specify calibration and range for each.
- Specify number of each and identify locations.
- Specify electric service characteristics if any component of the system, as a bag shaker, calls for power connections.
- Require that wiring and materials comply with National Electrical Code, NEMA standards, and local codes as applicable.

22-6. PACKAGE-TYPE CHEMICAL FEEDERS

(a) *General Notes*

- These units are used for injecting chemical feed into boiler feedwater systems, waste treatments, and where proportioning is needed in process work.
  The package consists of a tank, metering pump, strainer, relief valve, motor for driving the pump, and interconnecting piping. Mixing equipment and dissolving baskets are specified if needed, as extras.
- Specify kind of chemical to be injected, and for what purpose.

(b) *Chemical Tank*

- Specify the capacity of the tank, ________ gal.
- Specify construction, ________ gage steel, welded.
- Specify lining for tank, as (polyvinyl chloride) (polyethylene) (other). Thickness: ________ mils.
- Specify pump suction piping of ________________ (material) with a foot strainer in the bottom of the tank. Strainer to be of ________________ (material) with perforations ________ in. diam.
- Specify tank to have removable cover to permit easy inspection.
- Specify exterior finish of tank, as ________ coats of ________________ color paint.

(c) *Pump*

- Specify location of pump, as on a bracketed platform over the tank.
- Specify pump capacity in gph, as zero up to 1.1 or more, as standardized by the manufacturer.
- Specify pressure against which pump must discharge, ________ psig. *Note:* This may be as high as 500 psig.
- Specify type of pump, as plunger-metering, driven by electric motor for ________ volt, ________ phase, ________ cycle service.
- Specify pump construction materials, as steel, bronze, stainless steel, etc.
- Require bidders to submit drawings and complete data including dimensions, weight, diameter, and stroke of plunger, ________ in., number of strokes per minute, ________. Also type of motor, as open, single or three phase, motor speed, ________ rpm, overload protection, horsepower.

- Require bidders to submit pump performance curve showing discharge rate, ________ gph, versus discharge head, ________ ft, and hp, ________.

(d) *Piping, Valves, and Fittings*

- Specify materials of piping, and working pressure, ________ psig.
- Specify safety relief valve and pipe discharge back into tank. Valve to meet ASME requirements.
- Specify vacuum breaker, or other automatic fitting to avoid creation of a vacuum inside the tank.

(e) *Electrical and Instruments*

- Specify that all electrical materials and arrangements must comply with National Electrical Code, NEMA standards, and local codes as applicable.
- Specify pump operation, as manual, automatic intermittent, or otherwise controlled.
- Specify location of control switches or push buttons.
- Require provision of vacuum breaker or other means to prevent the creation of a vacuum in the tank.
- Specify number, calibration range, and placement of pressure gages and thermometers.
- For details of instrument construction, refer to Sec. 4-21.

22-7. SNOW-MELTING INSTALLATIONS

(a) *General Arrangement*

- Refer to drawings for arrangement of buried coils, showing if design is in the form of a grid or a sinuous, continuous coil, or some of both. Drawing to show also the distance between pipes on centers, ________ in., the size of the pipes, as (1½ in.) (2 in.), etc. Depth of burial: ________ in.
- Provide arrangement drawing of equipment showing relative positions of the heat source, as (hot water boiler) (steam boiler) (converter) (other); the heat exchanger; the expansion tank for the pumped solution; the solution circulating pump; the hot water circulating pump if necessary, or the condensate recovery system from the steam heated converter; and the interconnecting piping, valves, and controls.

(b) *Solution*

- Specify the kind of solution, as (glycol) (dowtherm) (other). Also specify the concentration of the solution, ________ per cent.

- Require bidders to submit characteristic curves of the solution as follows:
  1. Curve to show relationships between the temperature of solution freezing, _______ F, and the per cent solution by volume.
  2. Curve showing the densities of the solution by _______ per cent weight and _______ pounds per gal, over a temperature range of _______ F freezing to _______ F maximum.
  3. Curves showing the specific heat of the solution, Btu per pound per deg F, over a temperature range of _______ F minimum to _______ F maximum, and at concentrations by volume of 10 per cent, 20 per cent, 50 per cent, etc., as desired.
  4. Curves showing coefficient of expansion for various concentrations of _______ percentages by volume, over a temperature range of _______ F minimum to _______ F maximum.
- Specify that when chemical is mixed with water to provide the desired solution, _______ per cent of concentration, the mixture shall be circulated continuously through the system for at least 24 hours to make sure of complete mixing.
- Specify freezing point, _______ F, for the solution.
- Specify that inhibiter in the solution shall be of a nature to prevent reaction with the piping, valves, fittings, tubing, and other metals in the system.
- Specify that water used to mix the solution shall be soft water. Specify that the maximum acceptable level of hardness shall be not more than _______ ppm in terms of calcium carbonate.

(c) *Piping, Valves, and Fittings*

- Specify kind of piping: (wrought iron) (steel) (alloy) (other).
- Specify working pressure, _______ psig.
- Specify Schedule No. _______ of the piping.
- Specify materials and kinds of valves, as iron body, bronze mounted, outside stem, inside stem, etc., with threaded ends, welded ends, other.
- Specify material and ends of fittings, as malleable iron, screwed, etc.
- Specify test pressure on system of _______ psig and time test pressure is to be held _______ hours.

(d) *Solution Pump*

- Specify kind of pump; centrifugal, rotary, piston, etc.
- Specify materials, as iron casing, bronze rotor, steel shaft; all-bronze pump, etc., to suit nature of the solution to be pumped. Give solution temperature, _______ F.
- Specify the delivery pressure, _______ psig, _______ gpm capacity.
- Specify the pump drive, as electric motor for _______ volt, _______ phase, _______ cycle service.
- Connected to pump by (belt) (coupling), etc.
- Bidders to submit pump operating data, as pump speed, _______ rpm; motor speed, _______ rpm. Also a pump curve showing relationship between pump discharge pressure, _______ psig, and the motor hp, _______, and capacity, _______ gpm.
- Specify if pump and motor are to be set on a common baseplate and common foundation.
- Specify the solution density and temperature, _______ F, at which the pump performance is to be guaranteed.

(e) *Heat Exchanger*

- Specify kind of heat exchanger, as shell and tube, U-tube arrangement, with _______ passes, vertical or horizontal setting.
- Specify characteristics of heat medium. For steam, give pressure, _______ psig, and temperature, _______ F. For hot water, give temperature, _______ F, and rate of flow through the exchanger, _______ gpm.
- Specify capacity of the exchanger in terms of raising the temperature of a _______ density solution from _______ F to _______ F when flowing through the exchanger at a rate of _______ gpm.
- Specify the materials in the heat exchanger to best withstand the effects of the solution. Tubes to be of ____________ (material), shell of (steel) (other), tube sheet of ____________ (material), liquid boxes of ____________ (material).
- Require bidders to submit drawings showing dimensions, weight, and internal arrangement of the exchanger being offered. Also size and locations of connections.
- Specify maximum pressure drop acceptable, _______ psig, in solution being pumped through exchanger.
- For steam heating, specify kind of condensate trap under

the exchanger. Also trap capacity in ______ pounds per hour delivered against a back pressure of ______ psig.

- Specify kind and thickness of insulation on the exchanger and on the piping.

(f) *Solution Expansion Tank*

- Specify capacity of expansion tank, ______ cu ft.
- Specify that expansion tank shall comply with requirements of ASME "Code for Unfired Pressure Vessels."
- Specify a gage glass suitable for a working pressure of ______ psig, for checking level of solution in tank.
- Specify a makeup charging valve for the solution, and ditto for air; the latter on the tank, the former in the connecting piping as convenient.

(g) *System Controls*

- Specify the circulating temperature of the solution, as ______ F leaving the exchanger and ______ F returning to the exchanger.
- Specify the method of operation control. If by outside thermostat, specify temperature of air, ______ F, at which system is energized. If by thermostatic bulb buried in the ground, also specify temperature, ______ F, which energizes system.

  If the weight of snow falling on a plate is used to depress the plate and close electric contacts, provide information on size of plate, the depth of snow on the plate, ______ in., required to close the contacts, and other details of the device desired. Include a thermostat that prevents closing of the contactors if the outside temperature stays above 35 F or ______ F as selected.

  Closing of the contacts may open the steam or hot water admission valve to the heat exchanger, start the solution circulator, and do whatever else is necessary to energize the system.

  A return solution thermostat is sometimes provided to maintain circulation of the solution through the coils so long as the return temperature remains below a predetermined ______ F. This continuing circulation of warm solution keeps ground temperatures above freezing and there is no delay in the prompt melting of any snow that happens to fall.
- For steam admission valve to the heat exchanger, refer to Sec. 12-3(c).

- For hot water admission valve, specify the size and type of valve, and give its capacity, ________ gpm. Specify if valve is to respond to a thermostatic bulb immersed in the solution leaving the exchanger, or is to be actuated from some other source.
- Specify a manually operated switch or push-button station that may be used to start and stop the solution circulator at any time, regardless of the position of the automatic controls.

(h) *Instruments*

- Prepare a list of required instruments, as pressure gages, thermometers, thermostats, humidistats, etc., giving the quantity required of each size and type, and stating its point of location in the system. Specify calibrations in range and kind.
- For details of instrument construction, see Sec. 4-21.

22-8. DEHUMIDIFICATION EQUIPMENT

(a) *General Notes*

Dehumidifiers are available that perform on the basis of mechanical refrigeration, and also through the functions of a desiccant, either liquid or solid. When the desiccant is changed physically or chemically by the process of accepting moisture from the air or gas being dried, it is termed an "absorbent." Calcium chloride is an illustration of absorbent properties; the acceptance of moisture converts this material into a hydrate, and if the exposure to moisture continues long enough, the material actually dissolves into the water.

By contrast, an "adsorbent" stores the moisture in countless crevices and voids until these are filled, after which the moisture may be driven out by the application of heat. At no time, during either the storing or the expelling of moisture, is there any change in the physical or chemical properties of the material. Silica gel, activated alumina, and activated charcoal, among other materials, are examples of adsorbent materials.

Mechanical refrigeration reduces the humidity in the air by lowering the temperature below the dew point. Most domestic-type dehumidifiers are of this kind. If higher temperatures are needed, reheating of the saturated air becomes necessary.

(b) *Domestic Dehumidifiers, Mechanical Refrigeration*

- Specify the volume of air to be dried: ______ cfm. *Note:* 200 cfm is a common size.
- Specify the condition of the air entering the machine, giving temperature dry bulb, ______ F, and temperature wet bulb, ______ F.
- Specify wet bulb temperature of air leaving, ______ F.
- Specify ultimate desired temperature of dried air, ______ F dry bulb, and relative humidity, ______ per cent.
- If reheating is necessary, specify method, as (electric resistance heater) (hot water or steam coils), etc.
- Specify electrical connections to meet National Electrical Code, NEMA standards, and local codes as applicable.
- Specify electric service available, as ______ volts, ______ phases, ______ cycle.
- Specify if delivery of air is to be at a pressure of ______ in. W.G., or is a free atmospheric delivery.
- If noise is a factor, specify maximum level permitted: ______ decibels.
- If a filter is required, specify kind, (throwaway) (washable), etc. Specify velocity of air through the filter, ______ fpm.
- Specify general construction of machine, as casing of ______ gage steel, finished in ____________ color paint, and kinds of fans and blowers: (propeller) (centrifugal).
- Specify method of removing condensate. If by connection to a drain, specify kind and size of drain piping, with particular reference to corrosion-resistant materials. If condensate accumulates in a pot, pan, or other container, specify capacity per 24 hours, as ______ qt, ______ gal, as suitable.
- Specify materials of condensate container to be noncorrosive.
- Require bidders to submit information on unit dimensions and weights. Also to provide data on performance, compressor speed, ______ rpm; motor speed, ______ rpm. Also condensing coil fan speed, ______ rpm; motor speed, ______ rpm. Air blower speed, ______ rpm. Also horsepower of each motor, full-load amperes and locked rotor amperes.
- Require information on the cooling coil and the con-

densing coil, area of each in sq ft, size and materials of tubing in the coils.

- If unit is located in a duct system, so state, and specify the friction loss, ________ in. W.G., that fans must overcome. Provide sketch showing duct and where dehumidifiers are located.
- Specify method of control, as by humidistat for automatic operation. Also a manual switch or push button.

(c) ***Liquid Dehumidifiers, Adsorbent Type***

- Specify volume of air to be dried, ________ cfm.
- Specify the air entering conditions, giving dry and wet bulb temperatures, ________ F.
- Specify moisture removal rate, ________ grains per hour.
- Specify desired ultimate conditions, temperatures dry bulb and wet bulb, ________ F.
- Specify if air is to be cooled or reheated to attain the ultimate temperature, dry bulb ________ F and wet bulb ________ F.
- Specify the kind of adsorber to be used: ________ ________ (name).
- Specify general construction of machine, as casing of ________ gage steel, painted ________________ color outside and finished in anticorrosion materials inside: ________________ (name kind).
- Specify number, size, and placements of the spray nozzles, also specifying materials.
- Specify piping, giving materials, working pressure, ________ psig.
- Require bidders to furnish drawings and data showing dimensions, weights, internal arrangements, etc.
- Require information from bidders concerning the adsorbent pump, including capacity, ________ gpm, at a discharge head of ________ ft when handling solution at a temperature of ________ F. Also pump speed, ________ rpm, motor speed, ________ rpm.
- Specify or require bidders to describe methods used for reactivation of the adsorbent, as steam coils, electric heaters, etc. Specify volume of reactivation air, ________ cfm.
- Specify if reactivation shall be continuous or periodic. If the latter, specify length of continuous runs between reactivations, ________ hours.
- Provide a sketch showing how warm air from the re-

activation process is discharged to atmosphere, giving dimensions and lengths of ducts. Also material and construction of ducts.

- If a bypass around the dehumidifier is required to control the final temperatures of the dried air, provide sketch showing location of bypass, method of dampering and damper control; also size, construction, and painted finish of the bypass duct.
- If cooling is required for the discharge air, specify means, as mechanical with direct expansion coils, chilled water coils, or other. Specify who furnishes the cooling equipment, as the dehumidifying contractor or the owner. Specify cooling capacity, _______ tons of refrigeration.
- For cooling of the adsorbent solution, specify means: (city water) (chilled water) (cooling tower water), etc.
- Specify temperature of the water, _______ F, wherever used for cooling. Also specify the available rate of flow, _______ gpm, and pressure, _______ psig.
- Require bidders to state the quantity of water required by their cooling coils in each application, _______ gpm.
- Require information on means provided to make up losses of the adsorbent solution by evaporation. Also on the concentration of solution, _______ per cent, for proper performance.
- If heat for regeneration is provided by steam coils, specify pressure, _______ psig, of steam, and temperature, _______ F. For hot water, specify temperature, _______ F. For direct-fired regenerators, with gas, specify kind of gas and calorific value, _______ Btu per cu ft.
- Specify system of controls, as humidistat in space requiring the dried air, and another humidistat in the final air discharge stream. The former indicates, or records, if desirable, the continuing conditions in the subject space; the latter instrument gives evidence of the performance of the dehumidifier as a whole. The space humidistat likewise starts, stops, and maintains the continuous operation of the dehumidifier as necessary to satisfy conditions desired.

  Control interlocks between the outlet air, the cooling or heating coils, bypass dampers, reactivation procedures, spray pumps, and other components of the installation, shall be specified as set up by the system design engineer.

- Specify master switch or push-button station, giving location.
- Require bidders to submit curves showing performance of adsorbent, expressed in terms of ________ grains of moisture per pound of dry air, with outlet air temperatures of 50 F, 70 F, 90 F, 110 F, and in between as desired, and inlet air conditions of ________ grains of moisture per pound of dry air.
- For specifying refrigeration processes, if these are used in connection with the drying requirements, refer to Divisions 15 and 16, as applicable.
- If air filter is to be used, specify kind of filter and give size of particles, ________ microns, that are to be removed from the airstream.
- Specify electric service available, ________ volt, ________ phase, ________ cycle.

(d) *Solid Adsorbent Dehumidifiers*

- Specify air conditions entering and leaving, as suggested in the preceding section for liquid adsorbents.
- Specify if reactivation is to be continuous or periodic. If periodic, specify times.
- Specify kind of dehumidifying machine, as stationary cabinet or rotary.
- For stationary cabinet, specify construction, as ________ gage steel, finished inside and outside, with coats of ________________ paint, anticorrosive in nature.
- Require bidders to submit drawings showing internal arrangement of dehumidifier, thickness of adsorbent, air movement, dimensions, and maximum operating weight.
- Specify manner of reactivation, as steam, hot water, gas, etc.
- For steam, specify working pressure, ________ psig; for hot water, specify temperature, ________ F.
- For gas, direct-fired or through a heat exchanger, specify calorific value of the gas, ________ Btu per cu ft.
- Refer to preceding section on liquid adsorbents, and incorporate appropriate information given and required into this section dealing with solid adsorbents.
- For rotary dehumidifiers, require bidders to furnish drawings and complete data, stating among other facts the speed of rotation, as ________ hours, ________ minutes, per revolution of the drum.
- Require information on the motors used for driving fans

and the drum, including horsepower for each; kind of motor; temperature rise, ______ C, above an ambient of ______ C; motor speed, ______ rpm; protection against overload. Also method of drive, whether direct-connected or through a belt.

- Specify kind of bearings desired and methods of lubrication.
- Specify kind of adsorbent: ____________ (name).
- Specify material of rotating drum, as noncorrosive, stainless steel, etc.

(e) *Dehumidification Instruments*

- Make a list of all instruments needed for the installation, including humidistats, pressure gages, thermometers, remote indicators and recorders, flowmeters, etc., giving number of each size, calibration range, units of calibration, general design, as straight stem or angle stem, and points of installation in the system.
- For details of instrument construction to be specified, refer to Sec. 4-21.

Note: Most of these instruments will be required by all kinds of adsorbent systems; rotary, stationary, liquid. Fewer are needed by the domestic refrigeration units, particularly those which are portable.

22-9. PLASTIC-LINED PIPING

(a) *Service and Operating Conditions*

- Specify service in which the lined piping is to be placed, as (chemicals) (paper) (plating) (fertilizers) (waste disposal) (petrochemical) (other).
- Specify the operating pressure, ______ psig, and the range of temperatures, from ______ F minimum to ______ F maximum.
- State if lining is to be joined to cast iron, steel, wrought iron, alloy, etc.

(b) *Pipe Characteristics*

- Specify the thickness of the lining, ______ in.
- Specify that plastic liner and host pipe must be so joined as to expand and contract monolithically and without any separation of plastic from metal.
- Specify the tensile strength of the liner, ______ psi, ultimate.
- Specify elongation, ______ per cent.

- Specify coefficient of thermal expansion, _______ in. per in. per deg Fahrenheit.
- Specify the thermal conductivity, _______ Btu per hour per sq ft per deg F.
- Specify kind of abrasion to which liner may be subject, if any.
- Specify maximum acceptable pressure drop, _______ psi per 100 ft, at a flow of _______ gpm and a temperature of _______ F, in the specified service.

(c) *Erection Practices*

- Refer to drawings for pipe locations and general arrangements.
- State that sizes are to be as shown on the drawings.
- Specify that lined pipe shall not be bent in the field and that all changes in direction shall be accomplished by means of taper spacers or by lined fittings.
- Specify spacing of hangers or supports, _______ feet center to center.
- Specify lengths in which pipe is to be delivered.
- Specify kind of pipe joints, as flanged, threaded.
- Specify how pipe ends are to be protected during shipment, as thread protectors for threaded ends and flanges for flanged ends.
- Specify if threaded ends are to be threaded in the factory or in the field. Field threading must comply exactly with instructions of pipe manufacturer.
- Specify kind of gaskets to be used for flanged joints. Specify kind of pipe dope to be used in making threaded joints.

(d) *Plastic-lined Valves*

- Specify kind of plastic desired, using a proprietary identification with the words, "or approved alternate."
- As with pipe, specify service, pressure, _______ psig, and temperature, _______ F.
- Specify kind of valve ends; threaded or flanged.
- Specify ferrous material of valve body, as cast iron or steel.
- Specify rising or nonrising stem, as desired.
- If necessary, specify a chain-operated handwheel.
- If valves are to be operated by power, specify whether pneumatic or electric. For former, specify air pressure, psig, and state if source of compressed air is existing or must be provided by a separate compressor. For elec-

tric motors, specify available service, _______ volts, _______ phase, _______ cycle.

- Specify that all electric components, wiring, etc., must comply with National Electrical Code, NEMA standards, and local codes as applicable.
- Specify nature of liquid or slurry passing through valves, as acid, abrasive, etc. Indicate if corrosive attack is severe or mild.
- For diaphragm valves, specify if diaphragm is to be formed of a rubber, neoprene, etc.
- Specify size of valve, _______ in.
- If considered important, specify maximum acceptable pressure drop through valve, _______ psi.
- If plug-type valve, specify whether two-way or three-way.
- If check valve, specify horizontal or vertical.

(e) ***Plastic-lined Fittings***

- Specify kind of fittings, as ell, short radius; ell, long radius; tee, standard; tee, reducing; concentric reducers; eccentric reducers; laterals, etc.
- Specify material of fittings, iron or steel.
- Specify working pressures, _______ psig.
- Specify temperature, _______ F maximum.
- Specify liquid passing through fittings, as acid corrosive, abrasive, etc.
- Specify fitting ends, threaded or flanged.

(f) ***Information from Bidders***

- Require bidders to furnish complete information on the kind and characteristics of the plastic to be used in the pipe, valve, and fitting linings.
- If bidder buys the ferrous valves and fittings from another supplier, require information on the source of these items, for approval.
- Require all piping to conform to the schedule number that is indicated by the working pressure of the system.
- Also require description of process by which plastic is joined to the inner surfaces of the pipe, valves, and fittings.
- If power operation of the valves is specified, call for information from bidders on characteristics of the valve operator motors; kind, speed, whether modulating or off and on, etc.

Index